Polymer Electrolytes for Energy Storage Devices

Volume I

Polymer Electrolytes for Energy Storage Devices

Volume I

Edited by
Prasanth Raghavan and Jabeen Fatima M. J.

CRC Press is an imprint of the
Taylor & Francis Group, an **informa** business

First edition published 2021
by CRC Press
6000 Broken Sound Parkway NW, Suite 300, Boca Raton, FL 33487-2742

and by CRC Press
2 Park Square, Milton Park, Abingdon, Oxon, OX14 4RN

Copyright ©2021 Prasanth Raghavan, Jabeen Fatima M. J.
CRC Press is an imprint of Taylor & Francis Group, LLC

Reasonable efforts have been made to publish reliable data and information, but the author and publisher cannot assume responsibility for the validity of all materials or the consequences of their use. The authors and publishers have attempted to trace the copyright holders of all material reproduced in this publication and apologize to copyright holders if permission to publish in this form has not been obtained. If any copyright material has not been acknowledged please write and let us know so we may rectify in any future reprint.

Except as permitted under U.S. Copyright Law, no part of this book may be reprinted, reproduced, transmitted, or utilized in any form by any electronic, mechanical, or other means, now known or hereafter invented, including photocopying, microfilming, and recording, or in any information storage or retrieval system, without written permission from the publishers.

For permission to photocopy or use material electronically from this work, access www.copyright.com or contact the Copyright Clearance Center, Inc. (CCC), 222 Rosewood Drive, Danvers, MA 01923, 978-750-8400. For works that are not available on CCC please contact mpkbookspermissions@tandf.co.uk

Trademark notice: Product or corporate names may be trademarks or registered trademarks and are used only for identification and explanation without intent to infringe.

Library of Congress Cataloging-in-Publication Data

Names: Raghavan, Prasanth, editor. | J, Jabeen Fatima M., editor.
Title: Polymer electrolytes for energy storage devices / edited by Prasanth
   Raghavan & Jabeen Fatima M.J.
Description: First edition | Boca Raton : CRC Press, 2021. | Includes
   bibliographical references and index.
Identifiers: LCCN 2020053414 | ISBN 9780367701451 (hbk) | ISBN
   9781003144793 (ebk)
Subjects: LCSH: Storage batteries--Materials. | Solid state
   batteries--Materials. | Polyelectrolytes.
Classification: LCC TK2945.P65 P65 2021 | DDC 621.31/24240284--dc23
LC record available at https://lccn.loc.gov/2020053414

ISBN: 9780367701451 (hbk)
ISBN: 9780367701536 (pbk)
ISBN: 9781003144793 (ebk)

Typeset in Times
by Deanta Global Publishing Services, Chennai, India

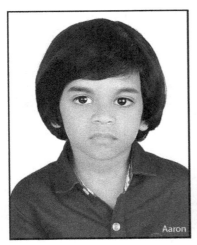

dedicated to
Aaron *and* **Ayaan**,
*who is never going to
read this book*

Contents

Foreword ix
Editors xi
Contributors xiii
Abbreviations xv

Chapter 1 Electrochemical Energy Storage Systems: The State-of-the-Art
Energy Technologies 1

*Abhijith P. P., Jishnu N. S., Neethu T. M. Balakrishnan,
Akhila Das, Jou-Hyeon Ahn, Jabeen Fatima M. J., and
Prasanth Raghavan*

Chapter 2 The Great Nobel Prize History of Lithium-Ion Batteries:
The New Era of Electrochemical Energy Storage Solutions 35

Prasanth Raghavan, Jabeen Fatima M. J., and Jou-Hyeon Ahn

Chapter 3 Polyethylene Oxide (PEO)-Based Solid Polymer Electrolytes
for Rechargeable Lithium-Ion Batteries 57

*Prasanth Raghavan, Abhijith P. P., Jishnu N. S., Akhila Das,
Neethu T. M. Balakrishnan, Jabeen Fatima M. J., and
Jou-Hyeon Ahn*

Chapter 4 Polymer Nanocomposite-Based Solid Electrolytes for
Lithium-Ion Batteries 81

Prasad V. Sarma, Jayesh Cherusseri, and Sreekanth J. Varma

Chapter 5 Poly(Vinylidene Fluoride) (PVdF)-Based Polymer Electrolytes
for Lithium-Ion Batteries 111

*Jishnu N. S., Neethu T. M. Balakrishnan, Akhila Das,
Jarin D. Joyner, Jou-Hyeon Ahn, Jabeen Fatima M. J., and
Prasanth Raghavan*

Chapter 6 Poly(Vinylidene Fluoride-*co*-Hexafluoropropylene)
(PVdF-*co*-HFP)-Based Gel Polymer Electrolyte for
Lithium-Ion Batteries 133

*Akhila Das, Neethu T. M. Balakrishnan, Jishnu N. S.,
Jarin D. Joyner, Jou-Hyeon Ahn, Jabeen Fatima M. J., and
Prasanth Raghavan*

Chapter 7	Polyacrylonitrile (PAN)-Based Polymer Electrolyte for Lithium-Ion Batteries	149

Neethu T. M. Balakrishnan, Akhila Das, Jishnu N. S., Jou-Hyeon Ahn, Jabeen Fatima M. J., and Prasanth Raghavan

Chapter 8	Polymer Blend Electrolytes for High-Performance Lithium-Ion Batteries	167

Jishnu N. S., Neethu T.M. Balakrishnan, Anjumole P. Thomas, Akhila Das, Jou-Hyeon Ahn, Jabeen Fatima M. J., and Prasanth Raghavan

Chapter 9	Polymer Clay Nanocomposite Electrolytes for Lithium-Ion Batteries	187

Jishnu N. S, Krishnan M. A., Akhila Das, Neethu T. M. Balakrishnan, Jou-Hyeon Ahn, Jabeen Fatima M. J., and Prasanth Raghavan

Chapter 10	Polymer Silica Nanocomposite Gel Electrolytes for Lithium-Ion Batteries	219

Akhila Das, Anjumole P. Thomas, Neethu T. M. Balakrishnan, Nikhil Medhavi, Jou-Hyeon Ahn, Jabeen Fatima M. J., and Prasanth Raghavan

Chapter 11	Polymer-Ionic Liquid Gel Electrolytes for Lithium-Ion Batteries	235

Jayesh Cherusseri

Chapter 12	Biopolymer Electrolytes for Energy Storage Applications	255

S. Jayanthi and M. Ulaganathan

Index 277

Foreword

Energy storage systems have emerged as an inevitable part of modern life ever since the commercialization of batteries began. An era of portable devices was initiated when lithium-ion batteries were commercialized by Sony in 1991, which led to a revolution of electronic gadgets such as mobile phones, laptops, palmtops, tablets, etc. In the current scenario, such storage devices have also been used commercially for electric vehicles, to reduce pollution and the exploitation of severely depleted stocks of fossil fuels. Energy storage devices capable of satisfying the present energy demands of portable devices consist mainly of lithium-ion batteries (LIBs), supercapacitors and fuel cells. The performance of these storage devices is assessed using two main parameters: the energy density and the power density. The first parameter defines the amount of energy that can be stored in a given volume or weight, whereas the second parameter describes the speed at which energy is stored in or discharged from the device. An ideal energy storage device should have high energy density and power density. The commercially available LIBs possess high energy density, whereas supercapacitors possess high power density. Present-day research and development of new and innovative component materials are progressing well to address the demands of super gadgets. The ideal energy storage devices for long-range applications are still in their infancy, and, hence, there is still plenty of room available in the exploratory domain of materials.

Even though LIBs have already been widely used in different areas, they are still facing a lot of challenges, including poor safety, short performance life, and relatively low specific energy. To address these issues, the new format of batteries, namely solid-state Li batteries, has been developed. Improving the efficiency of any storage device depends directly on the performance of its components, in particular, the behavior of its electrodes and electrolytes during charging and discharging. Material selection is the primary concern in developing advanced energy storage applications. The electrolyte is considered to be the heart of the energy storage device and its properties greatly affect the energy capacity, rate performance, cyclability and safety of these devices. Safety is a prime concern, since portability is the major requirement for these devices, demanding special electrolytic systems capable of replacing conventional liquid electrolyte. The organic chemicals used in the liquid electrolytes initiate the formation of a solid electrolyte interface at the anode of the LIBs, which, on continuous cycling, results in the growth of dendritic projections extending toward the cathode, leading to short circuiting and causing explosions in the devices.

The present book offers a detailed explanation of recent evidence of progress and challenges in electrolyte research for energy storage devices. The influences of electrolyte properties on the performances of different energy storage devices are discussed in detail. The detailed explanation has been classified under two major categories, which include a general introduction to energy storage devices and a history of lithium-ion batteries followed by a detailed investigation on polymer gel electrolytes for energy storage devices. In total, this volume consists of twelve chapters. The introduction chapter (Chapter 1) deals with types of energy storage systems and the working principles of these devices. A detailed view on the history of lithium-ion batteries has been carefully presented in Chapter 2. These chapters are followed by a detailed review of polymer-based gel and solid-state electrolytes for lithium-ion batteries, consisting of seven chapters based on different types of polymeric systems including PVdF, PVdF-*co*-HFP, PAN, blend polymeric systems, two chapters on composite polymeric systems (silica, clay etc.) and polymeric ionic liquid gel electrolytes for lithium-ion batteries. A detailed review on biopolymer electrolytes for energy storage applications was also included, as Chapter 12.

The present book covers a wide range of polymer-based electrolytes and includes useful information for the development of the various electrolytes. We truly believe that this book will be very useful, not only for researchers, but also for engineers in the development of next-generation energy storage devices. As the battery industry has grown so much over the past ten years, there have been many new entrants into the battery world from other industries. So, whether you are looking to learn something about one aspect of energy storage devices to reinforce your current knowledge, or are entirely neoteric and are looking to learn all the basics, this book will be a good manual to add to your dictionaries of science, engineering and technology.

Prof. (Dr.) Rachid Yazami
Draper Prize winner 2014 for the discovery of LIBs

Editors

Prasanth Raghavan, PhD, Professor, at the Department of Polymer Science and Rubber Technology, Cochin University of Science and Technology (CUSAT), Kerala, India, and Visiting Professor at the Department of Materials Engineering and Convergence Technology, Gyeongsang National University, Republic of Korea. Dr. Prasanth also holds the Associate/Adjunct Faculty position at the Inter-University Centre for Nanomaterials and Devices (IUCND), CUSAT. He earned his PhD in Engineering under the guidance of Prof. Jou-Hyeon Ahn, from the Department of Chemical and Biological Engineering, Geyongsang National University, Republic of Korea, in 2009, supported by a prestigious Brain Korea (BK21) Fellowship. Dr. Prasanth earned his B. Tech and M. Tech from CUSAT. After a couple of years of attachment as Project Scientist at the Indian Institute of Technology (IIT-D), New Delhi, he moved to the Republic of Korea for his PhD studies in 2007. His PhD research was focused on fabrication and investigation of nanoscale fibrous electrolytes for high-performance energy storage devices. He completed his Engineering doctoral degree in less than three years, still a record in the Republic of Korea. After his PhD, Dr. Prasanth joined the Nanyang Technological University (NTU), Singapore, as a Research Scientist, in collaboration with the Energy Research Institute at NTU (ERI@N) and TUM CREATE, a joint electromobility research centre between Germany's Technische Universität München (TUM) and NTU, where he was working with Prof. (Dr.) Rachid Yazami, who has successfully introduced graphitic carbon as an anode for commercial lithium-ion batteries, and received the Draper Prize, along with the Nobel laureates, Prof. (Dr.) John B. Goodenough and Prof. (Dr.) Akira Yoshino. After four years in Singapore, Dr. Prasanth moved to Rice University, USA as Research Scientist, where he worked with Prof. (Dr.) Pulickal M. Ajayan, the co-inventor of carbon nanotubes, and was fortunate to work with 2019 Chemistry Nobel Prize laureate, Prof. (Dr.) John B. Goodenough. Dr. Prasanth was selected for a Brain Korea Fellowship (2007), SAGE Research Foundation Fellowship, Brazil (2009), Estonian Science Foundation Fellowship, European Science Foundation Fellowship (2010), Faculty Recharge, University Grants Commission (UGC), Ministry of Higher Education, India (2015), etc. Dr. Prasanth has received many international awards, including Young Scientist award, Korean Electrochemical Society (2009), and was selected for the Bharat Vikas Yuva Ratna Award (2016) etc. He developed many products such as a high-performance breaking parachute, flex wheels for space shuttles, high-performance lithium-ion batteries for leading companies for portable electronic devices and electric automobiles etc. Dr. Prasanth has a general research interest in polymer synthesis and processing, nanomaterials, green/nanocomposites, and electrospinning. His current research focuses on nanoscale materials and polymer composites for printed and lightweight charge storage solutions, including high-temperature supercapacitors and batteries. Dr. Prasanth has published many research papers in high-impact factor journals, and a number of books/book chapters, and has more than 5000 citations and an h-index of 45 plus. Apart from science and technology, Dr. Prasanth is a poet, activist, and a columnist in online portals and printed media.

Jabeen Fatima M. J., PhD, is a research scientist at the Materials Science and NanoEngineering Lab (MSNE Lab), Department of Polymer Science and Rubber Technology (PSRT), Cochin University of Science and Technology (CUSAT), Kerala, India. Before joining the MSNE Lab, she worked as a Guest Assistant Professor at the Department of NanoScience and Technology, University of Calicut, India. Dr. Fatima earned her PhD in Nanoscience and Technology from the University of Calicut, India, in 2016, with the support of a prestigious National Fellowship CSIR JRF/SRF from the Council of Scientific and Industrial Research (CSIR), under the Ministry of Science and Technology, Government of India. Her research area was focused on the synthesis of nanostructures for photoelectrodes for photovoltaic applications, energy storage devices, photoelectrochemical water splitting, catalysis etc. She earned her MS degree in Applied Chemistry (University First Rank) after earning her BSc degree in Chemistry from Mahatma Gandhi University (MGU), Kottayam, India. Dr. Fatima has received many prestigious fellowships, including Junior/Senior Research fellowships (JRF/SRF) from the Centre for Science and Research, Department of Science and Technology, Ministry of India, a post-doctoral/research scientist fellowship from Kerala State Council for Science, Technology and Environment (KSCSTE), an InSc Research Excellence award, etc. Dr. Fatima has published many full-length research articles in peer-reviewed international journals, as well as chapters in books with international publishers. She serves as a reviewer for many STM journals published by Wiley International, Elsevier, Springer Nature etc. Dr. Fatima's current research areas of interest include the development of flexible and free-standing electrodes for printable and stretchable energy storage solutions and the development of novel nanostructured materials, ternary composite electrodes and electrolytes for sustainable energy applications, such as supercapacitors, fuel cells and lithium-ion batteries.

Contributors

Abhijith P. P.
Department of Nanoscience and Technology
University of Calicut
Kerala, India

and

Department of Polymer Science and Rubber Technology (PSRT)
Cochin University of Science and Technology (CUSAT)
Kerala, India

Jou-Hyeon Ahn
Department of Materials Engineering and Convergence Technology

and

Department of Chemical and Biological Engineering
Gyeongsang National University (GNU)
Jinju, Republic of Korea

Neethu T. M. Balakrishnan
Department of Polymer Science and Rubber Technology (PSRT)
Cochin University of Science and Technology (CUSAT)
Kerala, India

Jayesh Cherusseri
NanoScience Technology Center
University of Central Florida
Orlando, USA

Akhila Das
Department of Polymer Science and Rubber Technology (PSRT)
Cochin University of Science and Technology (CUSAT)
Kerala, India

Jabeen Fatima M. J.
Department of Polymer Science and Rubber Technology (PSRT)
Cochin University of Science and Technology (CUSAT)
Kerala, India

S. Jayanthi
Department of Physics
The Standard Fireworks Rajaratnam College for Women
Tamilnadu, India

Jishnu N. S.
Rubber Technology Centre
Indian Institute of Technology (IIT-KGP)
West Bengal, India

and

Leibniz Institute of Polymer Research Dresden e. V.
Dresden, Germany

and

Department of Polymer Science and Rubber Technology (PSRT)
Cochin University of Science and Technology (CUSAT)
Kerala, India

Jarin D. Joyner
Department of Materials Science and Nano Engineering

and

Department of Chemistry
Rice University
Houston, Texas, USA

Krishnan M. A.
Department of Polymer Science and
　Rubber Technology (PSRT)
Cochin University of Science and
　Technology (CUSAT)
Kerala, India

and

Department of Mechanical Engineering
Amrita Vishwa Vidyapeetham
Kerala, India

and

Department of Electrical Engineering
Pennsylvania State University
Pennsylvania, USA

Nikhil Medhavi
Department of Polymer Science and
　Rubber Technology (PSRT)
Cochin University of Science and
　Technology (CUSAT)
Kerala, India

Prasanth Raghavan
Department of Polymer Science and
　Rubber Technology (PSRT)
Cochin University of Science and
　Technology (CUSAT)
Kerala, India

and

Department of Materials Engineering
　and Convergence Technology
Gyeongsang National University (GNU)
Jinju, Republic of Korea

and

Department of Materials Science and
　Nano Engineering
Houston, Texas, USA

Prasad V. Sarma
Department of Physics
Sanatana Dharma College
Alappuzha, Kerala, India

Anjumole P. Thomas
Department of Polymer Science and
　Rubber Technology (PSRT)
Cochin University of Science and
　Technology (CUSAT)
Kerala, India

M. Ulaganathan
CSIR-Central Electrochemical
　Research Institute (CECRI)
Tamil Nadu, India

Sreekanth J. Varma
Department of Physics
Sanatana Dharma College
Alappuzha, Kerala, India

Abbreviations

([BMIM]BF$_4$)	1-butyl-3-methylimidazolium tetrafluoroborate
0D	Zero-dimensional
1D	One-dimensional
2D	Two-dimensional
AFC	Alkaline fuel cells
AFD	Average fiber diameter
Al	Aluminum
Al$_2$O$_3$	Alumina
Al-BTC	Aluminum benzenetricarboxylate
AN	Acrylonitrile
ATRP	Atom transfer radical polymerization
BDC	1,4-benzenedicarboxylate
BEV	Full-battery electric vehicles
BMIMBF$_4$	1-butyl-3-methylimidazolium tetrafluoroborate
BMITFSI	1-butyl-3-methylimidazolium bis(trifluoromethanesulfonyl)imide
BPEG	Triboron-based PEG
CA	Cellulose acetate
CMS	Chloromethylstyrene
CN	Cyanide
CNT	Carbon nanotube
CO$_2$	Carbon dioxide
CPEs	Composite polymer electrolytes
CS	Corn starch
DD	Deacetylation
DDS	Dimethyldichlorosilane
DMAc	Dimethylacetamide
DMC	Dimethyl carbonate
DMF	Dimethyl fluoride
DMFC	Direct methanol fuel cell
DMP	Dimethyl phthalate
DMSO	Dimethyl sulfoxide
DSC	Differential scanning calorimetry
DVIMBr	1,4-di(vinylimidazolium)butane bis bromide
EC	Ethyl carbonate
EDLC	Electrical Double Layer Capacitance
EES	Electrical energy storage
EMC	Ethyl methyl carbonate
EMF	Electro motive force
EO	Ethylene oxide
ePPO	Elastomer poly(propylene oxide)

et al.	Et alia
GO	Graphene oxide
GPE	Gel polymer electrolytes
HBPAGS	Hyper branched polyamine ester grafted nano-silica
HEMA	Hydroxyethyl methacrylate
HEV	Hybrid electric vehicle
HFP	Hexafluoropropylene
HMPP	2-hydroxy-2-methylpropiophenone
HNT	Halloysite nanotubes
ILGE	Liquid-based gel polymer electrolyte
IR	Internal resistance
KI	Potassium iodide
LAGP	Lithium aluminum germanium phosphate
LATP	Lithium aluminum titanium phosphate
LCO	Lithium cobalt oxide
LED	Light emitting diode
LFP	Lithium iron phosphate
Li	Lithium
LiBOB	Lithium bis(oxalato)borate
LIBs	Lithium-ion batteries
LiClO$_4$	Lithium perchlorate
LiDFOB	Lithium difluoro(oxalate)borate
LiFePO$_4$	Lithium iron phosphate
Li-FeS$_2$	Lithium iron disulfide
Li$^+$-ion	Lithium-ion
Li-M	Lithium manganese dioxide
LiRAP	Li-rich anti-perovskites
Li-S	Lithium sulfur
LiSnZr(PO$_4$)	Li-tin-zirconium phosphate
LiTFSI	Lithium bis(trifluoromethane sulfonyl)imide
LLTO	Lithium lanthanum titanate
LLZTO	Lithium lanthanum zirconium thallium oxide
LMO	Lithium manganese oxide
LTC	Lithium-thionyl chloride
MA	Methyl acetate
MC	Methylcellulose
MCFC	Molten carbonate fuel cell
MFC	Microbial fuel cell
MH	Metal hydride
MIBs	Magnesium-ion batteries
MIL-53	1,4-benzenedicarboxylate
MIL-53(Al)	Al, 1,4-benzenedicarboxylate
MMA	Methyl methacrylate
MOF	Metal-organic framework
MPE	Microporous polymer electrolytes

NASICON	Sodium superionic conductor
NiCd	Nickel-cadmium
NiMH	Nickel–metallic hydride
NiO	Nickel oxide
NiOOH	Nickel oxide hydroxide
NMP	*N*-methyl pyrrolidone
NPs	Nanoparticles
NWs	Nanowires
OLEs	Organic liquid electrolyte
PAA	Polyacrylic acid
PAFC	Phosphoric acid fuel cell
PAN	Polyacrylonitrile
PC	Propylene carbonate
PCL	Polycaprolactone
PDADMA	Poly(diallyldimethylammonium)
PDAD-MATFSI	Poly[diallyldimethylammonium] bis-trifluoromethane sulfonimide
PDMS	Polydimethyl siloxane
PE	Polymer electrolytes
PEG	Polyethylene glycol
PEGDA	M-UiO-66-NH-PEG diacrylate
PEMFC	Polymeric electrolyte membrane fuel cells
PEO	Polyethylene oxide
PET/PETE	Polyethylene terephthalate
PFPE	Perfluoropolyether
PFPE-diol	Hydroxy-terminated perfluoropolyether
PHEA-*co*-AN	Polyhydroxy ethyl acrylate-*co*-acrylonitrile
PHEMO	(poly(3-{2-[2-(2-hydroxyethoxy) ethoxy] ethoxy} methyl-3′-methyloxetane))
PIL	Polymer ionic liquid
PILGEs	Polymer ionic liquid gel polymer electrolytes
PLL	Polyethylene oxide-lithium bis(trifluoromethane sulfonyl) imide-lithium lanthanum zirconium thallium oxide
PMMA	Polymethyl-methacrylate
PNSE	Polymer nanocomposite-based solid-state electrolyte
POEM	Poly (oligo-oxyethylene) methacrylate
PPC	Poly(propylene carbonate)
PPCl	1-methyl-1-propyl piperidinium chloride
PPME	Poly(ethylene glycol) methyl ether methacrylate
PPTA	Poly(*para*-phenylene terephthalamide) (PPTA)
PrTrif	Praseodymium (III) trifluoromethanesulfonate
PS	Polystyrene
PS	Potato starch
PVA	Polyvinyl alcohol
PVAc	Polyvinyl acetate

PVC	Polyvinyl chloride
PVdF	Polyvinylidene difluoride
PVdF-*co*-CTFE	Polyvinylidene fluoride-*co*-chlorotrifluoroethylene
PVdF-*co*-HFP	Polyvinylidene fluoride-*co*-hexafluoro propylene
PVdF-*co*-HFP-g-PPEGMA	Polyvinylidene fluoride-*co*-hexafluoro propylene-grafted-poly(polyethylene glycol) methyl ether methacrylate)
PVP	Polyvinylpyrrolidone
PYR14TFSI	*N*-methyl *N*-butyl pyrrolidinium bis(trifluoromethanesulfonyl) imide
RAM	Random-access memory
RFID	Radio-frequency identification
RS	Rice starch
RTIL	Room-temperature ionic liquid
SBS	Styrene-butadiene-styrene
SEM	Scanning electron microscopy
SEs	Solid electrolytes
SiO$_2$	Silicon dioxide
SN	Succinonitrile
SPE	Solid-state polymer electrolyte
SS	Sago starch
Ta	Tantalum
TEM	Transmission electron microscope
TEOS	Tetra ethoxy silane
T$_g$	Glass transition temperature
TGTG'	*Trans-gauche-trans-gauche'*
THF	Tetrahydro furan
TiO$_2$	Titanium dioxide
TPFPB	Tris(penta-fluorophenyl)borane
TPU	Thermoplastic urethane
UV	Ultraviolet
VDF	Vinyl difluoride
vs.	*Versus*
wt.%	Weight percentage
XRD	X-ray diffraction

SYMBOLS

α	Alpha
A	Area of the electrode
β	Beta
E$_{cell}$	Cell terminal voltage
I	Current density
δ	Delta
S	Deterioration rate

Abbreviations

F	Faraday constant
γ	Gamma
$\geq$	Greater than or equal to
:	Is to
$\leq$	Less than or equal to
n	Number of electrons transferred
Ω	Ohm
I_p	Peak current density
ΔE_p	Peak to peak separation voltage
%	Percentage
I_F/I_R	Ratio of forward-to-backward current
J_f/J_b	Ratio of forward-to-backward current density
®	Registered trademark
ν	Scan rate
$E°_{anode}$	Standard anode potential
$E°_{cathode}$	Standard cathode potential
Γ^*	Surface concentration
T	Temperature
t	Time
R	Universal gas constant

UNITS

$A\ g^{-1}$	Ampere per gram
cm	Centimeter
°C	Degree celsius/Degree centigrade
$F\ g^{-1}$	Farad per gram
$F\ cm^{-3}$	Faraday per centimetre cubed
GHz	Gigahertz
GPa	Gigapascal
g	Gram
h	Hour
$ions\ cm^{-2}$	Ions per centimetre squared
K	Kelvin
keV	Kilo electron volts
kV	Kilo Volt
$KW\ kg^{-1}$	Kilowatt per kilogram
$kWh\ L^{-1}$	Kilowatt hour per litre
MHz	Megahertz
MPa	Megapascal
$m^2\ g^{-1}$	Meter squared per gram
$\mu Wh\ cm^{-2}$	Micro watt hour per centimetre squared
μm	Micrometer
$mA\ mg^{-1}$	Milliampere per milligram
$mAh\ g^{-1}$	Milliampere hour per gram

mW cm^{-2}	Milliwatt per centimetre squared
mF cm^{-2}	Millifarad per centimetre squared
mS cm^{-1}	Millisiemens per centimetre
mA	Milliampere
mA h	Milliampere hour
mV	Millivolt
mV s^{-1}	Millivolt per second
M	Molar
nm	Nanometer
%	Percentage
S	Siemens
S cm^{-1}	Siemens per centimetre
V	Voltage

1 Electrochemical Energy Storage Systems
The State-of-the-Art Energy Technologies

Abhijith P. P., Jishnu N. S., Neethu T. M. Balakrishnan, Akhila Das, Jou-Hyeon Ahn, Jabeen Fatima M. J., and Prasanth Raghavan

CONTENTS

1.1 Introduction 2
1.2 Types of Electrochemical Energy Storage Devices 2
1.3 Batteries and Their Classification 4
    1.3.1 Primary (Non-Rechargeable) Batteries 5
        1.3.1.1 Types of Primary Batteries 6
    1.3.2 Secondary (Rechargeable) Batteries 10
        1.3.2.1 Lead-Acid Batteries 10
        1.3.2.2 Nickel-Cadmium Batteries 10
        1.3.2.3 Ni-Metal Hydride Batteries 11
        1.3.2.4 Lithium-Ion Batteries 12
        1.3.2.5 Magnesium-Ion Batteries 12
        1.3.2.6 Fluoride-Ion Batteries 12
        1.3.2.7 Sodium-Ion Batteries 13
        1.3.2.8 Ion-Ion Batteries 13
1.4 Principles and Types of Lithium-Ion Batteries 13
    1.4.1 Lithium Iodide Battery 15
    1.4.2 Lithium Air Battery 16
    1.4.3 Lithium Redox Flow Battery 17
    1.4.4 Lithium Sulfur Battery 18
1.5 Supercapacitors 19
1.6 Fuel Cells 21
    1.6.1 Basic Structure of Fuel Cell 22
    1.6.2 Classification of Fuel Cells 23
        1.6.2.1 Polymeric Electrolyte Membrane Fuel Cells (PEMFC) 23
        1.6.2.2 Direct Methanol Fuel Cells (DMFCs) 25
        1.6.2.3 Alkaline Fuel Cells (AFCs) 26

          1.6.2.4   Phosphoric Acid Fuel Cell (PAFC)..27
          1.6.2.5   Molten Carbonate Fuel Cells (MCFC) 27
          1.6.2.6   Microbial Fuel Cells (MFC) 28
1.7   Conclusion 29
Acknowledgment 29
References..29

1.1 INTRODUCTION

Environmental challenges emanating from the use of fossil fuels (petroleum, natural gas, and coal) [1], harmful emissions of greenhouse gases (which adversely affect human health as well as contributing to global warming), dependence of industrial nations on oil that led to the oil crisis and accelerated depletion of fossil fuels over a few decades, basically due to high demand and, in some cases, extravagant consumption, have forced mankind to move away from using fossil fuels as the main global energy source. It is widely reported that global energy consumption is expected to increase by 28% by 2040 [2]. Hence, future energy demands cannot be satisfied by current energy technologies. Over the past decade, the whole world has been stepping toward the use of automation and on-road electric/hybrid vehicles having zero emissions, and increasing demands from the portable power market and the telecommunication industry have seen moves toward more advanced and environmentally friendly energy technologies, rather than fossil fuels. This ever-increasing demand for energy with time and growing population, has forced mankind to depend increasingly on alternative energy sources, where energy from renewable resources, such as solar, hydroelectric, thermal, wind, and tidal energy, will eventually replace traditional energy sources; however, most of these renewable energy sources are typically periodic or intermittent in nature. Hence, the efficient storage and effective release of energy from these renewable sources need energy storage devices; this increasing demand has stimulated the development of various electrochemical energy storage devices, such as supercapacitors, batteries, and fuel cells [3]. The plot of specific power against specific energy (Ragone plot) for various electrochemical energy storage devices is shown in Figure 1.1. [4].

1.2 TYPES OF ELECTROCHEMICAL ENERGY STORAGE DEVICES

The basic principles of electrochemical energy conversion were discovered by Alessandro Volta in the late 1700s [5]. Electrochemical energy storage devices play a vital role in the development of robust and sustainable energy technologies. In this mature technology, the electrical energy is stored in the form of chemical energy. This novel storage technique benefits from the fact that both electrical and chemical energy share the same carrier, the electron. This common point allows losses to be limited due to the energy conversion from one form to another. Based on the electrochemical reaction or chemistry behind the working principle, the electrochemical energy storage devices are broadly classified as batteries, supercapacitors, or fuel cells (Figure 1.2).

Electrochemical Energy Storage Systems

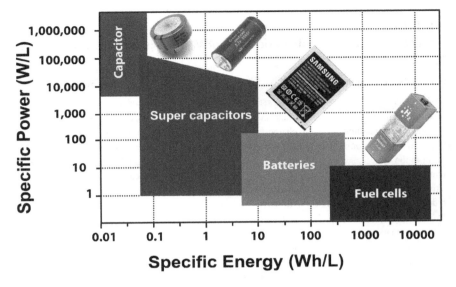

FIGURE 1.1 Ragone plot showing the specific power relationship with specific energy for various electrochemical energy conversion systems. Adapted and reproduced with permission from Ref. [3, 4]. Copyright © 2004 American Chemical Society.

FIGURE 1.2 Different types of electrochemical energy storage devices.

All three systems show "electrochemical similarities" even though energy storage and conservation mechanisms are different. Structurally, all these electrochemical devices have three common essential components: an anode and a cathode, separated by a membrane (porous membrane or ion-exchange membrane), and an electrolyte. The characteristics and working principles of the various energy storage solutions

TABLE 1.1
Characteristics and Working Principles of Various Energy Storage Devices [3, 5–12]

Batteries/Supercapacitors	Fuel Cells	Supercapacitors
Closed systems	Open system	Closed system
Anode and cathode act as charge-transfer medium	Anode and cathode act as charge-transfer medium	Anode and cathode may not act as charge-transfer medium
Energy storage and conversion occur in the same compartment	Energy storage and energy conversion are locally separated	Formation and release of electrolyte ions take place at electrolyte/electrolyte interface
Active masses undergoing redox reaction are delivered from anode and cathode	Active masses undergoing redox reaction are delivered from outside the cell	Electrolyte ions orient at electrolyte/electrolyte interface (EDLC)
Established market position	Niche markets	Development stage
Mobile phones, laptops automobiles	Memory protection in several electronic devices, transportation	Space and automobile applications

are listed in Table 1.1. The generation of electrical energy is by conversion of chemical energy *via* redox reactions at the anode and cathode in the case of both fuel cells and batteries. The terms "negative" and "positive" electrode are used, as reactions at the anode usually take place at lower electrode potentials than at the cathode [3]. A common feature is the process of energy provision taking place at the phase boundary of the electrode/electrolyte interface. Transport of electrons and ions involve two electrodes in contact with an electrolyte solution [3] (Figure 1.3).

1.3 BATTERIES AND THEIR CLASSIFICATION

A battery is a collective arrangement of electrochemical cells in which energy can typically be stored electrochemically *via* conversion of chemical energy into electrical energy, and *vice versa*, taking place between two electrodes (anode and cathode) and the electrolyte by means of an electrochemical redox reaction [13], acting as a source to power an electronic device, electrical device or an electric vehicle [14]. The first battery was developed by Volta in the year 1800. After more than 200 years of development, battery technology has reached a point where batteries can be made in any size, ranging from macro- to nanosized, in shapes ranging from cylindrical to prismatic or even paper batteries, and with fabrication techniques from roll-to-roll printing [15] to paintable batteries [16] for a wide spectrum of applications. Depending on the number of usage cycles, batteries are broadly classified as disposable/non-rechargeable (primary) or rechargeable (secondary) batteries.

Electrochemical Energy Storage Systems

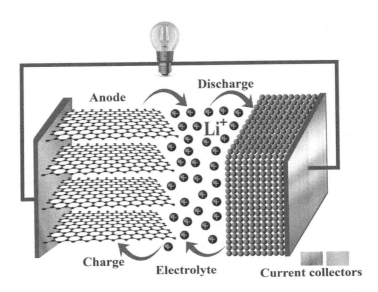

FIGURE 1.3 Schematic illustration on the structure and operating principles of lithium-ion batteries, including the movement of ions between electrodes during charge (forward arrow) and discharge (backward arrow) states.

1.3.1 Primary (Non-Rechargeable) Batteries

Primary or non-rechargeable batteries, commonly referred to as dry cells, are basically electrochemical devices that are discarded once used and cannot be recharged with electricity. The electrochemical reaction occurring in the cell is not reversible, rendering the cell non-rechargeable. Upon the use of a primary cell, electrochemical chemical reactions in the battery consume the chemicals and generate the power until these chemicals are completely consumed. They have the advantages of low unit cost and ease of operation, with the disadvantage of costing extra over the long term. Generally, these batteries have a better capacity and preliminary voltage than rechargeable batteries, and a sloping discharge curve. Primary batteries make up about 90% of the $50 billion battery market share. Globally, 15 billion batteries are thrown away each year, virtually all ending up in landfills. Because of the toxic metals and strong acids which they contain, primary batteries represent hazardous waste. The main advantages of primary batteries include:

- High power density because no design compromises are needed to allow for recharging.
- Best for low-drain applications, which include watches or hearing aids.
- Meet the apparent demand for single-use programs, along with guided missiles and army ordnance.
- Low initial price.
- Convenient.
- Wide availability of products with different characteristics.

These kinds of batteries are not suitable for excessive drain applications, because of short operating time and the cost of continuous maintenance. For usual energy performances, one-time use, disposable, primary batteries are an extremely uneconomical power source, seeing that they produce at best approximately 2% of the energy used in their manufacture. Owing to their high pollutant content, compared to their low energy density, primary batteries are considered to be the product of a wasteful, environmentally unfriendly technology, since they produce much more waste than rechargeable batteries. All primary batteries produce a small amount of hydrogen gas on discharge, and battery-powered devices must make provision for venting. Pressure buildup in the cell can rupture the seal and cause corrosion. This is visible in the form of a feathery crystalline structure that can develop and spread to neighboring parts in the device and cause damage.

1.3.1.1 Types of Primary Batteries

The primary battery or dry cell is a modified version of the Leclanché cell and was successfully introduced in 1888 by Gassner. The first dry cell was a zinc-carbon or Leclanché battery, which was one of the earliest and least expensive primary batteries, and they deliver an output voltage of 1.5 V. The first zinc-carbon battery was invented by Georges Leclanché in 1859, but it was a wet battery, in which the electrolyte used was in liquid form. However, the dry cell is not really dry, hence the term "dry cell" is a misnomer because a completely dry battery will not function under ordinary temperatures. Hence, the term "dry" only conveys that the constituent materials of the battery are non-spillable or flowable in any battery position even under pressure or in motion, which is achieved by employing the electrolyte in the form of a gel or as some absorbent material to achieve non-spillability.

Primary dry cells are distinguished from sealed non-spillable rechargeable or secondary batteries. Some of the rechargeable batteries remain dry until they are activated for use or are all-solid secondary batteries, which are assembled with a solid-state electrolyte (ceramic electrolyte or solid polymer electrolyte); however, they function only at high temperatures (typically above 60°C). Table 1.2 lists the potential anode and cathode materials for primary batteries and their typical electrochemical properties. Based on the chemicals packed or the anode/cathode combination in the cell for the electrochemical reaction, primary batteries are classified as follows.

1.3.1.1.1 Alkaline-Manganese Batteries

The alkaline-manganese battery, commonly known as an alkaline battery, which is an improved version of the zinc-carbon battery or Leclanché battery, was invented in 1949 by Lewis Urry, while working with the Eveready Battery Company laboratory in Ohio, USA. Alkaline batteries deliver an output voltage of 1.5 V, delivering more energy at higher load currents than zinc-carbon batteries, as well as having little self-discharge. Alkaline batteries do not leak electrolyte when depleted, as the old zinc-carbon ones do, but it is not totally leak-proof. Furthermore, a regular household alkaline battery provides about 40% more energy than the average Li^+-ion battery, but an alkaline battery is not as strong as a Li^+-ion battery on loading.

TABLE 1.2
Potential of Anode and Cathode Materials for Primary Batteries and Their Typical Electrochemical Properties

Electrode type	Cathode Material	Open Circuit Potential Developed (V) Acid	Open Circuit Potential Developed (V) Alkali	Weight Capacity (mAh g^{-1})	Volume Capacity (Ah mL^{-1})
Anode	MnO$_2$	+0.80	+0.29	307	1.54
	PbO$_2$	+1.46	+0.25	248	2.26
	HgO	+0.85	+0.10	248	2.76
	CuO	+0.34	−0.36	670	4.32
	Ag$_2$O	+0.80	+0.35	232	1.67
	Ag$_2$O$_3$	+1.90	+0.57	432	1.92
	AgCl	+0.22	+0.22	187	1.64
Cathode	Mg	−2.35	−2.69	2204	3.84
	Al	−1.87	−2.35	2982	8.05
	Zn	−0.76	−1.25	820	5.68
	Fe	−0.44	−0.88	960	7.55
	Cd	−0.40	−0.81	477	4.13
	Sn	−0.14	−0.91	452	3.30
	Pb	−0.13	−0.54	259	2.90

1.3.1.1.2 Silver Oxide Batteries

Silver oxide batteries exist in two forms: one has a cathode of monovalent silver oxide (Ag$_2$O), whereas the other form uses a divalent silver oxide (AgO) cathode [1]. The second form has a better theoretical potential because there is further chemical reduction from AgO to Ag$_2$O. The surface may be handled to reduce AgO conversion to Ag$_2$O. In order to gain voltage stability, a "dual oxide" machine can be adopted. Higher raw material costs suggest that silver oxide cells are more expensive than their mercury equivalents.

The silver oxide battery consists of a depolarizing silver oxide cathode, a zinc anode of high surface area, and an alkaline electrolyte. The electrolyte is potassium hydroxide in hearing-aid batteries, to obtain the highest energy density. The electrolyte in watch batteries may be either sodium hydroxide or potassium hydroxide. The sodium hydroxide electrolyte, which has a lower conductivity than the potassium hydroxide electrolyte, is frequently used, as it has a lower tendency to "creep" on the seal [16]. The negative terminal of the cell consists of a plated steel cell top. The zinc anode is an amalgamated zinc powder and the cathode is a packed pellet of silver oxide, plus graphite for conductivity. A permeable cushion of a non-woven regular material holds the alkaline electrolyte, which is an unequivocally basic potassium hydroxide arrangement, and the separator is a synthetic ion-permeable membrane. A fixing grommet seals the cell and protects the positive and negative terminals. The plated steel can act as a cell holder and as the positive terminal of the cell.

The internal surface of the cell is of a metal, which is electrochemically compatible with zinc, to limit inefficient erosion and destructive gas development, and the cell can is nickel-plated steel, which is exceptionally impervious to the electrolyte.

1.3.1.1.3 Mercury Batteries

Mercury batteries have a substantially better energy-to-weight ratio than do carbon-zinc batteries, on account of the very high power density of the substances used in their production. Thus, mercury batteries are one-third the size of traditional dry batteries of the same potential. The anode is fashioned from cylinders or pellets of powdered high-purity amalgamated zinc, or a gelled mixture of electrolyte and zinc. The depolarizing cathode is compressed mercuric oxide-manganese dioxide in a sleeve or pellet form (cell voltage 1.4 V) or pure mercuric oxide (mobile voltage 1.35 V), whereas the electrolyte, which does not participate in the reaction, is concentrated aqueous sodium or potassium hydroxide. The cathode is separated from the anode with the aid of an ion-permeable barrier. In operation, this aggregate produces metal mercury, which does not inhibit the current drift inside the cell. Zinc is thermodynamically unstable with respect to water and, in strongly alkaline solutions, will tend to self-discharge, with the evolution of hydrogen.

1.3.1.1.4 Lithium Primary Batteries

Lithium battery technology provides greater energy density, greater energy per volume, longer cycle life, and improved reliability. Lithium is a lightweight metal with great electrochemical potential and provides the highest specific energy per unit weight. Lithium batteries have a wide range of applications in various portable electronic devices, such as laptops, computers, mobile phones, telecommunication devices, etc. Lithium-metal batteries are non-rechargeable and are collectively known as lithium primary batteries. For a lithium battery, the anode material used is metallic lithium and the cathode material is manganese dioxide, thionyl chloride ($SOCl_2$), or iron disulfide (FeS_2), among others, with the electrolyte being a salt of lithium dissolved in an organic solvent, usually composed of propylene carbonate and a low-viscosity solvent such as dimethoxyethane.

1.3.1.1.4.1 Lithium Iron Disulfide (Li-FeS$_2$) Batteries Lithium-metal batteries can replace ordinary alkaline batteries in many portable devices, such as calculators, pacemakers, cameras etc. The lithium iron disulfide (Li-FeS$_2$) battery is the most-recently developed battery to be added to the primary battery family, and offers an operating voltage of 1.5 V, greater stability, higher operating voltage, and improved overall performance, compared with an alkaline battery. Compared with alkaline batteries, the lithium iron disulfide battery has higher discharge capacity, higher energy density, lower internal resistance (IR), and weighs less. The lithium iron disulfide battery also has certain other advantages, like good performance at both high and low temperatures, superior leak resistance, and low self-discharge (1% per year), allowing 15 years of storage at ambient temperatures. Also, they are friendly to the environment (no added mercury, cadmium, or lead). The major disadvantages of the Li-FeS$_2$ technology are a higher price and transportation issues, due to the lithium metal content in the anode.

1.3.1.1.4.2 Lithium-Thionyl Chloride (LiSOCl₂ or LTC) Batteries Lithium-thionyl chloride (LiSOCl$_2$ or LTC) is one of the most powerful and potent battery chemistries. They are the most rugged of the lithium-metal or lithium-ion batteries, and are capable of withstanding high temperatures and strong vibrations, so that they can be used to power horizontal drilling, also known as fracking. Due to the safety issues, LTC is not used in consumer devices and is used only by trained workers. Some LTCs could safely operate and perform well over a wide range of temperatures, from 0 to 200°C (32–392°F). Compared with lithium-ion technology, LTC offers twice the capacity with a specific energy of over 500 Wh kg^{-1}, and the cells show a nominal voltage of 3.60 V, with the end-of-discharge cutoff voltage of 3.00 V. However, the runtime of the LTC is based not only on capacity but also on thermal conditions and load pattern. Constant current is more enduring than pulsed load, a phenomenon that applies to most of the battery technologies. Similar to the alkaline battery technology, LTC has a relatively high internal resistance and can be used only for moderate discharge loads. However, LTC has low self-discharge and long shelf life due to the formation of a passivation layer between the lithium anode and the carbon-based cathode, which could dissipate when applying a load. The extended temperature range of LiSOCl$_2$ batteries makes it ideal for use in radio-frequency identification (RFID) asset-tracking tags that monitor the location and status of medical equipment throughout a hospital, nursing home, or medical research facility. Fitted with this battery, the instrument can undergo high temperature autoclave sterilization cycles of up to 125°C without having to remove the battery from the RFID tracking device, providing an uninterrupted data stream. These robust batteries can also be modified to operate in the medical cold chain, where wireless sensors must endure temperatures as low as −80°C to continually monitor the safe transport of tissue samples, transplant organs, or pharmaceuticals.

1.3.1.1.4.3 Lithium Manganese Dioxide (LiMnO₂ or Li-M) The lithium manganese dioxide (LiMnO$_2$ or Li-M) battery is similar to the LTC battery, comprising metallic lithium (metal powder or foil) as the anode and solid MnO$_2$ as the cathode, but it has a lower specific capacity and is safe for public use. Upon operation, the Li-M battery delivers a discharge voltage of 3.0–3.3 V, with a specific energy of about 280 Wh kg^{-1}. Compared to the LTC battery, the Li-M battery is economically priced, has a long life, allows moderate loads, and can deliver high pulse currents. Unlike the LTC battery, the Li-M battery can operate at much lower temperature ranges, from −20°C to 60°C (−22°F to 140°F). It has low self-discharge (less than 1% per year at room temperature), superior shelf life compared with the LTC battery, no memory effect, superior operational life, high capacity for a longer running time, no need of expensive safety electronics and is friendly to the environment (no added mercury, cadmium, or lead).

1.3.1.1.4.4 Lithium Sulfur Dioxide The lithium sulfur dioxide (LiSO$_2$) battery is a primary battery with a voltage of 2.8 V and an energy density of up to 330 Wh kg^{-1}. It offers a wide operational temperature range of −54 to 71°C (−65 to 160°F), with a projected shelf life of nearly 10 years at room temperature. LiSO$_2$ is inexpensive to make and is commonly used by the military.

1.3.2 Secondary (Rechargeable) Batteries

A secondary battery is a type of electric battery which may be charged, discharged right into a load, and recharged repeatedly, instead of a disposable or primary battery, which is furnished fully charged and discarded after use. It consists of one or more electrochemical cells. The term "accumulator" is used because it accumulates and stores electricity through a reversible electrochemical response. Rechargeable batteries are produced in many specific sizes and styles, ranging from button cells to megawatt systems designed to stabilize an electrical distribution community. Several different mixtures of electrode materials and electrolytes are used, such as lead-acid, zinc-air, nickel-cadmium (NiCd), nickel-metallic hydride (NiMH), lithium-ion (Li^+-ion), and lithium-ion polymer (Li^+-ion polymer).

1.3.2.1 Lead-Acid Batteries

Lead-acid batteries are the initial type of rechargeable battery, using a well-understood technology invented by Gaston Plantae in 1859. These are one kind of galvanic cells that convert the chemical energy to electrical energy by using sponge lead at the anode and lead oxide at the cathode. Sulfuric acid serves as the electrolyte. The lead-acid battery has many advantages, including low price and high availability of lead, good reliability, high output cell voltage (2 V), high electrochemical efficiency, a cycle life from several hundred to thousands, and ease of recycling, and represents an economical solution to a broad range of applications. The very low energy-to-weight ratio and consequently the lower specific energy in the range of 30–50 Wh kg^{-1} are the major disadvantages. Also, they are heavy and their lifetime is reduced at high temperatures.

When the electrolyte splits into ions, the protons reach the PbO_2 plate, forming PbO from the hydrogen atom formed, whereas the sulfate ions form lead sulfate which will cause an imbalance in charges, leading to a current flow in the external circuit that is discharging, whereas reversing the electrodes will lead to the charging of the circuit [17]. The overall chemical reaction during charging (the reverse direction) and discharging (the forward direction) is given in Equation 1.1:

$$PbO_2 + Pb + 2H_2SO_4 \rightarrow 2PbSO_4 + 2H_2O \quad E° = +2.048 \text{ V} \quad (1.1)$$

The main applications of lead-acid batteries are in power stations and substations. Lead-acid batteries are low-cost batteries, which makes them attractive for use in motor vehicles to provide the high current required by starter motors. Large lead-acid batteries are also used to power the electric motors in diesel-electric watercraft and are also used as an emergency power source on nuclear submarines.

1.3.2.2 Nickel-Cadmium Batteries

Nickel-cadmium batteries are the main representative of batteries with a positive nickel electrode, that were conceptualized in the late 1800s. Other possible systems could be Ni-Fe, Ni-Zn, $Ni-H_2$, or Ni-MH. It is a type of rechargeable battery in which nickel hydroxide (positive electrode or cathode) and metallic cadmium (negative

Electrochemical Energy Storage Systems

electrode or anode) are used as electrodes, with potassium hydroxide solution as the electrolyte. Ni-Cd batteries can be made as portable sealed types, interchangeable with carbon-zinc dry cells, as well as large ventilated cells used for standby power and motive power, which means these batteries can be made in a wide range of sizes and capacities. In comparison with other types of rechargeable cells, they show good performance at a low temperature and at high discharge rates, so that these cells work better. In a fully discharged Ni-Cd battery, [Ni(OH)$_2$]–nickel hydroxide acts as the cathode and [Cd(OH)$_2$]–cadmium hydroxide as the anode [16]. The overall reaction is given in Equation 1.2. A typical Ni-Cd battery has a potential of 12 V and is used in emergency lighting as well as in stand-by power. When the battery discharges the process gets reversed.

$$2Ni(OH)_2 + Cd(OH)_2 \rightarrow 2NiOOH + Cd + 2H_2O \quad E^°_{Charge} = +1.3 \text{ V} \quad (1.2)$$

It is notable that the amount of water in the electrolyte falls during discharge. Ni-Cd batteries are designated as positive limited, utilizing the oxygen cycle. The oxygen which evolves at the positive electrode during charge diffuses to the negative electrode and reacts with cadmium to form Cd(OH)$_2$. In addition, CO$_2$ in the air can react with KOH-electrolyte to form K$_2$CO$_3$, and CdCO$_3$ can be formed on the negative plate. Both of these compounds increase the internal resistance and lower the capacity of the battery.

These batteries have a long shelf life, long cycle life, overcharge capability, high rate capability, almost constant discharge voltage, and high power delivery for a short duration in low- temperature applications. The disadvantages of this battery type include high self-discharge rate, especially at high temperatures, memory effect, complexity in charging, and not being economically viable for recycling. The cost of cadmium is several times higher than that of lead and the cost of the Ni-Cd cell construction is higher than that of the lead-acid battery. There is also a problem with the disposal of toxic cadmium. But the low maintenance costs and the good reliability of the Ni-Cd battery have made it an ideal candidate for a number of power supply applications, ranging from portable electronic gadgets to aircraft and space shuttle power systems. Depending on construction, Ni-Cd cells have energy densities in the range of 40–60 Wh kg^{-1} (50–150 Wh dm^{-3}) and cycle life values ranging from several hundred for sealed cells to several thousand for vented cells.

1.3.2.3 Ni-Metal Hydride Batteries

Ni-MH batteries are similar to Ni-Cd batteries, although a metal alloy is employed as an anode that can absorb and store large amounts of hydrogen during the discharge process and is Cd-free, which makes them a more environmentally friendly power source. The positive electrode is NiOOH (nickel oxide hydroxide) and concentrated KOH is used as the electrolyte. A hydrophilic polypropylene separator is used. The anode can absorb over one thousand times its own volume of hydrogen. The alloy usually consists of two metals, the first one absorbing hydrogen exothermically, the second one, endothermically. They serve as a catalyst for the dissociative adsorption of atomic hydrogen into the alloy lattice. The commonly used metals include Pd, V, Ti, Ni, Cr, Co, Sn or Fe. The overall reaction during discharge is given in Equation 1.3.

$$NiOOH + MH \rightarrow Ni(OH)_2 + M \quad (1.3)$$

These batteries are employed to power hybrid electric vehicles, portable electronic and electric devices, medical instruments, and a number of other applications, with long cycle life. The energy density is 25% higher than a Ni-Cd cell (80 Wh kg^{-1}), its power density is around 200 Wh kg^{-1}, it is less tolerant of overcharging than Ni-Cd cells, and self-discharge is high, up to 4–5% per day. This is caused especially by the hydrogen dissolved in the electrolyte, that reacts with the cathode. Also, metal hydride batteries are expensive, suffer from a memory effect, poor charge retention (high self-discharge) and low energy density, compared with lithium-ion based technologies.

1.3.2.4 Lithium-Ion Batteries
The lithium-ion battery is a type of rechargeable battery. Lithium-ion batteries are commonly used for portable electronics and electric vehicles and are growing in popularity for military and aerospace applications. These batteries have a high energy density and no memory effect. They are generally much lighter than other types of rechargeable batteries of the same size. The electrodes of a lithium-ion battery are made of lightweight lithium and carbon. Lithium is also a highly reactive element, meaning that a lot of energy can be stored in its atomic bonds. A typical lithium-ion battery can store 150 watt-hours of electricity in a 1-kg battery. The absence of a memory effect means that you do not have to completely discharge them before recharging, as with some other battery chemistries. Lithium-ion batteries can handle hundreds of charge/discharge cycles [8].

1.3.2.5 Magnesium-Ion Batteries
Magnesium-ion batteries (MIBs) utilize magnesium cations as the active charge transporting agent in solution and as the elemental anode of an electrochemical cell. MIBs are one of the most promising candidates, because of the ideal features of Mg to be used as the metal anode. Each magnesium atom releases two electrons during the battery discharge phase, when compared with lithium. The metal Mg exhibits a low reduction potential, a high theoretical volumetric capacity, and a high abundance in the Earth's crust, the latter characteristic reducing the battery's costs. The Mg anode material increases the energy density and enhances safety [18]. This gives the potential to deliver nearly twice the electrical energy of a lithium cell.

1.3.2.6 Fluoride-Ion Batteries
Fluoride-ion batteries are the next-generation electrochemical storage devices, which offer particularly high energy density. The higher energy density property helps it last eight times longer than other types of battery. The working of the fluoride-ion battery is slightly different from the others, in that it uses negative fluoride ions to generate the current. The reason that these batteries have such high energy density

Electrochemical Energy Storage Systems 13

is due to the molecular structure of metal fluorides. The ratio of fluorine atoms to metal atoms is high in a metal fluoride. They are also lighter than the lithium-ion batteries. These batteries are safer than lithium-ion batteries as they do not suffer from overheating issues [19]

1.3.2.7 Sodium-Ion Batteries

Sodium-ion batteries are a type of rechargeable battery using Na⁺ (sodium-ion) as the charge carrier. It has sodium-containing materials as the cathode, whereas the anode is made of hydrocarbons. The electrolyte is either an aqueous solution (such as Na_2SO_4 solution) or non-aqueous. During charging, Na⁺-ions extracted from the cathode migrate towards the anode whereas balancing electrons move from the cathode through the external circuit. During discharging, the process is reversed and Na⁺ ions travel back to the cathode. It has a voltage of 3.6 V [16].

1.3.2.8 Ion-Ion Batteries

This is the most promising and portable of the batteries available on the market. The anode used is lithium-intercalated graphite (LiC_6) and the cathode is lithium-intercalated lithium cobalt oxide ($LiCoO_2$). A liquid electrolyte of $LiPF_6$ and an alkyl carbonate mixture are used. During discharge, lithium ions from the anode go to the cathode and electrons are transferred in the circuit; when an electromotive force (EMF) is applied, lithium ions from the cathode are pushed back to the anode side until the lithium ions are saturated there [20]. It has around 3000 life cycles and is less toxic than other batteries. The main applications of ion-ion batteries are in automobiles.

1.4 PRINCIPLES AND TYPES OF LITHIUM-ION BATTERIES

The global lithium-ion battery (LIB) market is forecast to grow from USD 36.20 billion in 2018 to USD 109.72 billion by 2026, at a compound annual growth rate (CAGR) of 13.4%, during the forecast period. In recent years, lithium-ion batteries have increasingly been used as the power source for hybrid (HEV) and full-battery electric vehicles (BEV). Over the past couple of years, most of the sales of electric vehicles have been accounted for by China, the USA, and the European region, primarily as high-end electric vehicles. Roughly 1.6 million electric cars were on the roads in China in 2018, followed by 810,000 in the United States. By March 2018, BEV production and sales in China reached 27,673 and 24,127 units, respectively, rising by 88.35 and 69.21% year-on-year, respectively, and the corresponding figures for plug-in hybrid electric vehicles (PHEVs) were 11,210 and 11,171 units, respectively, rising 291.21 and 201.47% year-on-year, respectively [21]. The industry produced about 660 million units of cylindrical lithium-ion cells in 2012; the 18650 size is by far the most popular one for cylindrical cells. Tesla's Model S electric SUVs under 40,000 USD have a 85 kWh battery, using 7,104 lithium-ion cells. A 2014 study projected that the Model S alone would use almost 40 percent of the estimated global cylindrical battery production during 2014 [22]. Production of the cell was gradually shifted to the higher-capacity 3,000+ mAh cells [23].

Based on the user type, LIBs are categorized as primary LIBs or secondary or rechargeable LIBs. A primary battery is a one-direction galvanic device, designed to be used once and discarded when it is fully discharged, and not recharged with electricity and reused, like a secondary or rechargeable battery, i.e. the electrochemical reaction occurring in the cell is not reversible or it has only a discharging process. A lithium primary battery has metallic lithium as the anode. Hence this type of battery is also referred to as lithium-metal batteries. Lithium primary batteries presently represent the primary electrical energy storage (EES) systems, with a production of greater than 100 million cells/month and about 1500 tons/month of electrode materials. Lithium-manganese dioxide, lithium-iron disulfide, lithium thionyl chloride, and lithium iodine batteries are the most common lithium primary batteries. Among the various lithium primary batteries, lithium thionyl chloride batteries have the highest energy density of all lithium type cells, and have a service life of 15 to 20 years, whereas lithium iodine batteries provide excellent safety and long service life. In batteries, during discharging, reduction happens on the cathode, gaining electrons, and oxidation happens on the anode, which loses electrons, as per the electrochemical reaction shown in Equations 1.4, 1.5, and 1.6 [24].

$$\text{Cathode}: MS_2 + Li^+ + e^- \xrightarrow{\text{discharge}} LiMS_2 \tag{1.4}$$

$$\text{Anode}: Li^+ \xrightarrow{\text{discharge}} Li^+ + e^- \tag{1.5}$$

$$\text{Full cell}: Li + MS_2 \xrightarrow{\text{discharge}} LiMS_2 \tag{1.6}$$
$$(M = Ti \text{ or } Mo)$$

In contrast to lithium primary batteries, lithium secondary batteries, also referred to as lithium-ion batteries, are rechargeable batteries in which lithium ions move from the negative electrode to the positive electrode during discharge, with the opposite action happening during charging. Research on LIBs started in the early 1980s, and the principle of the current LIB was completed in 1985, then first commercialized in 1991 by Sony. Most of the technological developments to date have been directed toward the needs of portable electronics, but now the focus tends to be on the performance demands of medium- and large-scale applications. As shown in Figure 1.3, typically, lithium-ion batteries consist of three layers: (i) cathode or positive electrode which commonly consists of $LiCoO_2$ [25, 26], $LiNiO_2$ [27] or $LiMn_2O_4$ [28], etc.; (ii) anode or negative electrode, consisting of graphitic carbon [29], TiO_2 [30] or Fe_2O_3/Fe_3O_4 [31], etc.; and (iii) a separator-cum-electrolyte called a gel polymer electrolyte (GPE), which is permeable to the ions and the electrolyte (e.g., $LiPF_6$ in an organic solvent). GPEs are prepared by immobilization of organic liquid electrolytes, e.g., 1 M solution of $LiPF_6$, $LiClO_4$, or LiTFSI in polymer structures [32–34]. Polymers, such as polyethylene oxide (PEO) [35, 36], polyacrylonitrile (PAN) [37, 38], polyvinylidene difluoride (PVdF) [39, 40], and its co-polymer polyvinylidene fluoride-*co*-hexafluoro propylene (PVdF-*co*-HFP) [41, 42], and polymethyl-methacrylate (PMMA) [43, 44] are among the best-studied materials.

As the name implies, the working of a lithium-ion battery mainly relies on the repeated transfer of lithium ions between the anode and the cathode. The electrochemical properties of the electrodes are strongly influenced by the physical and chemical properties of the electrode active material, such as particle size, homogeneity, morphology, and surface area. Lithium-ion polymer batteries (LiPo battery) are by far the most common commercial secondary cell polymer battery, with the leading technology among other types of metal-ion polymer batteries. A LiPo battery is a rechargeable battery with lithium-ion technology, using a polymer electrolyte instead of a liquid electrolyte. LIBs are able to supply continuous energy due to the spontaneous oxidation-reduction reactions occurring at the electrodes. During the charging process ("de-lithiation"), Li$^+$-ions are extracted from the cathodic material by supplying energy from an external source. The extracted Li$^+$-ions diffuse through the electrolyte and enter the anodic material (according to the reaction $C_x + LiMO_2 \rightarrow Li_{1-y}MO_2 + C_xLi_y$) in the case of a traditional LIB [45, 46], while electrons are simultaneously transferred to the positive electrode through the external circuit. In the discharge process ("lithiation"), the opposite process takes place (i.e., $Li_{1-y}MO_2 + C_xLi_y \rightarrow C_x + LiMO_2$, in the considered example), i.e., Li$^+$-ions, extracted from the anodic material, are re-inserted into the cathodic material, and the cell provides energy. The oxidation and reduction process occurred at the two electrodes in the lithium rechargeable batteries as shown below [47, 48]:

$$\text{Cathode}: LiMn_2O_4 \rightarrow Li_{1+x}Mn_2O_4 + x Li^+ + x e^- \tag{1.7}$$

$$\text{Anode}: x Li^+ + x e^- + C_6 \rightarrow Li_xC_6 \tag{1.8}$$

$$\text{Full Cell}: LiMn_2O_4 + C_6 \rightarrow LiC_6 + LiMn_2O_4 \tag{1.9}$$

Each combination of the aforementioned materials and compounds will slightly influence cost, voltage, cycle durability, and other characteristics of the LiPo batteries. The secondary lithium-ion batteries, in general, operate at 3.7 V and demonstrate a capacity of 150 mAh g^{-1} [49].

1.4.1 LITHIUM IODIDE BATTERY

Lithium iodide batteries are the major energy storage for medical implants, such as pacemakers. These batteries are included in the primary energy storage device category, as they are impossible to recharge. The lithium iodine primary battery was introduced in 1972, by Moser [50], who patented the first solid-state energy storage device. Based on this solid-state battery, the first attempt at implanting a lithium iodide battery – a cardiac pacemaker – was achieved in the same year [51]. The anode of the battery is lithium metal, the cathode is iodine-poly-2-vinyl pyridine and, at the contact layer, lithium iodide is formed, which conducts lithium ions performing as the electrolyte [52]. Equations 1.10 and 1.11 represent the anodic and cathodic reactions taking place in the lithium iodide cell.

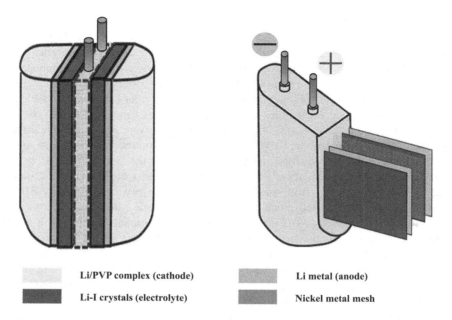

FIGURE 1.4 Schematic representation of the structure of lithium iodide battery used in pacemakers. Adapted and reproduced with permission from [54]. Copyright © 2015 Springer Nature.

$$\text{Anode} : \text{Li}_x \rightarrow x\text{Li}^+ + xe^- \quad (1.10)$$

$$\text{Cathode} : x\text{I}_2 + 2xe^- \rightarrow x\text{I}^- \quad (1.11)$$

A lithium iodide battery with enhanced performance was reported with iodine/iodide redox couples in an aqueous cathode along with an organic electrolyte to obtain a capacity of 200 mAh g^{-1} and a specific energy density of 0.33 kWh kg^{-1} [53]. The basic principle of the lithium iodide battery is the same within all examples, but the reagents and the phase of the system differs from battery to battery. A typical schematic image of lithium iodide battery used in pacemakers is shown in Figure 1.4 [54].

1.4.2 LITHIUM AIR BATTERY

The lithium air battery utilized the electrochemical reaction of lithium and oxygen, and has a high theoretical energy density of 3500 Wh kg^{-1}, which can substantiate the current demand for energy. A new type of energy storage device was first introduced in 1987 by Semkow and Sammells [55], in which they used a lithium alloy with the general formula Li$_x$FeSi$_2$, which has been immersed in a ternary molten salt (LiF, LiCl, and Li$_2$O), stabilized zirconia as the solid electrolyte, and La$_{0.89}$Sr$_{0.10}$MnO$_3$ as

Electrochemical Energy Storage Systems

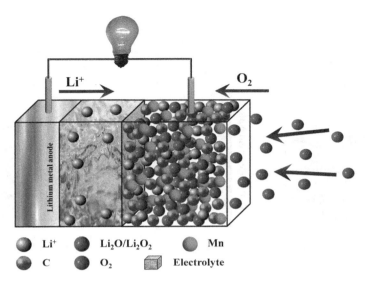

FIGURE 1.5 Schematic illustration of the structure and operating principles of typical lithium air batteries.

the oxygen electrode [55]. The overall reaction occurring in the typical cell is presented in Equation 1.12 [55].

$$\text{Li}_x\text{FeSi}_2 + \text{O}_2 \rightleftarrows \text{LiO}_2 + \text{FeSi}_2 \tag{1.12}$$

The major cathode materials reported for lithium oxygen batteries are activated carbon, carbon nanotubes, carbon nanofibers, graphene oxide, platinum oxide, palladium oxide, ruthenium oxide, copper, cobalt, metal alloys, transitional metal carbides, or chalcogenides. [56, 57]. The anodes mainly employed are based on lithium, such as lithium metal, alloys, etc. Lithium salt-based electrolytes in various organic solvents have been tried for lithium oxygen batteries [58]. A schematic of a lithium air battery is shown in Figure 1.5.

1.4.3 LITHIUM REDOX FLOW BATTERY

The lithium redox flow battery is a secondary battery, similar to other rechargeable batteries. Usual batteries function by storing chemical energy, whereas, in redox flow batteries, the electrodes are chemical components stored in separate reservoirs, which are circulated continuously. These circulating media are separated by an ion-selective selectively permeable membrane. Both reservoirs contain redox moieties, which are active materials dispersed in aqueous electrolyte (Figure 1.6) [59]. During the discharging process, the redox moiety present in the anolyte (anodic electrolyte) undergoes oxidation to form Li$^+$-ions, whereas, in the catholyte (cathodic electrolyte), the redox moiety undergoes reduction in the redox ion. In a typical charging

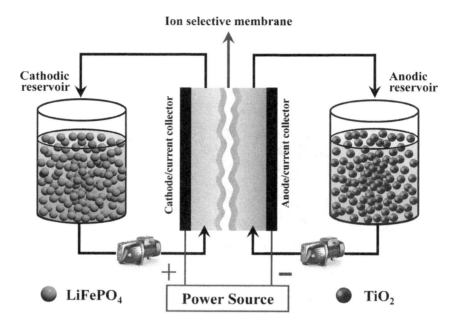

FIGURE 1.6 Schematic illustration of the structure and operating principles of redox flow batteries, including the circulation system for the electrode/electrolyte slurry.

process, the anolyte undergoes a reduction from lithium ion to lithium metal, whereas the catholyte undergoes an oxidation reaction. The charging reaction is given in Equations 1.13 and 1.14.

$$\text{Cathode}: RC^{x+} + ye^- \rightarrow RC^{(x-y)+} \tag{1.13}$$

$$\text{Anode}: zLi^+ + ze^- \rightarrow zLi \tag{1.14}$$

1.4.4 LITHIUM SULFUR BATTERY

The lithium-sulfur battery is another major energy storage device under study. The high theoretical specific capacity (1,675 mAh g^{-1}) of sulfur makes it more attractive in the area of energy storage devices. It also possesses a theoretical energy density of ~2,600 Wh kg^{-1} which is much higher than that with other storage systems. In a typical Li-S battery, lithium is used as the anode and sulfur as the cathode, with an electrolyte capable of conducting the sulfur ions. The overall reaction is depicted in Equation 1.15. Even though the capacity is much higher, they lack in the experimental performance owing to the poor conductivity of the sulfides used in the battery, as well as the volume expansion occurring in the sulfur during cycling. Hence, several attempts have been made to improve the properties of the cathode materials used in the Li-S battery.

$$16\,Li + S_8 \rightleftharpoons 8\,Li_2S \tag{1.15}$$

Electrochemical Energy Storage Systems 19

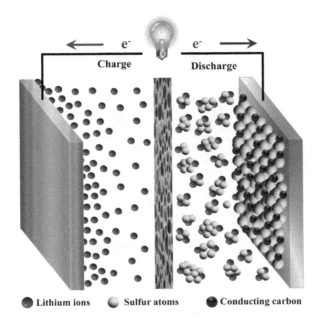

FIGURE 1.7 Schematic representation of the working principle of lithium-sulfur batteries, including the movement of ions between electrodes during charge (backward arrow) and discharge (forward arrow) states.

The electrolyte employed in the preliminary stages of Li-S battery development was a liquid electrolyte, which was subsequently replaced by solid-state, gel or polymer electrolytes [60]. Lithium polysulfide reactions can be catalyzed by using several metals, such as platinum, cobalt etc., and metal oxides, such as MnO_2, VO_2, and CeO_2. The polysulfide shuttle taking place in the lithium-sulfur battery is the backbone of the enhanced performance of these systems [61]. A schematic representation of the lithium-sulfur battery is given in Figure 1.7.

1.5 SUPERCAPACITORS

Supercapacitors are a class of electrochemical energy-storage devices that can rapidly store electrical energy through double-layer charging, faradic process, or a combination of both, and release energy instantaneously. They are also known as ultracapacitors and electrochemical double-layer capacitors. Supercapacitors can store hundreds or thousands of times more charge than conventional capacitors. According to the intrinsic principles of charge storage and discharge in supercapacitors, there are two kinds of capacitance: double-layer and pseudocapacitance. The latter involves a faradaic process whereas the former is non-faradaic. They can provide fast charge/discharge processes (in seconds) and high specific power (10 kW kg^{-1}), while maintaining a long cycle life (>10^5) [3, 62–66]. The role of supercapacitors becomes important because their parameters complement the deficiencies

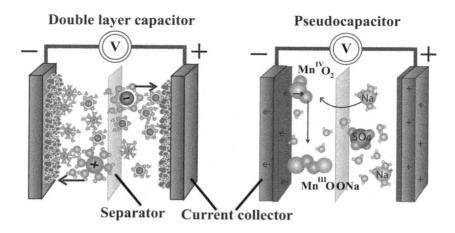

FIGURE 1.8 Schematic representation of a typical electrochemical double-layer capacitor and pseudocapacitor. Adapted and reproduced with permission from Ref. [69]. Copyright © 2014 Royal Society of Chemistry.

of other electrochemical power sources, such as batteries and fuel cells. The corresponding Ragone plot, as shown in Figure 1.1, illustrates that batteries suffer from slower power delivery than supercapacitors. This indicates that supercapacitors can be used when fast storage, coupled with high power, is required from the energy storage systems [66–68]. Components of a supercapacitor are similar to those of batteries, which contain an anode, cathode, and an electrolyte sandwiched between these electrodes, as shown in Figure 1.3. The charge storage and transfer determine the performance of a supercapacitor, which is governed by the components employed in the fabrication of the device.

Based on the electrode components, supercapacitors are classified into EDLC (electrical double-layer capacitance), pseudocapacitors, and hybrid capacitors, as shown in Figure 1.8. Based on their mode of energy storage, supercapacitors are divided into two main areas: i) redox supercapacitor or ii) electrochemical double-layer capacitor (EDLCs). Redox supercapacitors are also called as pseudocapacitors as the capacitance in this case is associated with a fast and reversible oxidation/reduction or faradaic charge transfer reactions of electro-active species on the electrode surface [70–76]. This type of supercapacitor behaves somewhat like a battery, as the charge storage process is based on a redox reaction. Commonly used electrode materials for this class of supercapacitors are transition metal oxides and conducting polymers. For such supercapacitors, the capacitance is related to the electrode potential, as $C=dQ/dV$, where C represents the capacitance of the pseudocapacitor, Q is the quantity of charge and V is the electrode potential (Figure 1.9).

The EDLCs are also like a battery, where there are two electrodes immersed in an electrolyte. During the charging process, the positive electrode attracts anions from the electrolyte and, similarly, the negative electrode attracts cations. However, unlike the batteries, the electrolyte ions do not react with the electrode material [77–80] as

Electrochemical Energy Storage Systems

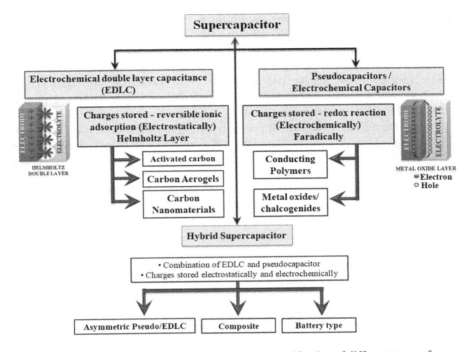

FIGURE 1.9 The flow chart representing the basic classification of different types of supercapacitors based on the electrochemical reaction/mechanism.

shown in Figure 1.8a. For such supercapacitors, the double layer capacitance at each electrode surface can be related to the effective surface area of the electrode and to the effective thickness of the double layer, and can be given by $C = \varepsilon A/d$, where ε is the dielectric constant of the double layer and A is the surface area of the electrode. Here, d represents the effective thickness of the double layer. Because of this kind of surface dependence for charge storage, optimization of pore size and structure, surface properties, and conductivity of electrode materials becomes very important.

1.6 FUEL CELLS

Fuel cells are electrochemical devices that use hydrogen (H_2), or H_2-rich fuels, together with oxygen from the air, to provide electricity and heat. However, there are many variations of this fundamental process, depending on the fuel-cell type and the fuel which is used. The major applications of fuel cells include the areas, such as stationary power generators, distributed power generators, portable power generators for transportation, military projects, the automotive market, micropower generators, and auxiliary power generators [81]. These are all applications used in a wide range of industries and environments on a global scale. The first fuel cell was demonstrated by Welsh scientist Sir William Grove in 1839, but the principle of the fuel cell was discovered by German scientist C. F. Schönbein. The first modification of the fuel cell was carried out by the chemist, W.T. Grubb, by using a sulfonated polystyrene ion-exchange membrane as the electrolyte in 1955 [82, 83].

The first commercial application of fuel cells was for the Gemini project carried out under NASA and further opened up a novel way of application of alkaline fuel cells for use in the Apollo space missions (1961–1972). More efficient forms of phosphoric acid fuel cell (PAFC) technology was developed in the 1970s, which was initiated by national governments as well as large companies, in order to overcome the energy shortages and higher oil prices, resulting in advances in PAFC, particularly in terms of stability and performance. The continuous technical development of PAFC suggests a bright future for the technology, mainly for stationary applications and buses.

The fuel cell industry has faced and keeps facing demanding situations, as it comes through a period of recession and completes the transition from R&D to commercialization. The achievement of some applications means that there has been a chance to consolidate particular technology into a popular reference design for a particular sort of fuel cell. This has brought about increasing development of fuel cells to provide scalable-strength solutions, capable of serving several kinds of market segments.

1.6.1 Basic Structure of Fuel Cell

The basic function of a fuel cell is to convert the chemical energy of a fuel and an oxidant into electrical energy. A basic single cell is developed in a manner which consists of an electrolyte layer in contact with a porous anode and cathode on either side. In the case of a typical fuel cell, the anode consists of a gaseous fuel, which is continuously fed into this particular part, and the cathode consists of an oxidant, which is oxygen from the air, which is continuously fed into the cathode compartment. An electric current is produced as a result of electrochemical reactions taking place at each electrode. A simple schematic of the fuel cell is shown in Figure 1.10 [3]. The anodic and cathodic reactions of a fuel cell with an acid electrolyte are given in Equations 1.16, 1.17, and 1.18.

$$\text{Anodic reaction}: H_2 \rightarrow 2H^+ + 2e^- \qquad (1.16)$$

$$\text{Cathodic reaction}: \tfrac{1}{2}O_2 + 2H^+ + 2e^- \rightarrow H_2O \qquad (1.17)$$

$$\text{Overall reaction}: H_2 + \tfrac{1}{2}O_2 \rightarrow H_2O + \text{heat} \qquad (1.18)$$

Both fuel cells and batteries have similar components and characteristics but differ in many aspects. The available energy in a battery is determined by the chemical reactants, which are stored in the battery. The battery will be discarded once the stored chemical reactants are consumed. The fuel cells are a type of secondary battery and, for them, the reactants are being continuously taken from an external source very similar to redox flow batteries. Thus, fuel cells are capable of producing electrical energy as long as the oxidant and fuel are being given to the electrodes [84], and they

Electrochemical Energy Storage Systems

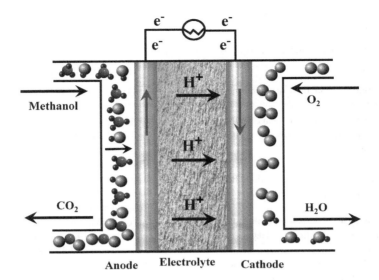

FIGURE 1.10 Schematic representation of the structure and working mechanism of a typical fuel cell (methanol fuel cell).

have their own advantages over other electrochemical energy storage devices. The main advantages and disadvantages of fuel cells are listed in Table 1.3.

1.6.2 Classification of Fuel Cells

Various types of fuel cells are available on the market now, and each fuel cell differs from the others with respect to the type of fuels which can be used, the operative temperature range, the type of catalyst used by the cell, and the efficiency ratio of the energy conversion. Primarily, the fuel cells are categorized as alkaline fuel cells (AFC), polymeric electrolyte membrane fuel cells (PEMFC), molten carbonate fuel cells (MCFC), direct methanol fuel cells (DMFC), phosphoric acid fuel cells (PAFC), and microbial fuel cells (MFC). Table 1.4 shows the classifications of fuel cells, based on their physicochemical working principles.

1.6.2.1 Polymeric Electrolyte Membrane Fuel Cells (PEMFC)

Polymer electrolyte membrane fuel cells (PEMFCs) are eco-friendly fuel cells, which convert energy efficiently, and thus will be required to assume a predominant role in future energy applications. These fuel cells are responsible for the conversion of the chemical energy of a fuel into electricity, utilizing a hydrogen oxidation reaction and an oxygen reduction reaction over a proper catalyst. Water and heat are the only by-products of this process.

The fuel, hydrogen, is separated from the oxidant by a polymer electrolyte membrane which allows the hydrogen ions to pass through it. The overall process takes place at a temperature range between 70°C and 100°C. A noble-metal catalyst, like

TABLE 1.3
Advantages and Disadvantages of Fuel Cells

Advantages	Disadvantages
• Noise: offers a much more quiet and smooth alternative to conventional energy production	• Fuel cells price for stationary electric powered technology are still too high and inappropriate for substitution of the technologies primarily based on fossil fuels
• Eco friendly, with zero emission of pollutants, greatly reduced CO_2 and harmful pollutant emissions	
• Low maintenance cost	• The degradation time and life cycles of many technologies of fuel cells, especially the high-temperature technologies, are still not completely known
• Fuel flexibility, modularity, and siting flexibility	
• Low thermal and chemical emissions	
• Higher efficiency conversion: while utilizing co-generation, fuel cells can achieve 80% energy efficiency.	• One of the main and primary fuels for fuel cell technology is hydrogen and the cost for hydrogen is very high; a network does not exist for its production and distribution.
• Design flexibility and size reduction; fuel cells are significantly lighter and more compact.	
• Good reliability: quantity of power provided does not degrade over time	• Expensive to manufacture due to the high cost of catalysts (palladium)
• High power density: high power density allows fuel cells to be a relatively compact source of electric power, which is beneficial in applications with space constraints	• Low energy density

TABLE 1.4
Classifications of Fuel Cell Based on Its Physicochemical Working Principles

Type of Fuel Cell	Anode	Electrolyte	Cathode
Polymer electrolyte membrane fuel cells (PEMFCs)	Fuel: H_2	PEMFC (80°C) $H^+ \rightarrow$	Input: O_2 (air) Product: H_2O
Solid oxide fuel cells (SOFCs)	Fuel: H_2 or CO Product: H_2O, CO_2	SOFC (500–1000°C) $O^{2-} \leftarrow$	Input: O_2 (air)
Alkaline fuel cells (AFCs)	Fuel: H_2 Product: H_2O	AFC (70°C) $OH^- \leftarrow$	Input: O_2 (air)
Phosphoric acid fuel cells (PAFCs)	Fuel: H_2	PAFC (200°C) $H^+ \rightarrow$	Input: O_2 (air) Product: H_2O
Molten carbonate fuel cells (MCFCs)	Fuel: H_2 or CO Product: H_2O, CO_2	MCFC (650°C) $CO_3^{2-} \leftarrow$	Input: O_2 (air) and CO_2

platinum, is used at low temperature to start the electrochemical process, which results in less wear on system components and leads to greater durability of the cells. The hydrogen will be oxidized at the anode and oxygen reduced at the cathode. The hydrogen ions produced as a result of oxidation pass through the membrane from anode to cathode and thus generate an electric current. Because PEMFC systems operate at low temperatures with no hazardous fluids, they have been suggested as a power source for zero-emission vehicles. The advantages of PEMFCs are high power density, ease of operation with a pressurized system, due to the high operative pressure difference between the anode and the cathode, lower operating temperature compared to phosphoric acid (PAFC) or molten carbonate fuel cell (MCFC) s, and economic viability for both fabrication and maintenance [85]. Disadvantages include:

- Due to the low operating temperature, PEMFC catalysts are susceptible to CO poisoning.
- The recovered heat can be utilized only as hot water.
- Issues are associated with water management of the membrane electrolyte

The most commonly used membrane for PEMFC is Nafion, which is a sulfonated tetra fluoro ethylene-based fluoro polymer-co-polymer discovered in the late 1960s by Walther Grot of DuPont, having a thickness between 50 and 175 µm. The Nafion membrane is fabricated by the addition of sulfonic acid groups onto the bulk of a polymer matrix of Teflon. The thinner membrane offers higher conductivity, whereas the thicker membrane reduces the conductivity. The principle role of this membrane is to act as a separator between hydrogen and oxygen. Sulfonated polyphosphazene-based membranes, polymer-zeolite nanocomposite proton-exchange membranes, and phosphoric acid-doped poly(bis-benzoxazole) [85] high-temperature ion-conducting membranes are the other types of membranes that are being researched. Among all these membranes, Nafion is considered to be the best membrane because of its high chemical stability; as a consequence, all other membranes are compared with Nafion [84, 86, 87]. The characteristics of Nafion membrane are:

- Intermixing of the hydrogen and oxygen can be restricted.
- Offers high chemical resistance as well as high mechanical resistance because of the Teflon.
- Acts as a good proton conductor.
- Allows high water uptake.

1.6.2.2 Direct Methanol Fuel Cells (DMFCs)

Direct methanol fuel cells are a kind of fuel cell, that consists of a polymer electrolyte membrane and operates at a temperature range between 70 and 100°C. The fuel used in DMFCs is liquid methanol, which is dissolved in water. Similar to PEMFCs, the DMFCs are also eco-friendly fuel cells. The polymer electrolyte ion-exchange membrane is the main part of DMFC, which is in direct contact with both anode and cathode, each of which consists of a three-layer structure consisting of a catalytic layer, a diffusion layer, and a backing layer [84]. The combination of platinum at the cathode and a platinum-ruthenium alloy at the anode, along with an ionomer, combines to form the catalytic layer. The ionomer membrane is composed of a perfluoro sulfonic acid polymer. The mixture of carbon and Teflon combines to form the diffusion layer which allows both the transportation of oxygen molecules to the catalyst layer of the cathode and the escape of CO_2 molecules from the anode.

DMFCs are simple in design and quite quiet, and consist of environmentally friendly technology without the hazard of explosions, making them appropriate for portable applications and generators. DMFCs also have applications in both civil and military environments, although DMFCs show the lowest efficiency, at 35%, of the fuel cells; this is the only drawback associated with DMFCs.

1.6.2.3 Alkaline Fuel Cells (AFCs)

Alkaline fuel cells can work under different operative temperatures, ranging from 30°C to 250°C, which makes them useful in various areas like the space sector. The electrolyte used in the alkaline fuel cell is a solution of potassium hydroxide, with pure oxygen as the oxidant. Depending on the design and ability of electrolytes, each cell generates a voltage between 0.5 V to 0.9 V and an efficiency of up to 65% [84]. In addition, the AFCs have a simple cell structure, can perform at different operative temperatures, have high electrical efficiency, and require less maintenance.

AFC consists of a porous anode and a porous cathode, made with a nickel or silver catalyst, and these electrodes are separated by a liquid KOH solution, which acts as the electrolyte. In an AFC cell, oxygen is continuously fed into the cathode part, and hydrogen is continuously fed into the anode part. An electric current is developed when the ions are transported between the cathode and the anode [88]. Potassium titanate, ceria, asbestos, and zirconium phosphate gel are also being used in the microporous separator for AFCs, although asbestos is carcinogenic and it is not being used nowadays [88]. Based on the type of electrolyte, AFCs are classified as follows:

(i) Mobile electrolyte AFCs: In a mobile electrolyte AFC, the electrolyte is pumped inside the cell via an external circuit. The anode and cathode parts are hydrogen and air, respectively. The reaction between KOH and CO_2 in the air presents the main challenges facing mobile AFCs. This problem can affect the overall efficiency of an AFC.
(ii) Static electrolyte AFCs: In the case of a static electrolyte AFC, the anode and cathode are separated by an electrolyte that is held inside the asbestos. Injection of pure oxygen inside the cathode is necessary for the working of the AFC.
(iii) Dissolved fuel AFCs: The electrolyte which separates the anode and cathode is combined with a fuel like hydrazine or ammonia; as a result, this type of AFC cannot be used for large power generators. Also, it uses hydrazine, which is toxic, carcinogenic, and explosive.

1.6.2.4 Phosphoric Acid Fuel Cell (PAFC)

Phosphoric acid fuel cells are the first type of fuel cell to be used and commercialized inside power applications. The development and growth of PAFCs occurred in the period 1960–1970. As the name of the fuel cell indicates, phosphoric acid is the electrolyte which is used inside the PAFC. For better function of PAFCs, the temperature must be maintained between 150 to 200°C due to the poor ionic conductivity of phosphoric acid. Pure hydrogen is not required in PAFC, however, the process takes place at at a high-temperature range. PAFC offers an overall efficiency between 37% and 42% [86].

The overall structure is made of a ceramic matrix of thickness 0.1–0.2 mm in which the porous electrodes are separated by the electrolyte, phosphoric acid. A platinum nanocatalyst is used to develop the diffusion electrodes and is then functionalized by high surface carbon, dispersed inside a layer made of carbon bonded with polytetrafluoro ethylene (PTFE). The phosphoric acid electrolyte provides the high thermal, chemical, and electrochemical stability required in order to be actively used inside the cell. Furthermore, the phosphoric acid does not react with CO_2, minimizing the problems associated with carbon monoxide and carbonate [89]. The characteristics of PAFCs includes:

- The operative temperature is between 150°C and 220°C, with an operating life cycle of more than 65,000 hours
- PAFCs achieve an efficiency of up to 40% and are capable of extending the efficiency up to 60% by means of combined heat systems
- There is less chance of carbon monoxide poisoning
- The se of PAFC reduces the cost of power generation.

1.6.2.5 Molten Carbonate Fuel Cells (MCFC)

The base structure of MCFC is made of a ceramic matrix in which the porous anode electrode is fueled by hydrogen and the porous cathode is usually fueled by oxygen. Two mixtures of molten carbonate salts, such as a combination of either lithium carbonate and potassium carbonate or lithium carbonate and sodium carbonate could be used as electrolytes. Ceramic powder and fibers are used inside the matrix to enhance the entire mechanical strength. Usually, nickel is used as the catalyst instead of an expensive catalyst like platinum [90, 91]. The cell operates at a temperature range of 650°C, since it needs high operative temperatures. Upon reaching this temperature, carbonate salts begin to melt and become conductive by carbonate ions (CO_3^{2-}). These ions are then transported from the cathode to the anode and are collected at the anode part which leads to the generation of an electric current. The advantages of characteristics of the molten carbonate fuel cell are:

- MCFCs are capable of achieving an efficiency of up to 45%.
- Use of MCFCs reduces the costs of power generation
- Use of stainless steel and nickel-based alloys results in the reduction of the production cost of MCFCs

1.6.2.6 Microbial Fuel Cells (MFC)

Microbial fuel cells (MFC) are also known as biological fuel cells. They are bio-electrochemical devices that convert chemical energy to electrical energy through electrochemical reactions, which involve bio-enzymatic catalysis and biochemical pathways. Here, organic matter is converted into electricity, using bacteria as the catalyst, providing the possibility of using a wide range of microbe-degradable organic or inorganic matter such as organic waste and soil sediments. The microorganisms carry out glycolysis, citric acid cycle, etc. which generate electrons and protons that are used for the generation of electricity. Microbial fuel cells operate at ambient temperature and atmospheric pressure and are currently used for generating energy from organic matter [92].

An MFC consists of three major components, namely an anaerobic anode chamber, an aerobic cathode chamber, and a separator connecting the two chambers. A simple schematic representation of MFCs is presented in Figure 1.11. The growth and the electron extraction from microorganisms take place in the anode chamber. Protons and electrons are produced by means of oxidative microbial metabolism in this chamber. Electrons are transferred by microbes to the anode part and flow to the cathode through a resistor, leading to the production of electricity.

An unmediated MFC is a type of microbial fuel cell that was developed in the 1970s, and bacteria in this type of MFC have electrochemically active redox proteins,

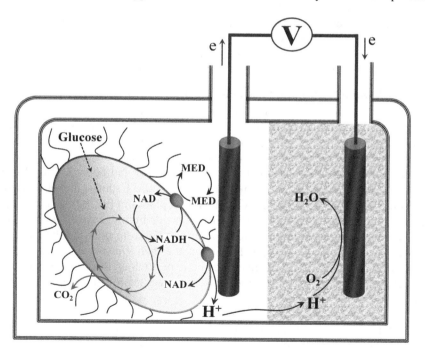

FIGURE 1.11 Structure and working principle (including the Krebs cycle) of microbial fuel cells.

such as cytochromes, on their outer membrane that can transfer electrons directly to the anode. Carbon dioxide, protons, and electrons are produced by organic electron donors in most of the MFCs. Sulfur compounds and hydrogen are the other reported electron donors. Since MFCs require very low power for action, this which makes them suitable for use in power generation applications as well as in wireless sensor networks, because MFCs use energy more efficiently than do standard internal combustion engines, which are limited by the Carnot cycle. Also, MFCs can be used in a biosensor for monitoring BOD [biological (biochemical) oxygen demand] values because an MFC- type BOD sensor can provide real-time BOD values.

1.7 CONCLUSION

From this review, it is clear that each energy storage system has its own strengths and weakness. High energy density and high power density are the two main properties required for an energy storage system. In most cases, supercapacitors provide low energy density and batteries provide low power densities. Rechargeable batteries are the best and most reliable sources for energy storage, of which lithium-ion batteries show the highest efficiency. Lithium-sulfur batteries are known to be the most efficient batteries capable of powering a smartphone for five continuous days. While comparing supercapacitors with batteries, the supercapacitors have a life expectancy of 10 to 15 years, whereas the batteries can last up to only 5–10 years. It has also been proved that supercapacitors can reach up to one million charge-discharge cycles, whereas typical batteries have only 500–1000 cycles, which makes supercapacitors suitable for use in various applications such as in automobiles, buses, trains, cranes, and elevators, where they are used for regenerative braking, short-term energy storage, or burst-mode power delivery. The simplicity of design and the cheap cost of the fuel cells is probably suitable for their applications in static power generation. Alkaline fuel cells are among the most efficient type of fuel cells, reaching up to 60% efficiency and up to 87% in combined heat and power generation. Of the fuel cells, polymer electrolyte membrane fuel cells (PEMFCs) and direct methanol fuel cells (DMFCs) have attracted the greatest interest regarding applications such as distributed power generation, portable applications, and all applications concerning the automotive and transportation sector.

ACKNOWLEDGMENT

Authors Dr. Jabeen Fatima M. J. and Dr. Prasanth Raghavan, would like to acknowledge Kerala State Council for Science, Technology and Environment (KSCSTE), Kerala, India for financial assistance.

REFERENCES

1. Kumar A, Kumar K, Kaushik N, et al. (2010) Renewable energy in India: Current status and future potentials. *Renew Sustain Energy Rev* 14:2434–2442. https://doi.org/10.1016/j.rser.2010.04.003

2. Kim J, Niedzicki L, Scheers J, et al. (2013) Characterization of N-butyl-N-methylpyrrolidinium bis (tri fl uoromethanesulfonyl) imide-based polymer electrolytes for high safety lithium batteries. *J Power Sources* 224:93–98. https://doi.org/10.1016/j.jpowsour.2012.09.029
3. Winter M, Brodd RJ (2004) What are batteries, fuel cells, and supercapacitors? *Chem Rev* 104:4245–4269. https://doi.org/10.1021/cr020730k
4. Stelbin Peter F, Raghavan P (2018) Graphene and carbon nanotubes for advanced lithium ion batteries. *Graphene Carbon Nanotub Adv Lithium Ion Batter* 781138. https://doi.org/10.1201/9780429434389
5. Haas O, Cairns EJ (1999) Electrochemical energy storage. *Annu Rep Prog Chem Sect C: Phys Chem* 95:163–198.
6. Ajayan PM, Schadler LS, Braun PV (2003) *Nanocomposite Science and Technology*. Weinheim: Wiley-VCH.
7. Hari Singh Nalwa (1999) *Handbook of Nanostructured Materials and Nanotechnology*, Academic Press. ISBN: 9780080533643
8. Aricò AS, Bruce P, Scrosati B, et al. (2005) Nanostructured materials for advanced energy conversion and storage devices. *Materials for Sustainable Energy*. pp. 148–159. ISBN: 978-981-4317-64-1
9. Tarascon JM, Armand M (2010) Issues and challenges facing rechargeable lithium batteries. In: *Materials for Sustainable Energy: A Collection of Peer-Reviewed Research and Review Articles from Nature Publishing Group*. pp. 171–179.
10. Landi BJ, Ganter MJ, Cress CD, et al. (2009) Carbon nanotubes for lithium ion batteries. *Energy Environ Sci* 2:638–654. https://doi.org/10.1039/b904116h
11. Balaish M, Kraytsberg A (2014) *Perspective*. https://doi.org/10.1039/c3cp54165g
12. Liu X, Zhang B, Ma P, Yuen MMF (2012) Carbon nanotube (CNT) -based composites as electrode material for rechargeable Li-ion batteries: A review author's personal copy Carbon nanotube (CNT) -based composites as electrode material for rechargeable Li-ion batteries: A review. https://doi.org/10.1016/j.compscitech.2011.11.019
13. Lim D-H, Haridas AK, Figerez SP, et al. (2018) Tailor-made electrospun multilayer composite polymer electrolytes for high-performance lithium polymer batteries. *J Nanosci Nanotechnol* 18:6499–6505. https://doi.org/10.1166/jnn.2018.15689
14. Raghavan P, Manuel J, Zhao X, et al. (2011) Preparation and electrochemical characterization of gel polymer electrolyte based on electrospun polyacrylonitrile nonwoven membranes for lithium batteries. *J Power Sources* 196:6742–6749. https://doi.org/10.1016/j.jpowsour.2010.10.089
15. Du C-F, Liang Q, Luo Y, et al. (2017) Recent advances in printable secondary batteries. *J Mater Chem A* 5:22442–22458. https://doi.org/10.1039/C7TA07856K
16. Singh N, Galande C, Miranda A, et al. (2012) Paintable battery. *Sci Rep* 2:6–10. https://doi.org/10.1038/srep00481
17. Zou C, Zhang L, Hu X, et al. (2018) Review article A review of fractional-order techniques applied to lithium-ion batteries , lead-acid batteries , and supercapacitors. *J Power Sources* 390:286–296. https://doi.org/10.1016/j.jpowsour.2018.04.033
18. Lai X, He L, Wang S, et al. (2020) Co-estimation of state of charge and state of power for lithium-ion batteries based on fractional variable-order model. *J Clean Prod* 255:10319–10329. https://doi.org/10.1016/j.jclepro.2020.120203
19. Zhao-karger Z, Fichtner M (2019) Beyond intercalation chemistry for rechargeable Mg batteries: A short review and perspective. 6:1–12. https://doi.org/10.3389/fchem.2018.00656
20. Wang P, Buchmeiser MR (2019) Rechargeable magnesium–sulfur battery technology: State of the art and key challenges. *Adv Funct Mater* 29. https://doi.org/10.1002/adfm.201905248

21. Lithium ion battery market to reach. www.marketsandmarkets.com/Market-Reports/lithium-ion-battery-market-49714593.html?gclid=Cj0KCQiAwf39BRCCARIsAL XWETxWaipZklUZEpuLKev8r3QJA1ezCT6krXGNBgQnUI3sXcC4Pft7K2waAs lLEALw_wcB
22. Fisher T (2013) Will Tesla alone double global demand for its battery cells? Greencarreports.com. Archived from the original on 18 October 2017. Retrieved 16 February 2014.
23. Reduced cell cost suggests the upcoming era of large capacity cells. (2014) www.energytrend.com/pricequotes/20130506-5180.html#:~:text=Reduced%20cell%20cost%20suggests%20the%20upcoming%20era%20of%20large%20capacity%20cells,-published%3A%202013%2D05&text=According%20to%20EnergyTrend%2C%20a%20research,cells%20back%20into%20the%20market
24. Bhutani A, Schiller JA, Zuo JL, et al. (2017) Combined computational and in situ experimental search for phases in an open ternary system, Ba-Ru-S. *Chem Mater* 29:5841–5849. https://doi.org/10.1021/acs.chemmater.7b00809
25. Whittingham MS (2012) History, evolution, and the future status of energy storage. *Proceedings of the IEEE* 100(special centennial issue):1518–1534. doi: 10.1109/JPROC.2012.2190170
26. Oswal M, Paul J, Zhao R (2010) A comparative study of lithium-ion batteries. AME 578 Project: University of South California. www.ehcar.net/library/rapport/rapport204.pdf
27. Haruyama J, Sodeyama K, Han L, et al. (2014) Space-charge layer effect at interface between oxide cathode and sulfide electrolyte in all-solid-state lithium-ion battery. *Chem Mater* 26:4248–4255. https://doi.org/10.1021/cm5016959
28. Kim Y (2012) Lithium nickel cobalt manganese oxide synthesized using alkali chloride flux: Morphology and performance as a cathode material for lithium ion batteries. *ACS Appl Mater Interfaces* 4:2329–2333. https://doi.org/10.1021/am300386j
29. Chen Y, Lu Z, Zhou L, et al. (2012) Triple-coaxial electrospun amorphous carbon nanotubes with hollow graphitic carbon nanospheres for high-performance Li ion batteries. *Energy Environ Sci* 5:7898–7902. https://doi.org/10.1039/c2ee22085g
30. Date I (2009) This document is downloaded from DR-NTU, Nanyang technological. *Security* 299:1719–1722. https://doi.org/10.1063/1.2978249
31. Wang J, Li L, Wong CL, et al. (2013) Controlled synthesis of α-FeOOH nanorods and their transformation to mesoporous α-Fe_2O_3, Fe_3O_4 @C nanorods as anodes for lithium ion batteries. *RSC Adv* 3:15316–15326. https://doi.org/10.1039/c3ra41886c
32. Chagnes A, Swiatowsk J (2012) Electrolyte and solid-electrolyte interphase layer in lithium-ion batteries. *Lithium Ion Batter - New Dev.* https://doi.org/10.5772/31112
33. Prabakaran P, Manimuthu RP, Gurusamy S, Sebasthiyan E (2017) Plasticized polymer electrolyte membranes based on PEO/PVdF-HFP for use as an effective electrolyte in lithium-ion batteries. *Chin J Polym Sci* 35:407–421.
34. Chapman N (2016) Spectroscopic measurements of ionic association in common lithium salts and carbonate electrolytes. Dissertations and Master's Theses (Campus Access). Paper AAI10189407. https://digitalcommons.uri.edu/dissertations/AAI10189407
35. Scrosati B, Croce F, Persi L (2000) Impedance spectroscopy study of PEO-based nanocomposite polymer electrolytes. *J Electrochem Soc* 147:1718–1721. https://doi.org/10.1149/1.1393423
36. Shin JH, Henderson WA, Passerini S (2005) PEO-based polymer electrolytes with ionic liquids and their use in lithium metal-polymer electrolyte batteries. *J Electrochem Soc* 152:978–983. https://doi.org/10.1149/1.1890701
37. Gopalan AI, Santhosh P, Manesh KM, et al. (2008) Development of electrospun PVdF-PAN membrane-based polymer electrolytes for lithium batteries. *J Memb Sci* 325:683–690. https://doi.org/10.1016/j.memsci.2008.08.047

38. Newcomb BA, Chae HG, Gulgunje PV., et al. (2014) Stress transfer in polyacrylonitrile/carbon nanotube composite fibers. *Polymer (Guildf)* 55:2734–2743. https://doi.org/10.1 016/j.polymer.2014.04.008
39. Kim JR, Choi SW, Jo SM, et al. (2004) Electrospun PVdF-based fibrous polymer electrolytes for lithium ion polymer batteries. *Electrochim Acta* 50:69–75. https://doi.org/10.1016/j.electacta.2004.07.014
40. Mohamed NS, Arof AK (2004) Investigation of electrical and electrochemical properties of PVDF-based polymer electrolytes. *J Power Sources* 132:229–234. https://doi.org/10.1016/j.jpowsour.2003.12.031
41. Angulakshmi N, Thomas S, Nahm KS, et al. (2011) Electrochemical and mechanical properties of nanochitin-incorporated PVDF-HFP-based polymer electrolytes for lithium batteries. *Ionics (Kiel)* 17:407–414. https://doi.org/10.1007/s11581-010-0517-z
42. Stephan AM, Nahm KS, Anbu Kulandainathan M, et al. (2006) Poly(vinylidene fluoride-hexafluoropropylene) (PVdF-HFP) based composite electrolytes for lithium batteries. *Eur Polym J* 42:1728–1734. https://doi.org/10.1016/j.eurpolymj.2006.02.006
43. Chiu CY, Yen YJ, Kuo SW, et al. (2007) Complicated phase behavior and ionic conductivities of PVP-co-PMMA-based polymer electrolytes. *Polymer (Guildf)* 48:1329–1342. https://doi.org/10.1016/j.polymer.2006.12.059
44. Rajendran S, Kannan R, Mahendran O (2001) An electrochemical investigation on PMMA/PVdF blend-based polymer electrolytes. *Mater Lett* 49:172–179. https://doi.org/10.1016/S0167-577X(00)00363-3
45. Doughty DH, Butler PC, Akhil AA, et al. (2010) Stationary electrical energy storage. *Electrochem Soc Interface* 19:49–53.
46. Corporation LB, Beach N (1999) Laub biochem. *Computer (Long Beach Calif)* 18:461–472. https://doi.org/10.1016/0025-5408(83)90138-1
47. Chitra S, Kalyani P, Mohan T, et al. (1999) Characterization and electrochemical studies of LiMn2O4 cathode materials prepared by combustion method. *J Electroceramics* 3:433–441. https://doi.org/10.1023/A:1009982301437
48. Li X, Cheng F, Guo B, Chen J (2005) Template-synthesized LiCoO2, LiMn2O4, and LiNi0.8Co0.2O2 nanotubes as the cathode materials of lithium ion batteries. *J Phys Chem B* 109:14017–14024. https://doi.org/10.1021/jp051900a
49. Tarascon JM, Armand M (2001) Issues and challenges facing rechargeable lithium batteries. *Nature* 414:359–67. https://doi.org/10.1038/35104644
50. Moser JR (1972) Solid state lithium-iodine primary battery. U.S. Patent 3,660,163.
51. Greatbatch W, Holmes CF (1992) The lithium/iodine battery: A historical perspective. *PACE - Pacing Clin Electrophysiol* 15:2034–2036. https://doi.org/10.1111/j.1540-8159.1992.tb03016.x
52. Mallela VS, Ilankumaran V, Rao SN (2004) Trends in cardiac pacemaker batteries. *Indian Pacing Electrophysiol J* 4:201–212.
53. Zhao Y, Wang L, Byon HR (2013) High-performance rechargeable lithium-iodine batteries using triiodide/iodide redox couples in an aqueous cathode. *Nat Commun* 4:1–7. https://doi.org/10.1038/ncomms2907
54. Khanna VK (2016) Batteries for implants. *Implantable Medical Electronics*. Springer, Cham. https://doi.org/10.1007/978-3-319-25448-7_9. ISBN: 978-3-319-25446-3
55. Semkow KW, Sammells AF (1987) A lithium oxygen secondary battery. *J Electrochem Soc* 134:2084–2085. https://doi.org/10.1149/1.2100826
56. Wang C, Xie Z, Zhou Z (2019) Lithium-air batteries: Challenges coexist with opportunities. *APL Mater* 7:. https://doi.org/10.1063/1.5091444
57. Jung KN, Kim J, Yamauchi Y, et al. (2016) Rechargeable lithium-air batteries: A perspective on the development of oxygen electrodes. *J Mater Chem A* 4:14050–14068. https://doi.org/10.1039/c6ta04510c

58. Abraham KM, Jiang Z (1996) A polymer electrolyte-based rechargeable lithium/oxygen battery. *J Electrochem Soc* 143:1–5.
59. (2014) Charles Stark Draper prize for engineering. www.nae.edu/105792/2014-Charles-Stark-Draper-Prize-for-Engineering-Recipients.
60. Lin Z, Liang C (2015) Lithium-sulfur batteries: From liquid to solid cells. *J Mater Chem A* 3:936–958. https://doi.org/10.1039/c4ta04727c
61. He J, Manthiram A (2019) A review on the status and challenges of electrocatalysts in lithium-sulfur batteries. *Energy Storage Mater* 20:55–70. https://doi.org/10.1016/j.ensm.2019.04.038
62. Burke A (2000) Ultracapacitors: Why, how, and where is the technology. *J Power Sources* 91:37–50.
63. Zhang Y, Feng H, Wu X, et al. (2009) Progress of electrochemical capacitor electrode materials: A review. *Int J Hydrogen Energy* 34:4889–4899. https://doi.org/10.1016/j.ijhydene.2009.04.005
64. Conway BE (1999) Electrochemical Supercapacitors Scientific Fundamentals and Technological Application. ISBN: 978-0-306-45736-4, Publisher: Springer US.
65. Simon P, Gogotsi Y (2008) Materials for electrochemical capacitors. https://doi.org/10.1038/nmat2297
66. Miller JR, Simon P (2008) Energy management. 321:651–652. https://doi.org/10.1126/science.1158736
67. Yuan C, Gao B, Zhang X (2007) Electrochemical capacitance of NiO / $Ru_{0.35}V_{0.65}O_2$ asymmetric electrochemical capacitor. 173:606–612. https://doi.org/10.1016/j.jpowsour.2007.04.034
68. Zheng JP, Soc JE, Zheng JP (2005) Theoretical energy density for electrochemical capacitors with intercalation electrodes theoretical energy density for electrochemical capacitors with intercalation electrodes. *J Electrochem Soc* 152:A1864. https://doi.org/10.1149/1.1997152
69. Jost K, Dion G, Gogotsi Y (2014) Textile energy storage in perspective. *J Mater Chem A* 2:10776–10787. https://doi.org/10.1039/c4ta00203b
70. Yang X, Wang Y, Xiong H, Xia Y (2007) Interfacial synthesis of porous MnO_2 and its application in electrochemical capacitor. 53:752–757. https://doi.org/10.1016/j.electacta.2007.07.043
71. Wu N-L (2002) Nanocrystalline oxide supercapacitors. *Mater Chem Phys* 75(1–3):6–11.
72. Wu M, Huang Y, Yang C, Jow J (2007) Electrodeposition of nanoporous nickel oxide film for electrochemical capacitors. 32:4153–4159. https://doi.org/10.1016/j.ijhydene.2007.06.001
73. Toupin M, Brousse T, Be D (2002) Influence of microstucture on the charge storage properties of chemically synthesized manganese dioxide. 3946–3952.
74. Lao ZJ, Konstantinov K, Tournaire Y, et al. (2006) Synthesis of vanadium pentoxide powders with enhanced surface-area for electrochemical capacitors. 162:1451–1454. https://doi.org/10.1016/j.jpowsour.2006.07.060
75. Kim I, Kim K (2006) Electrochemical characterization of hydrous ruthenium oxide thin-film electrodes for electrochemical capacitor. https://doi.org/10.1149/1.2147406
76. Chem JM (2013) Nanotubes with high performance in supercapacitors. 594–601. https://doi.org/10.1039/c2ta00055e
77. Pandolfo AG, Hollenkamp AF (2006) Carbon properties and their role in supercapacitors &. 157:11–27. https://doi.org/10.1016/j.jpowsour.2006.02.065
78. Kötz R, Carlen M (2000) Principles and applications of electrochemical capacitors. *Electrochim Acta* 45:2483–2498. https://doi.org/10.1016/S0013-4686(00)00354-6
79. Frackowiak E (2001) Carbon materials for the electrochemical storage of energy in capacitors. 39:937–950.

80. Zhang LL, Zhou R, Zhao XS (2009) Carbon-based materials as supercapacitor electrodes. *J Mater Chem* 38:2520–2531. https://doi.org/10.1039/c000417k
81. Hirschenhofer JH (2000) Fuel cell handbook. ISBN: 9780899343600
82. Phillits TR (1889) *Philosophical Magazine.* 21:0–3.
83. Swansea WRGEMA a a (2009) Philosophical Magazine Series 3 XXIV. On voltaic series and the combination of gases by platinum. 37–41. https://doi.org/10.1080/14786443908649684
84. Haile SM (2003) Fuel cell materials and components. 51:5981–6000. https://doi.org/10.1016/j.actamat.2003.08.004
85. Bose S, Kuila T, Nguyen TXH, et al. (2011) Polymer membranes for high temperature proton exchange membrane fuel cell: Recent advances and challenges. *Prog Polym Sci* 36:813–843. https://doi.org/10.1016/j.progpolymsci.2011.01.003
86. Moseley PT (2001) Fuel cell systems explained. *J Power Sources* 93:285. https://doi.org/10.1016/s0378-7753(00)00571-1
87. Asensio JA, Sánchez EM, Gómez-Romero P (2010) Proton-conducting membranes based on benzimidazole polymers for high-temperature PEM fuel cells. A chemical quest. *Chem Soc Rev* 39:3210–3239. https://doi.org/10.1039/b922650h
88. Watt ME, Outhred HR (2014) Strategies for the adoption of renewable energy technologies. *Encyclopedia of Life Support Systems (EOLSS)* 8:265–275.
89. Giorgi L, Leccese F (2013) Send orders of reprints at reprints@benthamscience.net Fuel cells: Technologies and applications. *Open Fuel Cells J* 6:1–20. https://doi.org/10.2174/1875932720130719001
90. Bove R, Moreno A, Mcphail S (2008) International status of Molten Carbonate Fuel Cell (MCFC) technology output voltage internal resistance. JRC Scientific and Technical Reports.
91. Kawamoto H (2008) Research and development trends in solid oxide fuel cell materials. *Sci Technol Trends* 52–70.
92. Mohan Y, Das D (2009) Effect of ionic strength, cation exchanger and inoculum age on the performance of microbial fuel cells. *Int J Hydrogen Energy* 34:7542–7546. https://doi.org/10.1016/j.ijhydene.2009.05.101

2 The Great Nobel Prize History of Lithium-Ion Batteries
The New Era of Electrochemical Energy Storage Solutions

Prasanth Raghavan, Jabeen Fatima M. J., and Jou-Hyeon Ahn

CONTENTS

2.1	Introduction	36
2.2	Development of Energy Storage Devices	36
2.3	Classification of Batteries	38
2.4	History of Lithium-Ion Batteries	40
2.5	Structure of Lithium-Ion Batteries	43
2.6	Principle of Lithium-Ion Batteries	44
2.7	Other Types of Battery Based on Lithium-Ion Technology	44
2.8	Challenges of Next-generation Lithium-Ion Batteries	45
2.9	The Nobel Prize: The New Era of Lithium-Ion Batteries	48
	2.9.1 Prof. John Bannister Goodenough	49
	2.9.2 Prof. Michael Stanley Whittingham	49
	2.9.3 Prof. Akira Yoshino	50
2.10	The Draper Prize and Lithium-Ion Battery (2014)	50
	2.10.1 Prof. Rachid Yazami	51
	2.10.2 Mr. Yoshio Nishi	52
2.11	Summary	52
Acknowledgment		52
References		52

2.1 INTRODUCTION

Energy plays an exceptional role in human life, and its demand and supply have always been crucial factors in the evolution of civilization. Energy demand has increased dramatically over time or with increasing population. Energy has always been the most essential resource for achieving the improvement of human life, for example, for heating and cooling, operating electronic and electrical appliances, for transportation, communication, and recreation, etc. In this modern era, one cannot imagine even one day without external supplies of energy [1]. Global energy consumption is expected to show an increase of 28% by 2040 [2], and it is widely reported that the future energy demands cannot be satisfied by current technologies. There are also predictions that the next world war may also be over energy [3]. As we are moving to a revolution of automation and of electric/hybrid vehicles having zero emissions, we need more advanced and environmentally friendly technologies. The present reserves of fossil fuel energy sources will be depleted in a few decades, due to high demand and, in some cases, extravagant consumption. Petroleum, natural gas, and coal are generally referred to as fossil fuels [1, 4].

Recently, the inevitable depletion of non-renewable fossil fuels, the move to zero-emission vehicles, and the dream of a clean environment, have forced mankind to transit away from using fossil fuels as the main global energy source. Green energy sources, such as solar, hydroelectric, thermal, wind, and tidal energy, will eventually replace traditional energy sources, although most of these renewable energy sources are typically periodic or intermittent in supply. Production of electricity from the aforementioned renewable sources needs electrochemical energy storage devices, such as batteries, supercapacitors, and fuel cells, which will play an important role in the efficient use of renewable energy during periods when supplies are depleted. Figure 2.1 shows the Ragone plot of the specific power against specific energy for various electrochemical energy conversion systems [5]. Amongst different electrochemical energy storage devices, batteries are crucial in solving these problems, as they can efficiently store electricity in the form of chemicals and subsequently release it, according to demand.

The battery is a collective arrangement of electrochemical cells in which chemical energy is converted into electricity and is used as a source of power by a chemical reaction, thereby storing the energy, which is subsequently released as electrons and ions [6]. The first battery, the Voltaic pile, consisting of a series of copper and zinc discs separated by cardboard moistened with a salt solution, was developed by Volta in 1800. With more than 200 years of subsequent development, battery technology has reached an era where batteries can be made in any size, ranging from macro to nano, with shapes ranging from cylindrical to prismatic or even paper batteries, with fabrication techniques from roll-to-roll printing to paintable batteries [7, 8] and are useful for a wide range of different applications.

2.2 DEVELOPMENT OF ENERGY STORAGE DEVICES

The importance of portable energy storage devices was highlighted by the introduction of batteries. Batteries are broadly classified as primary or secondary batteries.

History of Lithium-Ion Batteries

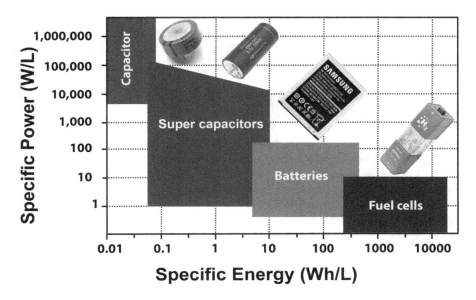

FIGURE 2.1 Ragone plot of specific power against specific energy for various electrochemical energy storage systems. Adapted and reproduced with permission from Ref. [5]. Copyright © 2004 American Chemical Society.

Primary batteries are irreversible batteries that are disposed of after being completely used, whereas secondary batteries are renewable and reversible battery systems, that can be charged and discharged a large number of times [1].

During the Parthian period (248 BC), a battery was constructed and was later displayed in Baghdad Museum. Wilhelm Konig investigated the details of this battery and it was termed the "Baghdad battery". The battery was a kind of primary voltaic cell with a copper compartment and a pointed iron rod. Addition of acidic electrolytes, such as vinegar or lime juice, would initiate the cell reaction [9]. Later, in 1749, Benjamin Franklin coined the name "battery" following his experiments on capacitors. But the discovery of the battery was achieved by Alessandro Volta, in 1800. The experiment was conducted with an aqueous salt membrane sandwiched between copper and zinc discs, and the connection produced a voltage of 0.76 V. The experiment is considered to be the initial experiment of electrochemistry. The cell so formed was known as the "galvanic cell" with two half-cells (zinc and copper). In 1836, John Frederic Daniel developed another type of electrochemical cell which generated about 1 V [10]. In 1859, the French physicist Gaston Planté introduced the first rechargeable battery, the "lead-acid battery" [11]. The battery was fabricated by rolling lead foils sandwiched between rubber strips. Georges-Lionel Leclanché introduced a new type of electrochemical cell in 1866, with carbon as the anode (negative electrode), and zinc as the cathode (positive electrode) in an electrolyte of ammonium chloride, generating a voltage of 1.4 V. The cell was termed the "Leclanché cell" [12]. A modified version of the Leclanché cell was subsequently commercialized as a dry cell with a carbon anode and zinc as the cathode. These cells were widely used in early telecommunication systems. A thermoelectric battery was introduced by Ernst Waldmar Jungner in 1869 [13]. In 1908, he also introduced "electrodes for reversible galvanic batteries". Nickel-iron, nickel-cadmium, etc. were used as electrodes for the fabrication of rechargeable batteries in an alkaline electrolytic medium.

The lead-acid battery is known to be the first commercial battery. These batteries possess comparatively very high efficiency (80–90%) but the capacity available decreases on the removal of input power [14]. Apart from the decrease in efficiency, the large size and particularly the heavy weight of these batteries, and leakage risks are some of the common disadvantages, which decreased the applications of these batteries as energy storage devices. Despite these problems, the system is still commercially used as a household energy storage system in combination with inverters, as well as in fossil fuel-powered transportation systems. Usually, aircraft powered by lead-acid batteries contain 6–12 batteries, connected in series to produce a voltage in the range 12–24 V [15]. Another major commercial battery is the nickel-cadmium (Ni-Cd) battery. Up to the past three decades, Ni-Cd batteries were the most-marketed energy storage device. These rechargeable energy storage systems were widely used in portable electronic devices like toys, AA-type batteries, AAA-type batteries, etc. Usually, the power generated was lower, so that, to enhance the power, two or more systems are connected in series with one another and sealed in a stainless-steel pack. The major drawback of these batteries is the "memory effect" [16]. The memory effect can be explained in terms of a memory of the initial point of the charging cycle, as a result of which a sudden potential drop is experienced at the same point. This drop affects the battery performance for potential applications. During the initial stages of battery commercialization, alkaline batteries were used as AA and AAA batteries. But, because these exhibited leakage issues, the basic components were replaced by nickel-cadmium, nickel-metal hydride, and lithium-ion batteries. Current energy storage depends largely on lithium-ion batteries.

2.3 CLASSIFICATION OF BATTERIES

The two mainstream classes of batteries are disposable/non-rechargeable (primary) and rechargeable (secondary) batteries. A primary battery is designed to be used once and then discarded, and not recharged with electricity. In general, primary batteries are assembled in a charged condition and the electrochemical reaction occurring in the cell is mostly irreversible, rendering the cell non-rechargeable. Leclanché, alkaline manganese dioxide, silver oxide, and zinc/air batteries are examples of primary batteries [5].

The secondary battery is a type of electrochemical device, in which the chemical reactions can be reversed by application of an external electrical energy source. Therefore, such a cell can be recharged many times by passing an electric current through it after it reached its fully discharged state, allowing it to be used for long periods. Generally, secondary batteries have a lower capacity and initial voltage, a flat discharge curve, higher self-discharge rates, and varying recharge life ratings. Secondary batteries usually have more active (less stable) chemistries that

TABLE 2.1
Common Battery Types Based on the Type of Materials and Their Rechargeability

Type of Battery	Type of Electrolyte in the Battery	
	Aqueous Electrolyte (Low Voltage Capacity)	Non-Aqueous Electrolyte (High Voltage Capacity)
Primary battery (disposable)	Manganese dry cell Alkaline dry cell Li-air battery	Li-metal battery
Secondary battery (rechargeable)	Lead-acid battery Ni-Cd battery Ni-MH battery Sodium-ion battery	Al-ion battery LIB Li-air battery

TABLE 2.2
The Characteristics and Performance of Commonly Used Rechargeable Batteries

Battery Type	Lead-Acid	Ni-Cd	Ni-MH	Li$^+$-Ion
Commercialization (year)	1970	1956	1990	1992
Nominal cell voltage (V)	2.1	1.2	1.2	3.6
Volumetric energy density (Wh L^{-1})	60–75	50–150	140–300	250–620
Gravimetric energy density (Wh kg^{-1})	30–50	40–60	60–120	100–250
Power density (W kg^{-1})	180	150	250–1000	250–340
Cycling stability	500–800	2000	500–1000	400–1200
Monthly self-discharge rate (%) at RT	3–20	10	30	5
Memory effect	No	Yes	No	No

need special handling, containment, and disposal. Aluminum-ion battery, lead-acid battery, lithium-ion battery, nickel-cadmium battery, and sodium-ion battery are examples of secondary batteries. According to the chemical reaction involved, rechargeable batteries can further be classified as lead-acid, nickel-metal hydride, zinc-air, sodium-sulfur, nickel-cadmium, lithium-ion, lithium-air batteries, etc.

Batteries may also be classified by the type of electrolyte employed, either aqueous or non-aqueous systems. Some common battery types are listed in Table 2.1 and the characteristics and performance of commonly used rechargeable batteries are shown in Table 2.2 in accordance with these classifications. Among the aforementioned rechargeable batteries, lithium-ion batteries (LIBs) have gained considerable interest in recent years in terms of the high specific energy and cell voltage, good capacity retention, and negligible self-discharge [6]. Figure 2.2 shows the projected

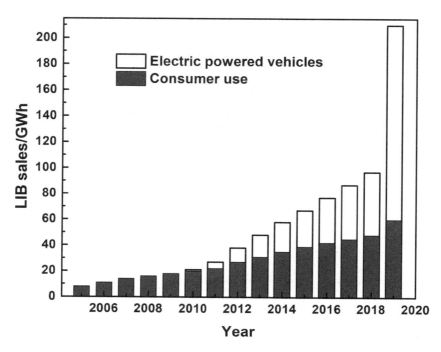

FIGURE 2.2 Predicted increase in demand for lithium-ion batteries from 2005 to 2019 (2019 value is approximated from [17]) for electric vehicle and consumer device applications.

increase in demand for LIBs in the present decade, and it is clear that the importance of LIBs in day-to-day life is increasing over time.

2.4 HISTORY OF LITHIUM-ION BATTERIES

Understanding the brief history of the development of LIBs is quite interesting. The first rechargeable LIBs were described by the British chemist M. Stanley Whittingham, a key figure in the history of the development of LIBs, while working at Exxon. He fabricated the first rechargeable LIB with layered titanium disulfide (TiS$_2$) as the cathode and metallic lithium as the anode in 1976 [18]. Exxon subsequently tried to commercialize the LIBs, but they were unsuccessful due to the problems of lithium-ion (Li$^+$-ion) dendrite formation and subsequent short circuiting, following extensive cycling and consequent safety concerns [19]. In addition, TiS$_2$ has to be synthesized under completely sealed conditions, and is quite expensive (~$ 1000 per kilogram for TiS$_2$ raw material in the 1970s). When exposed to air, TiS$_2$ reacts to form toxic hydrogen sulfide, which has an unpleasant odor, causing environmental issues. Lithium (Li) is the lightest metal and the lightest solid element, and is a highly reactive element; it burns under normal atmospheric conditions because of its spontaneous reactions with water and oxygen [20]. During the charging cycle, the tendency for Li to readily precipitate onto the negative electrode (anode) in the form of dendrites causes short circuiting. The high chemical reactivity of metallic Li and the tendency for dendrite formation results not only in poor battery characteristics, including inadequate cycling stability, because of side reactions with the electrolyte, but also poses an inherent risk of a thermal runaway reaction, which was an insoluble issue in terms of safety. As a result, the researchers focused on developing LIBs which employed only Li compounds which were capable of accepting and releasing Li$^+$-ions, instead of using metallic Li electrodes. As a result, reversible intercalation of Li$^+$-ions into graphite [21–23] and cathodic oxides [24–26] was reported in 1976 by J. O. Besenhard, who proposed its application as the anode and cathode in Li$^+$-ion cells [23, 26].

In 1977, Samar Basu demonstrated electrochemical intercalation of Li$^+$-ions into graphite, which led to the development of a workable Li$^+$-ion-intercalated graphite electrode (LiC$_6$) at Bell Labs to provide an alternative to the Li metal battery [27, 28]. In 1979, Ned A. Godshall et al. [29–31], and, in the following year, John Goodenough et al. [32–34] demonstrated a rechargeable Li$^+$-ion cell with a nominal voltage of 4 V, using layered LiCoO$_2$ as the high-energy and high-voltage material for the positive electrode and Li metal as the negative electrode, although layered LiCoO$_2$ did not attract much attention initially. In 1979, Godshall et al. demonstrated other ternary compound Li-transition metal-oxides, such as the spinel LiMn$_2$O$_4$, Li$_2$MnO$_3$, LiMnO$_2$, LiFeO$_2$, LiFe$_5$O$_8$, and LiFe$_5$O$_4$, as electrode materials other than LiCoO$_2$ for LIBs [29–31, 35], and Huggins was awarded a US patent in 1982 on the use of LiCoO$_2$ as cathodes in LIBs [36]. In 1983, Goodenough and colleagues also identified manganese spinel as a low-cost cathode material [37] and, in 1985, Godshall et al. identified Li-copper-oxide and Li-nickel-oxide as cathode materials for LIBs [35]. The lack of safe anode materials, however, limited the application of a layered oxide cathode of LiMO$_2$ (M=Mn, Ni, Co) in LIBs [38].

In 1978, Besenhard [23] and Basu [28], and, in 1980, Rachid Yazami [39, 40] demonstrated that graphite, also with a layered structure, could be a good candidate to reversibly store Li$^+$-ions by electrochemical intercalation/deintercalation of Li$^+$-ions in graphite. The publications of Yazami and Touzain [39, 40] are accepted as describing the world's first successful experimental demonstration of the electrochemical intercalation and the release of Li$^+$-ions in graphite. The organic electrolytes available for LIBs at the time decomposed during cell charging, when using graphite as an anode, slowing the commercialization of a practical rechargeable Li/graphite battery. In Yazami's studies, a solid electrolyte was used to demonstrate that Li could be reversibly intercalated in graphite through an electrochemical mechanism, and this experiment provided the scientific basis for the use of graphite as negative-electrode material, as is the standard in LIBs today. As of 2011, the graphite electrode developed by Yazami was the most commonly used electrode in commercial LIBs.

In 1985, Akira Yoshino assembled a prototype Li$^+$-ion cell, using a carbonaceous anode (polyacetylene, which is an electrically conductive polymer discovered by Prof. Hideki Shirakawa, who received the Nobel Prize in Chemistry in 2000) into which Li$^+$-ions could be inserted and discharged, with LiCoO$_2$ as the cathode [41, 42]. Both the carbon anode and the LiCoO$_2$ cathode are stable in air, which is highly beneficial from both an engineering and a manufacturing perspective. In addition, Yoshino [43] found that carbonaceous material with a certain crystalline structure provided greater capacity without causing decomposition of the propylene carbonate electrolyte solvent, as happened with the graphite electrode. This battery design, using materials without metallic Li, enabled industrial-scale manufacture, and proved to be the cornerstone of the current LIB.

After the successful demonstration of this prototype design of the LIB in 1986, Yoshino carried out the world's first safety tests on LIBs and proved that this LIB overcame the safety issues that had prevented commercialization of non-aqueous secondary batteries in the past. Because of the risk of ignition or even explosion during the safety test, Yoshino had to borrow a facility designed for testing explosives. In these tests, a lump of iron was dropped on to the batteries and the effect compared with that from a set of cells assembled using the Li metal electrode; test results showed that a violent ignition occurred with a metallic Li battery, whereas no ignition occurred with a LIB. According to Yoshino, this was a great relief, because, if ignition had occurred in this test, the LIB would not have been commercialized. This was the crucial turning point for the commercialization of the LIB. I consider the success of these tests to be "the moment when the lithium-ion battery was born" [43].

Eventually, Sony, the dominant manufacturer of personal electronic devices at the time, such as the Walkman and pocket cameras, commercialized LIBs in 1991, as did a joint venture between Asahi Kasei and Toshiba in 1992. Table 2.3 shows some milestones in the commercialization of LIBs. LIBs proved to be a tremendous success and facilitated a major reduction in the size and weight of the power supply for portable devices, thereby supporting the evolution of the portable electronics industry. Commercialization of the LIB made available an energy density, in terms of both weight and volume, of around twice what was possible with nickel-cadmium or nickel-metal hydride batteries, and providing an electromotive force of 4 V or more; in this way, the LIB made it possible to power a cell phone with a single cell. To acknowledge their pioneering contribution to the development of LIBs, John

TABLE 2.3
Historical Milestones of the Commercial Lithium-Ion Batteries

Year	Company	Historical Milestone
1991	Sony (Japan)	Commercialization of LIB
1994	Bellcore (USA)	Commercialization of Li polymer
1995	Group effort	Introduction of pouch cell, using Li polymer
1995	Duracell and Intel	Proposal of industry standard for SMBus*
1996	Moli Energy (Canada)	Introduction of Li+-ion with manganese cathode
1996	University of Texas (USA)	Identification of Li phosphate
2002	Group effort	Various patents filed on nanomaterials for batteries

*System management bus

History of Lithium-Ion Batteries

Bannister Goodenough, Rachid Yazami, Akira Yoshino, and Yoshio Nishi were awarded the 2012 IEEE Medal for Environmental and Safety Technologies and the Draper Prize in 2014. Later, in 2019, John Bannister Goodenough, Michael Stanley Wittingham, and Akira Yoshino received the Nobel Prize in Chemistry for the development of lithium-ion batteries, an important technology by which the world was able to move away from fossil fuels.

2.5 STRUCTURE OF LITHIUM-ION BATTERIES

Lithium-ion batteries are commercially available and are mostly marketed as portable batteries. Most of the next-generation electrical and electronic devices rely on this energy storage system. The components may vary from battery to battery, but the basic construction is the same. The size, shape, and components of the batteries varies, depending on the application. A LIB consists of four major parts – an anode, cathode, electrolyte, and a separator, as shown in Figure 2.3. The anode of a rechargeable battery is the positive electrode that receives electrons while charging, whereas, during discharging, the electrons flow from the anode toward the cathode. In the LIBs available on the market, the anode used is graphite rods. Other anode materials widely explored are silicon, lithium, alloys, ternary metal oxides, etc. The cathode is the negative electrode; while charging, the cathode provides electrons to the external circuit due to the de-intercalation of lithium ions. Commonly used cathode materials include lithium cobalt oxide ($LiCoO_2$, LCO), lithium iron phosphate ($LiFePO_4$, LFP), lithium manganese oxide ($LiMnO_4$, LMO), etc. The chemical

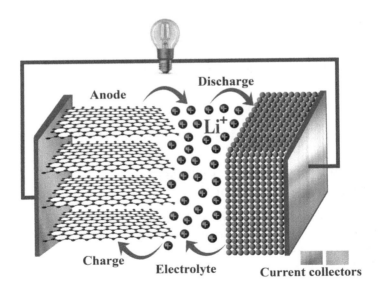

FIGURE 2.3 Schematic illustration on the structure and operating principles of lithium-ion batteries, including the movement of ions between electrodes during charge (forward arrow) and discharge (backward arrow) states.

reactions taking place at the cathode and anode in a typical LIB are given below (Equations 2.1 and 2.2):

$$\text{Anode (Graphite)}: C_n + xLi^+ + xe^- \rightarrow C_nLi_x \qquad (2.1)$$

$$\text{Cathode (LCO)}: LiCoO_2 \rightleftarrows Li_{1-x}CoO_2 + xLi^+ + xe^- \qquad (2.2)$$

2.6 PRINCIPLE OF LITHIUM-ION BATTERIES

A primary LIB is a one-direction device that has only a discharging process. During discharging, reduction happens at the cathode, gaining electrons, and oxidation occurs at the anode, losing electrons, as displayed in the following reaction (Equations 2.3, 2.4, and 2.5) [6]:

$$\text{Cathode}: MS_2 + Li^+ + e^- \xrightarrow{\text{discharge}} LiMS_2 \qquad (2.3)$$

$$\text{Anode}: Li^+ \xrightarrow{\text{discharge}} Li^+ + e^- \qquad (2.4)$$

$$(M = Ti \text{ or } Mo) \qquad \text{Full cell}: Li + MS_2 \xrightarrow{\text{discharge}} LiMS_2 \qquad (2.5)$$

In contrast with primary cell batteries, secondary cell LIBs are rechargeable. Figure 2.3 shows the schematic representation of the structure and working principle of LIBs. During the charge/discharge process, the oxidation and reduction processes occur at the two electrodes as shown below (Equations 2.6, 2.7, and 2.8). The secondary LIBs, in general, operate at 3.7 V and demonstrate a capacity of 150 mAh g^{-1}.

$$\text{Cathode}: LiMn_2O_4 \rightleftarrows Li_{1-x} + xMn_2O_4 + xLi^+ + xe^- \qquad (2.6)$$

$$\text{Anode}: xLi^+ + xe^- + C_6 \rightleftarrows Li_xC_6 \qquad (2.7)$$

$$\text{Full Cell}: LiMn_2O_4 + C_6 \rightleftarrows LiC_6 + LiMn_2O_4 \qquad (2.8)$$

2.7 OTHER TYPES OF BATTERY BASED ON LITHIUM-ION TECHNOLOGY

LIB is the most-marketed energy storage device worldwide. For powering from miniature devices to e-vehicles, LIBs vary in terms of the size and capacity of the battery. Batteries are the key energy suppliers for most of the portable devices and instruments. Both primary as well as secondary batteries, based on lithium, such as a lithium-iodide battery, or a lithium-manganese oxide battery, etc., have been employed chiefly as energy storage devices in implantable medical devices and instruments, like pacemakers, neurostimulators, and drug delivery systems, etc. Lithium-ion batteries are the main energy storage devices in laptops, palmtops, and mobile phones.

History of Lithium-Ion Batteries

Normal lithium-ion batteries are widely used in these portable devices. High-density batteries are required for electric vehicles. Lithium-ion batteries with polymer electrolytes are safer and more reliable power sources, and hence are employed in electric vehicles. Lithium-iron phosphate and lithium-manganese oxide are widely used cathode materials in commercial e-vehicle batteries. Several other energy storage devices based on lithium, other than standard LIBs, have been explored recently, such as the lithium-iodide battery, lithium-air battery, and the lithium-sulfur battery.

2.8 CHALLENGES OF NEXT-GENERATION LITHIUM-ION BATTERIES

As discussed at the beginning of this chapter, a move toward the electrification of road vehicles is becoming a societal goal of vital importance. In addition to factors such as finite fossil-fuel supplies, the need to curb global warming, etc., the issues associated with decreasing air quality have become most alarming, which made a major push toward the adoption of electric vehicles in densely populated cities a worldwide imperative, in order to improve the quality of air. As a mature technology, LIBs have aided the revolution in microelectronics and have become the power source-of-choice for portable electronic devices. Their success in the portable electronics market is due to the higher gravimetric and volumetric energy densities offered by LIBs, compared with other rechargeable systems; however, for powering electric vehicles, LIBs are still a nightmare. A quote featured in 1915 by the Washington Post summarized the then-current state of electric vehicles clearly: "Prices on electric cars will continue to drop until they're within the reach of the average family" [44]. Even a century later, the automobile industry is still facing the same cost issue. For portable electronics, the energy density is the most important factor. The current-day LIBs are sufficient to power portable electronic devices, but power density, cost, cycle life, and safety are also critical performance parameters, along with energy density (driving distance between charges), for electric vehicles. The adoption of electric vehicles is limited primarily due by the high cost, safety issues, and inadequate storage capacity of today's LIBs.

The performance parameters of LIBs are largely determined by the electrochemical properties and characteristics of the component materials used in fabricating the batteries, as well as the cell engineering and system integration involved. The characteristics of the materials employed rely on the underlying chemistry associated with the materials. Among present-day battery technologies, Li$^+$-ion technology is the best performing one, as a result of its delivered gravimetric energy densities of <250 Wh kg^{-1} and volumetric energy densities of <650 Wh L^{-1}, which exceeds any competing technology by a factor of at least 2.5 [45]. However, the adoption of LIBs for powering electric vehicles faces enormous challenges. If we really want to compete with gasoline, the energy delivered by a battery needs to increase by a factor of 12 to match that from one liter of gasoline (3000 Wh L^{-1}, taking into account corrections from Carnot's principle). Knowing that the energy density of batteries has increased by a factor of only five over the past two centuries, our chances of achieving a 10-fold increase over the next few years are very slim, in the absence of unexpected research breakthroughs. Fortunately, the automotive industry has set a more realistic target of doubling the present Li+-ion energy density (as high as ~500 Wh kg^{-1} and >1,000 Wh L^{-1}) in the next 10 years so that the autonomy of electric vehicles approaches 500 km. Accomplishing this goal is challenging; it will need innovations both in the component materials used in the cell and in the engineering involved in fabricating the cells. It should be recognized that the incremental improvements made in energy density since the first announcement of commercial LIBs in 1991 by Sony Corporation have been due largely to progress in engineering, as the component electrode materials still remain the same, with minor modifications.

The energy density of a battery is the product of its capacity and its potential, and is mainly governed by the capacity of the positive electrode. Simple calculations show that an increase in cell energy density of 57% could be achieved by doubling the capacity of the positive electrode, whereas one needs to increase the capacity of the negative electrode by a factor of 10 to get an overall cell energy density increase of 47% [46]. So, the chances of markedly improving the energy density of today's Li+-ion cells are mainly rooted in identifying better positive electrode materials, i.e. materials that could display either greater redox potentials (e.g., are highly oxidizing) or greater specific capacity (i.e., materials capable of reversibly inserting more than one electron per 3D metal orbital).

Many next-generation conversion-reaction cathodes, such as sulfur (or Li$_2$S) and oxygen (or Li$_2$O$_2$ or Li$_2$O), and anodes, such as Si, Sn, Sb, Ge, P, etc., are being actively pursued. The conversion reaction anodes offer much higher capacities than graphite, but they have higher operating voltages (which would lower the cell voltage) and suffer from large volume expansion and contraction (up to ~400%, depending on the anode and the Li content, compared to <10% for graphite) during charging and discharging, respectively [47]. The large volume changes during charge/discharge cycling, pulverization of the particles, continuous formation of a solid electrolyte interface (SEI), and the consequent trapping of active Li from the cathode in the anode SEI [48] are the major challenges associated with the conversion-reaction anodes. Many approaches have been pursued, such as reducing the particle size from micro- to nanosized, making composite electrodes with carbon nanotubes (CNT) or graphene, or deliberately leaving space within the active material architecture, but none of them have been successful enough to be practically viable at the present time [49, 50]. The above approaches drastically increase SEI formation and decrease volumetric energy density. The particle milling caused by volume changes results in a continuous formation of new surfaces during the charge/discharge process that further aggravates the formation of SEI, posing daunting challenges. It is well known that stable and reliable functioning of LIBs hinges on the formation of a stable electrode–electrolyte interface layer, which is conductive to Li+-ions, while being electronically insulating. The importance of the SEI layers for battery performance is widely acknowledged, but, despite more than four decades of research, the exact mechanisms by which formation of the SEI layer occurs remain poorly understood, and our understanding of the composition and properties of the SEI layer is limited. For the development of next-generation

LIBs, in-depth understanding of the reactions with the electrolyte, in light of the recent developments in high-capacity positive electrodes, is critically needed [51]. The concept of developing a systematic framework, which links electrochemistry and solid mechanics for LIBs, remains in its infancy. In this scenario, with the view of resolving the related issues, a recent report discussed in detail the electrochemical-mechanical coupling and their properties in the context of ion-conducting soft materials. The report also provided a comprehensive discussion on the various type of defects that can form in solid electrolyte membranes or at the interfaces [52]. Even though solid polymer electrolytes (SPEs) are well-studied electrolyte systems for the development of safer batteries, the development of a practical SPE having good electrochemical performance and thermomechanical and dimensional stability without compromising its Li$^+$-ion transport resistance and transport number still remains a considerable challenge.

Sulfur and oxygen (the so-called "Beyond Li$^+$-ion" chemistries) offer much higher capacities than the layered, spinel, and olivine-type cathodes. The oxygen-based cathodes suffer from clogging by insoluble products, catalytic decomposition of electrolytes, moisture from the air, and poor cycle life, making their practical viability extremely difficult, if not impossible. The critical challenges faced by sulfur-based cathodes are much less than those with oxygen-based cathodes, and remarkable progress has been made in recent years in increasing the active material content and loading, suppressing dissolved polysulfide migration between the electrodes, and reducing the electrolyte amount [53, 54]. However, the necessity for pairing a Li metal anode with a sulfur- or oxygen-based cathode poses formidable challenges, unless Li_2S and Li_2O_2 cathodes could be successfully paired with an anode like graphite or Si, or practical lithium-containing anodes, that could be paired with sulfur- or oxygen-based cathodes, could be developed. On the basis of simple faradaic calculations, it is reported that an electrolyte-to-sulfur ratio of between 11.1 and 4.0 is the point where an "idealized" lithium-sulfur (Li-S) battery will out-perform a traditional LIB on a specific energy (Wh kg^{-1}) and energy density (Wh L^{-1}) basis. Based on the close data analysis, McCloskey [55] makes a strong recommendation that all future Li-S battery studies focus on electrolyte-to-sulfur ratios of less than 11 mL g^{-1} or, more realistically, about 5. The literature shows that a practical Li-S battery can compete favorably with an LIB, comprising a Li metal anode coupled with an advanced intercalation cathode material, on the basis of volumetric energy density. To suppress dendrite formation and reduce parasitic reactions between Li and the electrolyte constituents, the practical Li-S battery requires a solid-state electrolyte or separator. To make a Li-S cell to be a cost-competitive alternative to LIBs, a cost target of ~$10 m^{-2} has been identified for the solid-state separator [56]. Practically, these targets are still daunting challenges to the battery technologists and will require some innovative solutions. In a nutshell, the challenges for the development of next-generation LIBs primarily focus on the following directions: increased energy density and safety, reduced cost, the achievement of sustainable and greener LIBs, and increased capacity, while ensuring sustainability and green storage.

2.9 THE NOBEL PRIZE: THE NEW ERA OF LITHIUM-ION BATTERIES

The Nobel Prize is the most prestigious award for scientists working in the fields of physics, chemistry, medicine, peace or literature around the globe. The prize is named after the Swedish scientist Alfred Nobel. He was a chemist as well as an engineer, inventor, and businessman. The scientist had left a will, in which the capital of his wealth was to be converted to safe securities and the interest obtained was to be utilized to establish annual prizes, which were later to be named the "Nobel prize". The will and testament included the lines "*the interest on which shall be annually distributed in the form of prizes to those who, during the preceding year, shall have conferred the greatest benefit on mankind*" (from the official website of the Nobel Prize) [57]. The first Nobel Prize was awarded for peace in 1901.

In 2019, the Nobel Prize in Physics was awarded for the establishment of a rechargeable world by the introduction of lithium-ion batteries. Three eminent scientists, who were behind the development and establishment of successful secondary lithium-ion batteries, were awarded the Nobel Prize in Chemistry: (i) Prof. John B. Goodenough, (ii) Prof. M. Stanley Whittingham, and (iii) Prof. Akira Yoshino (Figure 2.4) [58]. The prize was awarded for their work, that led to the commercialization of portable high-density batteries, which, in turn, made possible the practical use of mobile phones, laptops, palmtops, tablets, and even electric vehicles.

FIGURE 2.4 Images of Nobel laureates in Chemistry, 2019; from left to right: Prof. John. B. Goodenough, Prof. M. Stanley Whittingham, and Prof. Akira Yoshino. Adapted and reproduced with permission from The Royal Swedish Academy of Science Ref. [56]. Copyright © Nobel Media 2019. Photograph: Niklas Elmehed

2.9.1 Prof. John Bannister Goodenough

Prof. John Bannister Goodenough is a world-famous American scientist, who, at the age of 97, became the oldest winner of a Nobel prize. After graduating in mathematics, he made his reputation as a solid-state physicist. He completed his MS and PhD in physics from the University of Chicago, USA, and joined the Lincoln Laboratory at the famous Massachusetts Institute of Technology in 1952, where he developed the concept of cooperative orbital ordering in the transition metal compounds. This invention led to the discovery of random-access memory (RAM). Goodenough was also known for explaining electron transfer (super exchange) of electrons between the overlapped orbitals, which became known as the "Goodenough-Kanamori rules". In 1976, Goodenough joined the Inorganic Chemistry Laboratory of the University of Oxford, UK, as Head, where the major achievement in the battery industry, the introduction of lithium cobalt oxide as an intercalating compound of lithium, was introduced in 1981 [33]. This led to the commercialization of the wireless revolution with lithium-ion batteries. After ten years at the University of Oxford, Goodenough moved to the University of Texas, Austin, where he was appointed to the Virginia H. Cockrell Centennial Chair of Engineering in the Cockrell School of Engineering [57]. According to Scopus, the publication list of J. B. Goodenough has 771 articles published in peer-reviewed journals and 27 reviews, of which 663 publications were published during his time at the University of Texas, with a highest annual publication record of 43 in 2016. Goodenough has several awards and honors to his credit. He won the Japan Prize in 2001, the Enrico Fermi award in 2009, the National Medal of Science in 2011, the Charles Draper Prize in 2014, and the Eric and Sheila Samsun Prime Minister's Prize for Innovation in Alternative Fuels for Transportation in 2015. Goodenough also won the Fellow of the Electrochemistry Society award and the National Academy of Inventors award in 2016. The Welch Award in Chemistry was awarded in 2017 and the subsequent year he was awarded the Benjamin Franklin Award, before Goodenough was co-awarded the Nobel Prize in Chemistry for 2019 [57].

2.9.2 Prof. Michael Stanley Whittingham

Prof. Michael Stanley Whittingham is a British-American scientist, specializing in chemistry. He is currently a Distinguished Professor of Chemistry and Materials Science at Binghamton University, State University of New York, USA. He was the man behind the development of the intercalation of compounds. He also invented the first rechargeable battery, implementing the intercalation behavior of transitional metal chalcogenides. Whittingham was working for the Exxon research and development company, from where he developed and patented the first rechargeable lithium-ion battery in 1977 [33]. In his first battery, lithium metal was used as the anode and titanium disulfide as the cathode. This invention laid the foundation for rechargeable batteries, leading to the era of portable batteries. In 1988, he joined Binghamton University and continues his research in the field of energy [33]. In 2019, Whittingham was co-awarded the Nobel Prize in Chemistry.

2.9.3 Prof. Akira Yoshino

Prof. Akira Yoshino works at Meijo University in Nagoya, Japan, and is Honorary Fellow of the Asahi Kasei Corporation. He is a Japanese chemist who completed his MS in 1970 from the Department of Petrochemistry, Graduate School of Engineering, Kyoto University. Yoshino earned his doctorate from the Graduate School of Engineering, Osaka University in 2005. From 1972 onward, Yoshino has worked for the Asahi Kasei Corporation in various posts. In 1987, Yoshino patented the first secondary Li^+-ion batteries (the first rechargeable batteries) [42]. In 1999, Yoshino received the Fiscal 1998 Chemical Technology Prize from the Chemical Society of Japan and the award of the Electrochemical Society. In 2001, he received the Meritorious Achievement Prize, from the New Technology Development Foundation (Ichimura Foundation) and the Encouragement Prize of Invention of the Minister of Education, Culture, Sports, Science, and Technology, from the Japan Institute of Invention and Innovation. In 2011, Yoshino was awarded the Yamazaki-Teiichi Prize from the Foundation for the Promotion of Material Science and Technology of Japan and the C&C Prize from the NEC C&C Foundation. In 2012, he was awarded the Fellow of the Chemical Society of Japan award and the IEEE Medal for Environmental and Safety Technologies from the Institute of Electrical and Electronics Engineers. In 2013, he received The Global Energy Prize and The Kato Memorial Prize from the Kato Foundation for the Promotion of Science. In 2014, Yoshino was awarded the Medal of Honor with purple ribbon from the Government of Japan. In 2016 he received The NIMS Award 2016 from the National Institute for Materials Science. In 2018, he was awarded "The Japan Prize". In 2019, Yoshino, along with Whittingham and Goodenough, received the Nobel Prize in Chemistry, while he also received the European Inventor Award that year.

2.10 THE DRAPER PRIZE AND LITHIUM-ION BATTERY (2014)

The National Academy for Engineering, situated in Washington, D.C., USA, is a non-governmental organization established in 1964. The Academy established a prize in the name of Charles Stark Draper (1901–1987) in 1988. Charles Stark Draper was a well-known American scientist/engineer, known as "the father of inertial navigation". He was also the founder of the Massachusetts Institute of Technology Confidential Instrument Development Laboratory in 1932, the name of which was later changed to the Charles Stark Draper Laboratory. He was an expert in aerospace engineering and contributed to the development of several sensing and space craft instruments. He also worked as part of several national institutions, including the U.S. National Academy for Engineering. The Charles Stark Draper Prize was established as an endowment prize for the leading contributors in science, engineering, and technology. The prize consists of a hand-inscribed certificate, a $500,000 cash prize, and a gold medal. It is one of the three most prestigious awards in engineering. In 2014, four pioneers behind the development, establishment, and commercialization of lithium-ion batteries were awarded this prestigious award. Prof. John B. Goodenough, Mr. Yoshio Nishi, Prof. Rachid Yazami, and Prof. Akira Yoshino were

History of Lithium-Ion Batteries

FIGURE 2.5 Photograph of Draper Prize winners of 2014 for research on lithium-ion batteries. From left to right: Prof. Akira Yoshino, Prof. John B. Goodenough, Mr. Yoshio Nishi, and Prof. Rachid Yazami. Adapted and reprinted with permission from National Academy of Sciences for the National Academy of Engineering Ref. [57].

the four main people behind the lithium-ion battery (Figure 2.5) [59]. Brief biographies of Prof. John B. Goodenough and Prof. Akira Yoshino have been presented in Sections 2.9.1 and 2.9.2, and the biographies of Prof. Rachid Yazami and Mr. Yoshio Nishi are presented below.

2.10.1 Prof. Rachid Yazami

Prof. Rachid Yazami is the man behind the intercalation of lithium ions into graphite. He is a native of Morocco and completed his post-graduate studies in chemistry and chemical technology in 1978 in France. In 1983, the intercalation of alkali metal ions, such as lithium, into graphite was patented as part of his research work [60]. Yazami was co-inventor on this patent, along with Philippe Touzain and Jacques Maire. From 1985 onwards, he worked as a research associate at the French Centre National de la Recherche Scientifique (CNRS), becoming Research Director in 1998. In 2013, Yazami joined the Energy Research Institute (ERI) at Nanyang Technological University (NTU) in Singapore, as director of the energy storage program, Professor and adjunct Principal Scientist. According to the Scopus list, he has 162 publications in peer-reviewed journals to his credit and more than 150 patents. In 2019, Yazami earned a patent for an electrode fabricated with silicon, tin, and a non-alloy of a transition metal for a battery [61]. He has received several awards and honors, in addition to the Draper Prize of 2014. Yazami was awarded the Royal Wissam Malaki (Royal Medal) of Intellectual Competency from the King of Morocco, and Global Energy Prize in 2014. In 2016, he received the Medal of Chevalier de la Légion d'Honneur, and in 2018 he was the TAKREEM Awards Laureate for Science and Technological Achievement. In addition to these honors, Yazami has also received awards from the Institute of Electrical and Electronics Engineers (IEEE), the Japan Society for the Promotion of Science (JSPS), the North Atlantic Treaty Organization (NATO), the National Aeronautics and Space Administration (NASA), as well as the Marius Lavet Prize.

2.10.2 Mr. Yoshio Nishi

Mr. Yoshio Nishi is the retired senior vice-president and chief technology officer of the Sony Corporation. He was a specialist in chemistry, having graduated from the Faculty of Applied Chemistry of the Department of Technology at Keio University, Japan, in 1966. He joined Sony, where he was engaged in research and development on fuel cells and electrochemical cells with non-aqueous electrolytes. In 1991, Yoshio Nishi and Keizaburo Tozawa commercialized the first lithium-ion battery (LIB). He has several awards to his credit. A technical award was made to Nishi from the Electrochemical Society in 1994. He was also awarded the Kato Memorial Award from the Kato Foundation for the Promotion of Science (Japan) in 1998. Nishi was also awarded the Ichimura Award from the New Technology Development Foundation (Japan) in 2000 in recognition of his contributions to LIB technology.

2.11 SUMMARY

LIBs have clear fundamental advantages and decades of research history behind them, which have allowed the development of today's LIBs, with a high energy density, power density, cycling stability, rate capability, and Columbic efficiency. As energy demand is increasing from day to day, LIB research is continuing into high-performance batteries, which can be used efficiently even under less-than-ideal conditions. Because of the global effort taking to establish zero-emission automobiles, battery technologists are involved in hunting for new electrode materials to go beyond the current LIB boundaries of cost, energy density, power density, cycle life, rate capability, and safety, for heavy-duty applications, such as electric vehicles. This chapter narrates the history of the birth of LIBs and the important milestones up to the present day, as well as commemorating the eminent scientists responsible for the development and commercialization of practical LIBs. The chapter also provides an insight into the challenges faced and the opportunities offered by the next-generation LIBs.

ACKNOWLEDGMENT

Authors Dr. Jabeen Fatima M. J. and Dr. Prasanth Raghavan would like to acknowledge the Kerala State Council for Science, Technology and Environment (KSCSTE), Kerala, India, for financial assistance.

REFERENCES

1. Rinkesh 15+ amazing reasons why we should conserve energy. In: *Conserve Energy Future*. https://www.conserve-energy-future.com/whyconserveenergy.php

2. Administration UEI (2017) International energy outlook 2017. In: *IEO2017*. https://www.eia.gov/outlooks/ieo/pdf/0484(2017).pdf
3. Cobb K (2014) World war III: It's here and energy is largely behind it. In: *Oilprice.com*. https://oilprice.com/Energy/Energy-General/World-War-III-Its-Here-And-Energy-Is-Largely-Behind-It.html
4. Kumar A, Kumar K, Kaushik N, et al. (2010) Renewable energy in India: Current status and future potentials. *Renew Sustain Energy Rev* 14:2434–2442. https://doi.org/10.1016/j.rser.2010.04.003
5. Winter M, Brodd RJ (2004) What are batteries, fuel cells, and supercapacitors? *Chem Rev* 104:4245–4269. https://doi.org/10.1021/cr020730k
6. Lim D-H, Haridas AK, Figerez, SP, Raghavan, P, Matic, A, Ahn J-H (2018) Tailor-made electrospun multilayer composite polymer electrolytes for high-performance lithium polymer batteries. *J Nanosci Nanotechnol* 18:6499–6507. https://doi.org/10.1166/jnn.2018.15689
7. Singh N, Galande C, Miranda A, et al. (2012) Paintable battery. *Sci Rep* 2:6–10. https://doi.org/10.1038/srep00481
8. Hu L, Wu H, La Mantia F, et al. (2010) Thin, flexible secondary Li-Ion paper batteries. *ACS Nano* 4:5843–5848. https://doi.org/10.1021/nn1018158
9. Mills AA (2001) The "Baghdad battery". *Bull Sci Instrum Soc* 68:35–37.
10. Martins GF (1990) Why the Daniell cell works! *J Chem Educ* 67:482.
11. Ruftschi P (1977) Review on the lead-acid battery science and technology. *J Power Sources J* 2:3–24.
12. Leclanchè GL (1867) Improvement in combining generating and secondary or accumulating galvanic battery. Patent: Number 64113. 22–24. https://patentimages.storage.googleapis.com/bb/d3/e7/05646c7b9a032c/US64113.pdf
13. Jungner EW (1894) Thermoelectric battery. 2–4.
14. Pinnangudi B, Kuykendal M, Bhadra S (2017) *Smart Grid Energy Storage*. Elsevier Ltd.
15. Eismin TK (2014) *Study Guide for Aircraft Electricity and Electronics*. McGraw Hill Professional.
16. Berndt D (1997) Maintenance-free batteries: Lead-Acid, Nickel/Cadmium, Nickel/Metal hydride. *A Handbook of Battery Technology*. Research Studies Press. ISBN: 9780471970187
17. Bohlsen M (2019) A look at the top 5 lithium-ion battery manufacturers in 2019. In: *Seek. Alpha*. https://seekingalpha.com/article/4289626-look-top-5-lithium-ion-battery-manufacturers-in-2019
18. Whittingham MS (1976) Electrical energy storage and intercalation chemistry. *Science (80-)* 192.
19. Steve L (2010) The great battery race. In: *foreignpolicy.com*. https://foreignpolicy.com/2010/10/12/the-great-battery-race/
20. Kirchhoff B (1861) XXIV.—On chemical analysis by spectrum-observations. *Q J Chem Soc London* 13:270–289. https://doi.org/10.1039/QJ8611300270
21. Besenhard JO, Fritz HP (1974) Cathodic reduction of graphite in organic solutions of alkali and NR4+ salts. *J Electroanal Chem* 53:329–333. https://doi.org/10.1016/S0022-0728(74)80146-4
22. Besenhard JO (1976) The electrochemical preparation and properties of ionic alkali metal and NR^{4-}graphite intercalation compounds in organic electrolytes. *Carbon N Y* 14:111–115. https://doi.org/10.1016/0008-6223(76)90119-6
23. Eichinger G, Besenhard JO (1976) High energy density lithium cells: Part II. Cathodes and complete cells. *J Electroanal Chem Interfacial Electrochem* 72:1–31. https://doi.org/10.1016/S0022-0728(76)80072-1

24. Schöllhorn R, Kuhlmann R, Besenhard JO (1976) Topotactic redox reactions and ion exchange of layered MoO3 bronzes. *Mater Res Bull* 11:83–90. https://doi.org/10.1016/0025-5408(76)90218-X
25. Besenhard JO, Schöllhorn R (1976) The discharge reaction mechanism of the MoO3 electrode in organic electrolytes. *J Power Sources* 1:267–276. https://doi.org/10.1016/0378-7753(76)81004-X
26. (1976) Editorial board. *J Electroanal Chem Interfacial Electrochem* 72:iii. https://doi.org/10.1016/S0022-0728(76)80071-X
27. Zanini M, Basu S, Fischer JE (1978) Alternate synthesis and reflectivity spectrum of stage 1 lithium-graphite intercalation compound. *Carbon N Y* 16:211–212. https://doi.org/10.1016/0008-6223(78)90026-X
28. Basu S, Zeller C, Flanders PJ, et al. (1979) Synthesis and properties of lithium-graphite intercalation compounds. *Mater Sci Eng* 38:275–283. https://doi.org/10.1016/0025-5416(79)90132-0
29. Godshall NA, Raistrick ID, Huggins RA (1980) Thermodynamic investigations of ternary lithium-transition metal-oxygen cathode materials. *Mater Res Bull* 15:561–570. https://doi.org/10.1016/0025-5408(80)90135-X
30. Godshall NA (1979) Electrochemical and thermodynamic investigation of ternary lithium - transition metal-oxide cathode materials for lithium batteries: Li_2MnO_4 spinel, $LiCoO_2$, and $LiFeO_2$. In: *156th Meeting of the Electrochemical Society*, Los Angeles, CA.
31. Godshall NA, Raistrick ID, Huggins RA (1980) Thermodynamic investigations of ternary lithium-transition metal-oxygen cathode materials. *Mater Res Bull* 15:561–570. ISSN 0025-5408, https://doi.org/10.1016/0025-5408(80)90135-X
32. Mizushima K, Jones PC, Wiseman PJ, Goodenough JB (1980) Li_xCoO_2 (0<x<–1): A new cathode material for batteries of high energy density. *Mater Res Bull* 15:783–789. https://doi.org/10.1016/0025-5408(80)90012-4
33. Mizushima K, Jones PC, Wiseman PJ, Goodenough JB (1981) Li_xCoO_2 (0<x<1): A new cathode material for batteries of high energy density. *Solid State Ionics* 4:171–174.
34. Goodenough JB, Mizushima K, Wiseman PJ (1979) Li_xCoO_2 (0<x<1): A new cathode material for batteries of high energy density.
35. Godshall NA (1986) Lithium transport in ternary lithium-copper-oxygen cathode materials. *Solid State Ionics* 18–19:788–793. https://doi.org/10.1016/0167-2738(86)90263-8
36. Raistrick ID, Godshall NA, Huggins RA (1980) Ternary compound electrode for lithium cells. U.S. Patent US4340652A.
37. Thackeray MM, David WIF, Bruce PG, Goodenough JB (1983) Lithium insertion into manganese spinels. *Mater Res Bull* 18:461–472. https://doi.org/10.1016/0025-5408(83)90138-1
38. Bravo Diaz L, He X, Hu Z, Restuccia F, Marinescu M, Varela Barreras J, Patel Y, Offer G, Rein G (2020) Meta-review of fire safety of lithium-ion batteries: Industry challenges and research contributions. *J Electrochem Soc* 167:090559.
39. Yazami R, Touzain P, Bonnetain L (1983) Generateurs electrochimiques lithium/composes d'insertion du graphite avec $FeCl_3$, $CuCl_3$, $MnCl_2$ et $CoCl_2$. *Synth Met* 7:169–176.
40. Yazami R, Touzain P (1983) A reversible graphite-lithium negative electrode for electrochemical generators. *J Power Sources* 9:365–371. https://doi.org/10.1016/0378-7753(83)87040-2
41. Yoshino A, Sanechika K, Nakajima T (1985) Secondary batteries. Japanese Patent 1989293.
42. Yoshino A, Sanechika K, Nakajima T (1989) Secondary batteries. U.S. Patent US4668595A.

43. Yoshino A (2012) The birth of the lithium-ion battery. *Angew Chemie - Int Ed* 51:5798–5800.
44. Bryce R (2011) A million electric vehicles. In: *National Review* https://www.nationalreview.com/2011/06/million-electric-vehicles-robert-bryce/
45. Tarascon JM, Armand M (2010) Issues and challenges facing rechargeable lithium batteries. In: *Materials for Sustainable Energy: A Collection of Peer-Reviewed Research and Review Articles from Nature Publishing Group.* pp. 171–179.
46. Tarascon JM (2002) Vers des accumulateurs plus performants. Un problème de matériaux et d'interfaces. *L' Actual Chim* N251:130–137.
47. Li X, Gu M, Hu S, et al. (2014) Mesoporous silicon sponge as an anti-pulverization structure for high-performance lithium-ion battery anodes. *Nat Commun* 5:4105. https://doi.org/10.1038/ncomms5105
48. Pan L, Wang H, Gao D, et al. (2014) Facile synthesis of yolk–shell structured Si–C nanocomposites as anodes for lithium-ion batteries. *Chem Commun* 50:5878–5880. https://doi.org/10.1039/C4CC01728E
49. Chan CK, Peng H, Liu G, et al. (2008) High-performance lithium battery anodes using silicon nanowires. *Nat Nanotechnol* 3:31–35. https://doi.org/10.1038/nnano.2007.411
50. Yoon S, Manthiram A (2009) Sb-MOx-C (M = Al, Ti, or Mo) nanocomposite anodes for lithium-ion batteries. *Chem Mater* 21:3898–3904. https://doi.org/10.1021/cm901495h
51. Gauthier M, Carney TJ, Grimaud A, et al. (2015) Electrode–electrolyte interface in Li-Ion batteries: Current understanding and new insights. *J Phys Chem Lett* 6:4653–4672. https://doi.org/10.1021/acs.jpclett.5b01727
52. Kusoglu A, Weber AZ (2015) Electrochemical/mechanical coupling in ion-conducting soft matter. *J Phys Chem Lett* 6:4547–4552. https://doi.org/10.1021/acs.jpclett.5b01639
53. Manthiram A, Fu Y, Chung S-H, et al. (2014) Rechargeable lithium–sulfur batteries. *Chem Rev* 114:11751–11787. https://doi.org/10.1021/cr500062v
54. Chung S-H, Chang C-H, Manthiram A (2016) A carbon-cotton cathode with ultrahigh-loading capability for statically and dynamically stable lithium–sulfur batteries. *ACS Nano* 10:10462–10470. https://doi.org/10.1021/acsnano.6b06369
55. McCloskey BD (2015) Attainable gravimetric and volumetric energy density of Li–S and Li ion battery cells with solid separator-protected Li metal anodes. *J Phys Chem Lett* 6:4581–4588.
56. McCloskey BD (2015) Attainable gravimetric and volumetric energy density of Li–S and Li Ion battery cells with solid separator-protected Li metal anodes. *J Phys Chem Lett* 6:4581–4588. https://doi.org/10.1021/acs.jpclett.5b01814
57. Nobel Prize History. In: *Nobel Media.* www.nobelpeaceprize.org/History
58. Nobel Media (2019) Nobel Prize Chemistry 2019. www.nobelprize.org/prizes/chemistry/2019/summary
59. (2014) Charles stark draper prize for engineering.
60. Touzain P, Yazami R, Maire J (1986) Insertion compounds of graphite with improved performances and electrochemical applications of those compounds. U.S. Patent: US4584252A, 2–4.
61. Yazami R, Zhang W (2016) Electrode, battery cell and battery cell arrangement. U.S. Patent 15/738819.

3 Polyethylene Oxide (PEO)-Based Solid Polymer Electrolytes for Rechargeable Lithium-Ion Batteries

Prasanth Raghavan, Abhijith P. P., Jishnu N. S., Akhila Das, Neethu T. M. Balakrishnan, Jabeen Fatima M. J., and Jou-Hyeon Ahn

CONTENTS

3.1 Introduction 57
3.2 Preparation of PEO-Based Solid Polymer Electrolytes 58
3.3 Copolymer-Based PEO Solid Polymer Electrolytes 59
3.4 Conclusions 74
Acknowledgment 75
References 75

3.1 INTRODUCTION

Rechargeable batteries play a crucial role in the improved performance of portable electronic devices (smartphones, laptops, cameras, etc.), as well as electric/hybrid electric vehicles. A rechargeable battery is an electrochemical system with primarily three components: (i) the anode, which is oxidized, releasing electrons to the external circuit during the electrochemical reaction; (ii) the cathode, which is reduced, accepting electrons from the anode during the electrochemical reaction; and (iii) the electrolyte or the ionic conductor, which provides the medium for transfer of charge, as ions, inside the cell between the anode and the cathode. The electrolyte typically has a finite ionic conductivity and a negligible electronic conductivity. Among the different types of batteries, lithium-ion batteries (LIBs) have attracted much attention due to their characteristic electrochemical properties, such as high energy density, negligible self-discharge, long cycling stability, etc. Typically, lithium-ion batteries consist of a graphite negative electrode, organic liquid electrolyte, and lithium transition-metal oxide (LiCoO$_2$) positive electrode. LIBs were first commercialized in 1991, since when such batteries have been widely distributed all over the world as a power source for mobile electronic devices, such as cell phones, laptops and camcorders [1–3]. The conventional type of electrolyte, which was being used in batteries, was a liquid electrolyte; however, the use of liquid electrolytes may cause leakage in rechargeable batteries, leading to fire or explosions, which, in turn, cause serious safety issues [4]. In order to solve this problem, the liquid electrolyte had to be replaced by a polymer or ceramic-based solid electrolyte [5]. The solid electrolytes are not only capable of reducing the cathodic limit of the potential window and thus providing a high capacity [4] but can also help to prevent leakages, avoiding the risk of thermal runaway, fire or explosions.

Ceramic electrolytes are a type of electrolyte in which the ionic conduction takes place by the movement of ionic point defects. One of the main issues associated with ceramic electrolytes is their low room-temperature ionic conductivity, so that they are practically employed only in high-temperature batteries, operating above 70°C. The problems associated with liquid electrolytes (primarily leakage and associated safety issues) and ceramic electrolytes (mainly low-room-temperature ionic conductivity) can be solved by using solid polymer electrolytes, which possess high ionic conductivity with negligible electronic conductivity. These types of electrolytes also exhibit good thermal stability as well as chemical stability.

Based on the polymer matrix, there are various kinds of solid polymer electrolytes, among which the polymeric matrix polyethylene oxide (PEO) has been well studied and is an widely used electrolyte in LIBs. PEO is a polymer electrolyte matrix, which is widely used as a solid polymer electrolyte in rechargeable batteries. The most promising factors regarding PEO are its good interfacial stability and its good flexibility [6]. Since PEO exhibits a low glass transition temperature and shows excellent interfacial stability with lithium, this makes the formation of complexes with lithium salts easier. PEO shows low ionic conductivity at room temperature because it possesses a high degree of crystallinity at room temperature,which is the factor at which the commercialisation of PEO becomes complex.

3.2 PREPARATION OF PEO-BASED SOLID POLYMER ELECTROLYTES

The choices of polymer host and the metal salt play a key role in achieving a good complexed polymer-metal-salt system. PEO has a high molecular weight, and has been widely used as a polymer host for the preparation of solid polymer electrolytes because of its ability to dissolve high concentrations of a wide variety of metal salts. PEO and its derivatives have good chain flexibility, low glass transition temperatures ($T_g = -66°C$), low softening temperature (about 40°C), superior electrochemical stability to lithium metal, and excellent solvation properties with conductive lithium salts [7, 8]. Typically, the prototypical solid polymer electrolyte, PEO-LiX, can be simply prepared by mechanically blending PEO with a suitable lithium salt (LiX) [9, 10]. The transport of Li$^+$-ions in these electrolytes has been associated with the local relaxation and segmental motion of the amorphous regions (the "hopping" mechanism) in the PEO chains [11]. These polymers often show higher crystallinity at lower temperatures and the electrolytes exhibit low room-temperature ionic conductivity (typically of the order of 10^{-8}–10^{-6} S cm^{-1}) and inferior Li$^+$ transference numbers (0.2–0.3) at room temperature, which seriously affects the high rate capability of LIBs. This necessitates operation at higher temperatures (generally, ~70°C) for their successful utilisation in practical battery applications. For this reason, various methods have been adopted to synthesise PEO-based solid polymer electrolyte (SPEs) with good mechanical properties, superior electrochemical stability, and, in particular, high ionic conductivity [12, 13]. PEO-based solid polymer electrolytes are prepared by mechanical blending, hydraulic hot pressing [13] and solution casting. In the solution-casting method, the polymer and lithium salt (LiX) are first dissolved in a suitable solvent (methanol or Dimethylacetamide (DMAc)) to obtain a uniform viscous solution by mechanical stirring. Then, the degassed, bubble-free solution is cast on a mould to get a uniform film of sufficient thickness, and dried either at room temperature or in a vaccum oven. It was reported that, in the SPE, ionic conductivity increases and activation energy decreases with an increase in salt concentration. To improve the lithium ion conductivity, thermal stability, mechanical properties, dimensional stability and electrochemical performance of PEO-based solid polymer electrolytes, different approaches, such as blending with other polymers [14, 15], crosslinking [16, 17], incorporation of nano-sized ceramic fillers [18], mixing of plasticizers or room-temperature ionic liquids [19], etc., have been adopted.

Incorporation of various nanoparticles, such as TiO_2 [20], Al_2O_3 [21] and SiO_2 [22], helps to improve the thermal stability and reduce the crystallinity of the polymer [23]. The interfacial resistance between PEO-based solid electrolytes and electrodes can be controlled by using metal organic frameworks (MOFs), with an organic functional group as an active additive, which could also improve the conductivity of lithium ions [24, 25]. Blending of two or more polymers with PEO is also a promising way of improving the physical as well as the electrochemical properties of the polymer electrolyte. The blending of two or more polymers can combine the advantageous physical as well as electrochemical properties of different polymers, which could not be achieved with a single polymer [26, 27]. The various types of polymers which are being used for copolymerisation are PMMA, PVdF, and PAN, etc., which result in a solid polymer electrolyte with good ionic conductivity and thermal stability.

3.3 COPOLYMER-BASED PEO SOLID POLYMER ELECTROLYTES

The crystallinity of the polymer host for the preparation of polymer electrolytes, especially for solid polymer electrolytes, has a great influence on their room-temperature lithium-ion conductivity and ion transport number. Copolymerisation with another monomer can effectively improve the amorphous content (considered to be the main conducting phase of Li$^+$-ions) of the resulting copolymer [13, 28]. It is well known that, generally, copolymers show much lower tendency toward crystallisation compared with their individual counterparts, known as homopolymers. Because of the characteristic nature of the copolymers, many researchers have made attempts to obtain solid polyether-type solvents, in which short ethylene oxide (EO) segments are separated by other monomeric units that do not enter the crystalline lattice of PEO. The reaction products of PEG (nEG; n = 1, 2, 4, 9) with dimethyldichlorosilane (DDS) [29] (structure and equations, SE-01) or methylene bromide [30, 31] (SE-02) through the polycondensation reaction, as shown below, were the first fully amorphous EO-based linear copolymers employed for the preparation of solid polymer electrolytes.

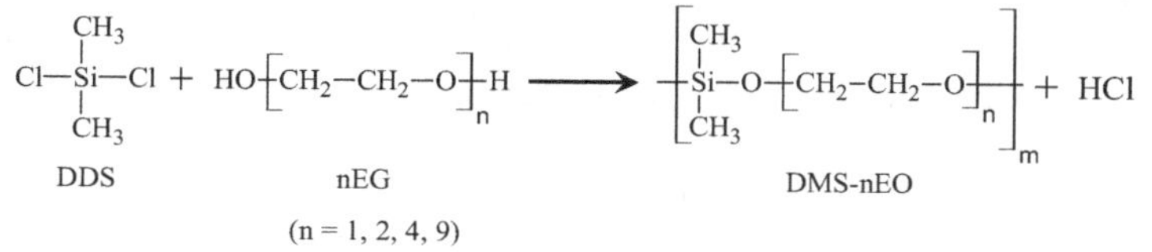

$$Cl-\underset{CH_3}{\underset{|}{Si}}-Cl + HO{\left[CH_2-CH_2-O\right]}_n H \longrightarrow {\left[\underset{CH_3}{\underset{|}{Si}}-O{\left[CH_2-CH_2-O\right]}_n\right]}_m + HCl \qquad SE\text{-}01$$

DDS   nEG   DMS-nEO

(n = 1, 2, 4, 9)

$$Br-\underset{H}{\underset{|}{\overset{H}{\overset{|}{C}}}}-Br + Na{\left[O-CH_2-CH_2-O\right]}_m Na \longrightarrow {\left[O{\left[CH_2-CH_2-O\right]}_m CH_2\right]}_n + 2NaBr \qquad SE\text{-}02$$

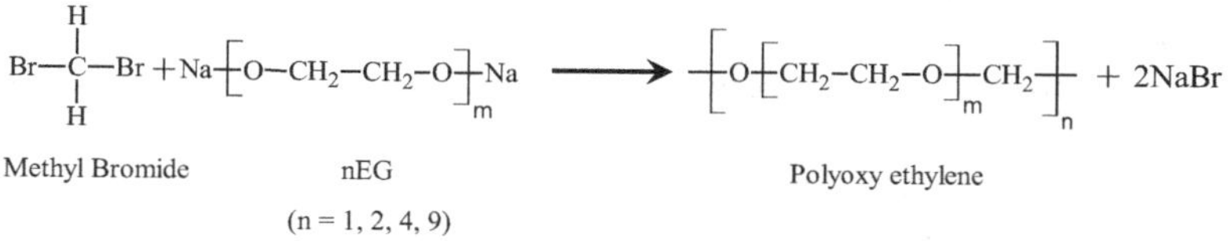

Methyl Bromide   nEG   Polyoxy ethylene

(n = 1, 2, 4, 9)

The preparation procedures explained in these studies show precise control over the chain length of EO monomeric units. The study shows the effect of chain length on the transport properties and lithium ion conductivity. Highest conductivities, in the order of 10^{-3} S cm^{-1}, were reported for the copolymer in which the length of EO monomeric units sequences was from five to seven, a finding which is attributed to the number of oxygen atoms in such chain segments, which is sufficient to dissociate the lithium salt and effectively complex lithium cations. In addition, the length of the chain segments is too short to form a crystalline phase in the polymer, so that the electrolyte exhibits the optimal amorphous content and provides the maximum soft segment solid phase for lithium transportation. The introduction of dimethyldichlorosilane units to PEO by copolymerization contributed to the increase in carrier mobility. The highest conductivity observed in the solution of poly(dimethylsiloxane-co-ethylene oxide) copolymer (at [LiClO$_4$]/[EO unit]= 0.03) was 1.5×10^{-4} S cm^{-1}. The increased conductivity was associated with the dissolution of LiClO$_4$ in the co-polymers, which showed that the dissociation of LiClO$_4$ and the higher conductivity were due to migration of the dissociated ions [29]. The solid-state electrolyte, prepared by blending the copolymer of optimized chain length with lithium salts exhibited alithium ion conductivity in the range of 1 to 5×10^{-5} S cm^{-1} at ambient temperature. The ionic conductivity of a mixture of polymer with LiCF$_3$SO$_3$ ([O]/[Li]=25) was about 5×10^{-3} and 3×10^{-3} S cm^{-1} at 25 and 20°C, respectively. This value of conductivity is similar to values reported for comb-shaped polymers containing short poly(oxyethylene) segments [32–35]. The copolymerisation of EO with epoxides catalysed with alkylalumoxanes, i.e. products of partial hydrolysis of trialkylaluminum compounds [36–38], resulted in random copolymers having the general structure as in SE-03, and is demonstrated to be a very effective method by which to synthesise amorphous matrices for the preparation of solid polymer electrolytes with good ionic conductivity.

$$\left[\left[-CH_2-CH_2-O\right]_m-CH_2-\underset{\underset{R}{|}}{CH}-O\right]_n \qquad \text{SE-03}$$

The copolymer polymerisation of propylene oxide (PO) with EO forms poly(ethylene oxide-*co*-propylene oxide) (SE-03), in which the EO/PO molar ratio was *ca.* 5, showed the most interesting ion transport properties. At this ratio, the random copolymer formed was completely amorphous, and the presence of methyl side groups led to internal plastification. Electrolytes obtained by mixing the copolymer poly(ethylene oxide-*co*-propylene oxide) with lithium salt exhibited a glass transition temperature below –70°C, compared with T_g values 30 to 40 degrees higher for analogous systems obtained with PEO. The fully amorphous nature of the material (i.e., the absence of the crystalline phase), high room-temperature flexibility and a glass transition temperature of the poly(ethylene oxide-*co*-propylene oxide) copolymer similar to that of natural rubber (*cis*-1,4 poly-isoprene) enabled fast lithium ion transport and a higher transference number. The appropriate complexing of the copolymer with a suitable lithium salt concentration resulted in an ionic conductivity of 1×10^{-4} S cm^{-1} at ambient temperature, which was three to four orders of magnitude higher than that for the PEO complexes [38]. The conductivity decreased in the following order with respect to the anion used: $CF_3SO_3^- > BF_4^- > CF_3COO^- > ClO_4^-$. It had been found previously that the amorphous electrolytes, prepared with high-molecular-weight poly(ethylene oxide-*co*-propylene oxide) containing 16–34 mol.% of PO, exhibited an ionic conductivity exceeding 10^{-5} S cm^{-1} at ambient temperature, which is about three orders of magnitude higher than that for analogous systems with PEO [37]. Other random copolymers, like PEO–epichlorohydrin and methyleneoxy–PEO, exhibited excellent ionic conductivity by increasing the EO ratio in the random co-polymer obtained. Nishimoto et al. [39] synthesized poly(ethylene oxide-*co*-2-(2-methoxyethoxy)ethyl glycidyl ether) P(EO-MEEGE) as a high-molecular-weight (>10^6) polyether comb polymer for the preparation of solid polymer electrolytes. The elastic polymer electrolyte films, having no chemically cross-linked structures, were prepared by mixing P(EO-MEEGE) (containing 9% MEEGE) with a sufficient concentration of lithium bis(trifluoromethylsulfonyl)imide (LiTFSI) at room temperature, and exhibited high ionic conductivities of 10^{-4} and 10^{-3} S cm^{-1} at 30 and 80°C, respectively. The improved ionic conductivity of the electrolyte can be attributed to the synergistic effects of lower crystallinity and higher number of highly mobile ether side chains, which could dissociate more lithium salt.

The solid phase solvents, such as poly(1,3-dioxolanes) [40] and linear EO co-polymers with sulfur dioxide or carbon dioxide [41–43] have been used for solubilizing lithium salts for the preparation of SPEs with high ionic conductivity. Unfortunately, the use of solid solvents or linear copolymers adversely affect the mechanical strength of the electrolytes and most of them are subject to creep under the pressure applied in electrochemical devices, leading to short-circuiting at temperatures above 40°C. Hence, more mechanically stable electrolytes were prepared by forming three-dimensional (3D) crosslinked network structures, either by radiation or chemical cross-linking, can greatly improve the mechanical properties [44] or prevent recrystallisation [45, 46]. The crosslinked network structures, based on polyethylene glycol (PEG) and trifunctional EO oligomers, has been extensively used to produce mechanically and dimensionally stable solid polymer electrolytes.

The most common approach was based on polyaddition reactions with isocyanates, leading to polyurethane structures [32, 47–51] or styrene-butadiene-styrene (SBS) block copolymers [50] with pendant short-chain poly(ethylene oxide) (PEO). The electrolyte prepared from a network of poly(dimethylsiloxane-grafted ethylene oxide) copolymer crosslinked by an aliphatic isocyanate (grafted PDMS) and containing 10 wt.% $LiClO_4$ [48] or the SBS block copolymer were combined with lithium trifluoromethanesulfonate [50] in different concentrations and exhibited room-temperature ionic conductivity of 10^{-5} S cm^{-1}. The electrolytes were mechanically stable up to 120°C [48–50] and were found to be suitable for battery applications at temperatures higher than 60°C. The conductivity exceeded 1.5×10^{-4} S cm^{-1} at a temperature of 80°C for the systems doped with $LiClO_4$. The immobilisation of EO segments in the network structure reduces the segmental mobility, which results in some sacrifice of conductivity. This sacrifice could be addressed to a great extent by reducing the degree of crosslinking or by using the polymeric chain segments with repeating units, which are very flexible. Among such different matrices, the most pronounced effect was achieved by using EO segments attached to the short polysiloxane chains as polyols (SE-04) [48].

$$H_3C-\underset{\underset{CH_3}{|}}{\overset{\overset{CH_3}{|}}{Si}}-O\left[\underset{\underset{CH_3}{|}}{\overset{\overset{CH_3}{|}}{Si}}-O\right]_x ---- \left[\underset{\underset{(CH_2)_3}{|}}{\overset{\overset{CH_3}{|}}{Si}}-O\right]_y ---- \underset{\underset{CH_3}{|}}{\overset{\overset{CH_3}{|}}{Si}}-CH_3 \quad \text{SE-04}$$

$$\left[H_2C-CH_2-O\right]_z H$$

where x is ~56 and y is ~16; unit PEO = -(CH_2- CH_2-O)$_{22}$-H, and average molecular weight is 22,000.

Floriańczyk et al. [52] synthesized a PEO–PAAM copolymer by the thermopolymerisation of high-molecular-weight PEO and methyl methacrylate (MMA). It was found that grafting of an MMA chain onto the backbone PEO chain can significantly reduce the T_g and crystallinity of PEO at room temperature. The PEO–PAAM–$LiClO_4$-based SPE, formed by combining the grafted polymer with $LiClO_4$, shows good mechanical properties and ionic conductivity of 6.6×10^{-5} and 10^{-3} S cm^{-1} at room temperature and 100°C, respectively. PAAM can also be a potential candidate to copolymerise with PEO, offering greater chain flexibility, lower crystallinity and higher polarity than the resulting copolymer. The strong polarity can promote the dissociation of more lithium salt, and the lower crystallinity facilitates the easy transport of the lithium ions, thereby improving the lithium-ion transference number. The PEO–PAAM-based electrolyte exhibited an optimal room-temperature ionic conductivity of 4×10^{-5} S cm^{-1} [53].

PEO-Based Solid Electrolytes 63

In addition to the random, block and grafted co-polymers, branched copolymers, having short ion coordinating segments, containing 5–22 EO monomeric units attached to the backbones of inert polymers, have also been demonstrated for the preparation of solid polymer electrolytes. From the earlier reports, it was found [32] that the phosphazene polymers are more attractive for introducing polar side chains than siloxanes. Linear polyphosphazenes consists of a backbone of alternating phosphorus and nitrogen atoms, with two side groups attached to each phosphorus atom [54]. Studies on amino-substituted ($NHCH_3$, $N(CH_3)_2$), short-chain alkoxy-substituted (OCH_2CF_3) and mixed-ligand (OC_6H_5, NHC_5H_4N) polyphosphazenes showed that these systems did not perform satisfactorily as electrolytes, which indicates complex formation between the polymer $(NP(OCH_2CF_3)_2)_n$ and the alkali salts. The alkali salt-doped polymer electrolyte systems exhibited relatively low ionic conductivities of $<10^{-6}$ S cm^{-1} at 100°C. The early studies with branched copolymers have been focused on comb-like derivatives of polyphosphazenes [32, 55] (SE-05), polymethacrylates [56, 57] (SE-06) and triblock styrene-butadiene-styrene (SBS) copolymers [50] (SE-07). The branch copolymers, having short side chains of ethylene oxide, doped with lithium salts, form amorphous electrolytes exhibiting ionic conductivities in the range of 10^{-5} to 10^{-6} S cm^{-1} at ambient temperature.

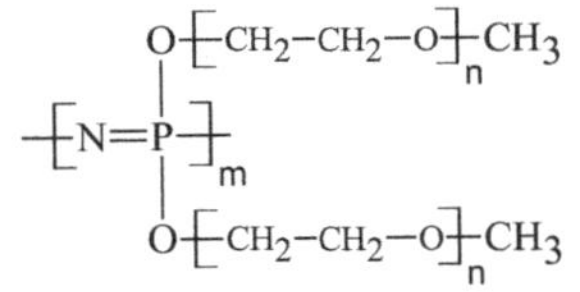

SE-05

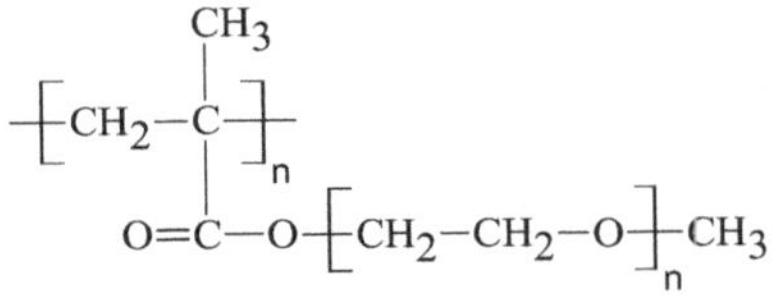

SE-06

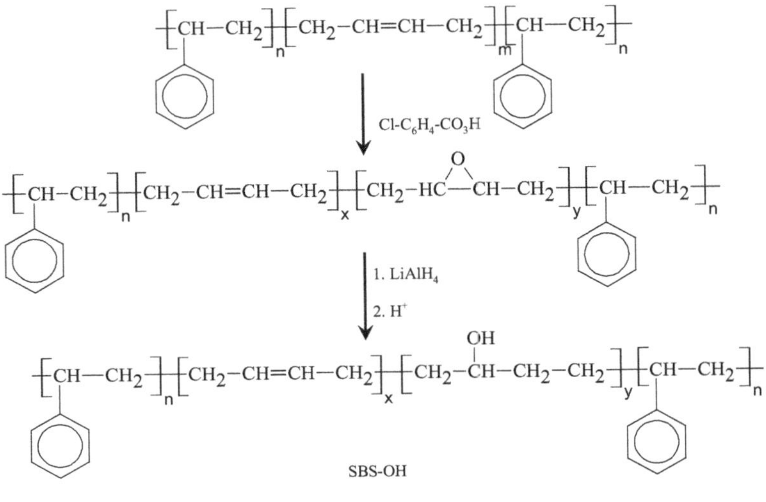

SE-07

SBS-OH

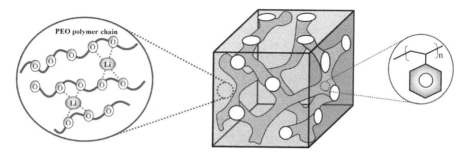

FIGURE 3.1 Schematic illustration of a novel nanostructure-controlled solid polymer electrolyte prepared from a block copolymer with a PEO moiety and a PS moiety by using the self-organised microphase-separation process.

To achieve a proper balance between ionic conductivity and mechanical properties, both di-block or tri-block copolymers have been investigated for the preparation of solid polymer electrolytes. Tri-block copolymers composed of stiff polystyrene [50, 58] or polymethyl methacryate [59] segments and elastic conducting segments derived from poly(ethylene glycol) methyl ether methacrylate (PPME) [59, 60] or SBS [50]. PEO–PMMA block copolymer [59], etc., have been reported. These copolymers exhibit microphase separation, with the hard segment domain and the ion-conducting continuous phase, as shown in Figure 3.1, respresenting the tri-block copolymer polystyrene-*block*-PPME-*block*-polystyrene (PS-*b*-PPME-*b*-PS) [60]

The first copolymers of this type were obtained by the living anionic polymerisation method [58] but the resultant copolymer failed to form membranes with satisfactory mechanical properties. Controlled radical polymerisation methods appeared to be more effective at obtaining the copolymer with improved mechanical properties, especially when applying ruthenium(II) complexes and bi-fuctional initiators [SE-08] [60, 61].

PEO-Based Solid Electrolytes

This method allows the block copolymer PS-*b*-PPME-*b*-PS of poly-ethylene glycol methyl ether methacrylate (PME) and polystyrene (PS) to be obtained, having a M_W greater than 100,000. It is difficult to prepare PS-*b*-PPME-*b*-PS copolymer by the anionic polymerisation technique, due to the restriction of the block polymerisable sequence. In addition, the high-molecular-weight copolymer (M_W >100,000) is difficult to prepare by the anionic polymerisation technique. Morphological studies on the products, obtained by means of the transmission electron microscopy (TEM) technique, shows microphase separation, with a continuous phase of PEO segments, forming a network structure when the PEO content exceeded 70%, as shown in Figure 3.1, which facilitates easy transportation of lithium ions, resulting in high ionic conductivity. When the concentration of the styrene moiety increased, the PEO phase could not form continuous pathways for ion conduction, and, consequently, the ionic conductivity decreased with decreasing PEO content.

These copolymers, doped with $LiClO_4$ (in a ratio of Li/EO = 0.03–0.08), exhibited high ionic conductivity of 2×10^{-4} S cm^{-1} at 30°C for the electrolyte with a 5 mol.% (Li/EO = 0.05) concentration of lithium salt. The conductivity decreased at lithium salt concentrations higher than 8 mol.%, due to the fact that the ion dissociation of $LiClO_4$ in the solid phase was gradually suppressed with increasing concentration of $LiClO_4$ in the PEO phase [62]. The ionic conductivity became higher with increasing PEO concentration, but, on the other hand, the ionic conductivity did not change with EO length; however, the increasing PEO concentration led to lower mechanical strength of the polymer electrolyte. The electrolyte forms free-standing films of tensile strength higher than 3 MPa [60]. The electrochemical evaluation in all-solid-state rechargeable lithium cells of $LiCoO_2$/SPE/Li showed excellent charge and discharge characteristics, while maintaining a discharge capacity of 100 mAh g^{-1} (against a theoretical 140 mAh g^{-1}, representing about 72% utilisation of the active material) after 100 cycles at room temperature at a discharge rate 0.1 C [60, 61]. The Coulombic efficiency of the first charge-discharge cycle was 93%, and it increased up to 99% at the fourth cycle. This high Coulombic efficiency was attributed to the electrochemical stability of the polymer electrolyte [61]. The cyclic voltammetry (CV) studies showed that the electrolyte was electrochemically stable up to 4.5 V, so that this solid-state electrolyte could be used in 4 V class lithium batteries.

Ghosh et al. [59] prepared a nanostructured LiBOB–PEO–*b*–(PMMA–*ran*–PAAMLi) dry solid-state thin-film polymer electrolyte membrane by copolymerisation of the PEO block and a random copolymer of methyl methacrylate (PEO-PMMA) with lithium methacrylic acid (MAALi), followed by combining with lithium bis(oxalate)borate ($LiBC_4O_8$, LiBOB). The resulting SPEs exhibited high room-temperature ionic conductivity (1.26×10^{-5} S cm^{-1} at 21°C) and tremendous Li$^+$-transference numbers, i.e., reaching up to 0.9 at room temperature. In the study, the PEO-based di-block copolymer is selected due to its ability to solvate alkali metal salts. The second block, which consists of a random copolymer of MMA and MAALi, was chosen for its ability to incorporate lithium ions within the microphase-separated spherical domains of the di-block copolymer PEO-*b*-(PMMA-*ran*-PMAALi) (as in Figure 3.2 and SE-09), creating a secondary lithium source. The room-temperature ionic conductivity of the LiBOB–PEO–*b*–(PMMA–*ran*–PAAMLi) electrolyte is

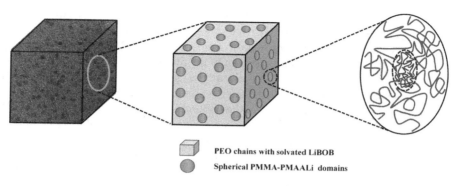

FIGURE 3.2 Schematic visualisation of the morphology of the di-block copolymer PEO-*b*-(PMMA-*ran*-PMAALi) solid-state electrolyte.

much higher than that measured (2.6×10^{-6} S cm^{-1}) for a PEO homopolymer of similar molecular weight (3.5 kDa) and molar composition of LiBOB. The value obtained for the di-block copolymer electrolyte is nearly two orders of magnitude higher than that shown by traditional high-molecular-weight PEO homopolymer electrolytes, in the absence of ceramic fillers or similar additives [4, 6].

SE-09

Lee et al. [63] synthesised a tri-block copolymer based on poly(ethylene oxide) (PEO), poly (2-naphthyl glycidyl ether)–*block*–poly(2-(2-(2-methoxyethoxy)ethoxy) ethyl glycidyl ether)-*block*-poly(2-naphthyl glycidyl ether)s (PNG–PTG–PNGs) by sequential ring-opening polymerisation, using a bi-directional initiator catalysed by a phosphazene base, as shown in Figure 3.3. The SPEs based on PNG–PTG–PNGs with LITFSI (Li/EO = 0.05) showed high lithium ion conductivity at room temperature, which could ascribed to the low T_g (about −65°C) and amorphous nature of the PTG blocks. It is worth noting that the lithium ion conductivity of the PNG$_{18}$–PTG$_{107}$–PNG$_{18}$ solid polymer electrolytes still exhibited anionic conductivity of 9.5×10^{-5} S cm^{-1}, even though the proportion of amorphous regions was reduced due to the higher content of crystalline PNG domains resulting from the formation of excellent Li$^+$-transport pathways formed by the microphase separation into soft PTG and hard PNG domains in PNG–PTG–PNGs [63]. Moreover, the degradation temperatures at 5 wt.% loss values (T_d, 5%) of PNG–PTG–PNGs were in the range 371–390 °C, implying that solid polymer electrolytes based on PNG–PTG–PNGs block polymers have a broad range of operating temperatures.

The methods of living anionic and/or controlled radical or sequential ring-opening polymerisation have also been utilised for the synthesis of star-branched EO

PEO-Based Solid Electrolytes

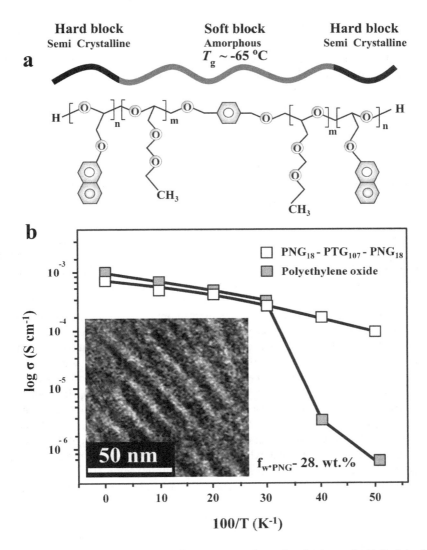

FIGURE 3.3 Chemical structure and temperature-dependent ionic conductivity based on poly(ethylene oxide) (PEO), poly (2-naphthyl glycidyl ether)–*block*–poly (2-(2-(2-methoxyethoxy)ethoxy) ethyl glycidyl ether)-*block*-poly(2-naphthyl glycidyl ether)s (PNG–PTG–PNGs) tri-block copolymers. Adapted and reproduced with permission from Ref. [63]. Copyright © 2018 American Chemical Society.

copolymers. The procedure developed by Professor Stanislaw Penczek's research group in the Centre for Molecular and Macromolecular Studies of the Polish Academy of Sciences is an example of a purely anionic process [64]. Based on this method, in the first stage, living EO oligomers are obtained in the classic anionic polymerisation, initiated with alkoxides, or in the reaction of metallic potassium with PEG mono ethers. In the next step, the living oligomers are subjected to the reaction with di-epoxides. The reaction between these reagents leads first to the formation of an A_2B_1-type reactive oligomer, which gradually increases the number of arms as a result of further addition reactions of primary and secondary alkoxide centres to epoxide groups (SE-10).

$$2\ R\text{-}[O\text{-}CH_2\text{-}CH_2]_n\text{-}O^- + \overset{O}{\triangle}\sim\overset{O}{\triangle}$$

$$\downarrow$$

$$R\text{-}[O\text{-}CH_2\text{-}CH_2]_n\text{-}O\text{-}CH_2\text{-}\underset{O^-}{CH}\sim\underset{O^-}{HC}\text{-}CH_2\text{-}O\text{-}[CH_2\text{-}CH_2\text{-}O]_n\text{-}R$$

SE-10

Based on the procedure developed by Penczek [64], Marzantowicz et al. prepared star-branched (A_xB_y)-type reactive star polymer matrices of poly(ethylene oxide) arms with a di-epoxide core by anionic polymerisation. The average number of poly(ethylene oxide) arms is between 20 (according to molecular weight measurements) and 26 (according to viscosity measurements). Each arm has a molecular weight of about 2,000 (conducting segments contain 20 or more EO monomeric units), and the morphology of these polymers depends essentially on the EO homo-sequence length. The EO arms are semi-crystalline in nature, but the crystalline phase disappears after doping with lithium salts. Because of the semi-crystalline nature, the mobility of ether segments in branched structures is lower than that of the amorphous phase in linear polymers, but selection of the appropriate salt and its concentration results in amorphous electrolytes, exhibiting a low-temperature ionic conductivity about 50 times higher than that of the analogous system with PEO [65]. The conductivity of the semi-crystalline electrolyte was more than 500 times lower than that of the amorphous one, which can be attributed to the dense-packed crystalline phase [65]. At temperatures below the melting temperature of electrolytes with the linear PEO matrix, the conductivity of electrolytes based on the branched PEO/lithium imide salt $LiN(CF_3SO_2)_2$ (LiTFSI) was much higher. The study has revealed two main disadvantages of PEO:LiTFSI electrolytes based on branched PEO. One issue is related to the higher values of the glass transition temperature in comparison with electrolytes based on linear PEO, and second problem is associated with mechanical properties and related practical applications in batteries. All of the electrolytes with LiTFSI salt were obtained in the form of a viscous liquid, which does not provide enough self-support to be a good membrane for a rechargeable battery [65].

A hyper-branched graft copolymer, consisting of a poly(methyl methacrylate) main chain and poly(ethylene glycol) methyl ether methacrylate side chains (poly(MMA-g-POEM)), was synthesised by atom transfer radical polymerisation (ATRP) of poly(oligo-oxyethylene) methacrylate (POEM), and a macro-initiator of poly(MMA-co-CMS) for ATRP is prepared by copolymerisation of methyl

PEO-Based Solid Electrolytes

methacrylate (MMA) and chloromethylstyrene (CMS). The hyper-branched graft copolymer of poly(MMA-*g*-POEM*x*), (*x* = 5 or 9), the chemical structure of which is shown in SE-11, and the structure of the hyper-branched graft copolymer is changed *via* (a) the average distance between side chains, (b) the side chain length, and (c) the branched chain length. The ionic conductivity of the SPE prepared from the graft copolymer, where the POEM content = 51 wt.% with lithium salt LiClO$_4$, was 2×10^{-5} S cm^{-1} at 30°C, and the maximum value was obtained when the [Li]/[EO] ratio was equal to 0.06. The ionic conductivity and tensile strength of the SPEs increase with increasing side chain length, branched chain length and/or average distance between the side chains, and the tensile strength exceeded 1 MPa [66].

SE-11

A series of other techniques are known for the preparation of hyperbranched or multi-armed copolymers, among which the anionic polymerisation technique is combined with that of controlled radical polymerisation [67], and condensation polymerisation of tri-functional monomers is used [68–70] for the preparation of PEO-based hyperbranched or multi-armed macromolecules (SE-12). As discussed in the aforementioned section (Figure 3.3), binary structures with amorphous and soft ion-conducting pathways containing EO segments can be generated in these systems, and SPE with LiClO$_4$ exhibited ambient temperature conductivities in the range 10^{-5}–10^{-4} S cm^{-1} [67–70]. The advantages of these hyperbranched electrolytes include low cost, high transparency, absence of crystallinity and good thermal stability (thermally stable above 150°C), even at high concentrations of salt, which implies that they can be used over a wide temperature range [70]. In addition, the ionic conductivity of multi-armed polymer–Li$^+$ complexes could be tuned by controlling the length and the number of oxyethylene arms or by balancing of the inter- and intramolecular complexes in the polymer matrix. Studies on the transport properties displayed that the ionic conductivity could be improved by changing from comb- to multi-armed copolymers. The long oxyethylene arms attached to the core facilitate high segmental mobility and high carrier ion concentrations there, to surprisingly improve the ionic conductivity [69].

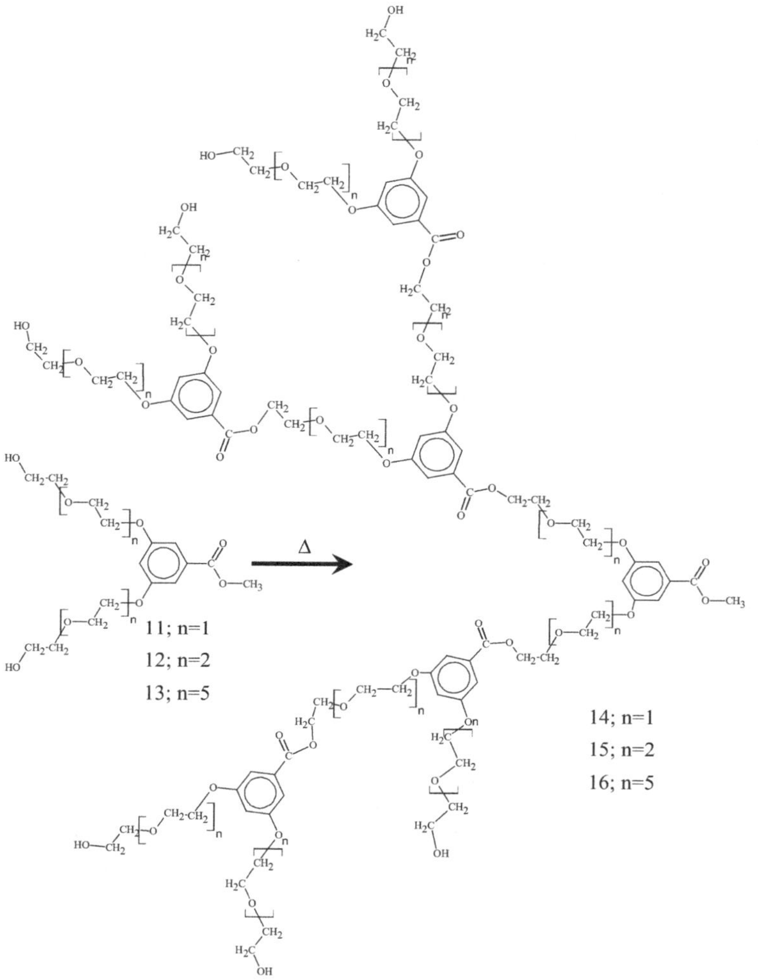

11; n=1
12; n=2
13; n=5

14; n=1
15; n=2
16; n=5

SE-12

With a view to creating low-impedance pathways for the lithium ion conduction and inhibiting ion aggregates, several amphiphilic polymers were synthesised, in which short EO segments were separated by monomeric units bearing long aliphatic side chains [71–74]. Systems based on poly(2,5,8,11,14-pentaoxapentadecam ethylene-(5-alkyloxy-1,3-phenylene)) have been the most extensively studied ones. The studies adapted the extended helical crystalline structures of PEO–alkali salt complexes [75–77] to synthesise organised low-dimensional polymer complexes [71, 72, 78–80] with the amphiphilic polymer poly(2,5,8,11,14-pentaoxapentadecamethyl ene-[5-alkyloxy-1,3-phenylene]) (SE-13). When these molecules are fully organised, the side groups R, which are generally long n-alkyl chains, condense together to create an ionophobic layer, causing the polyether segments to generate a helical substructure. The Langmuir–Blodgett (LB) films of low-dimensional polymer

PEO-Based Solid Electrolytes

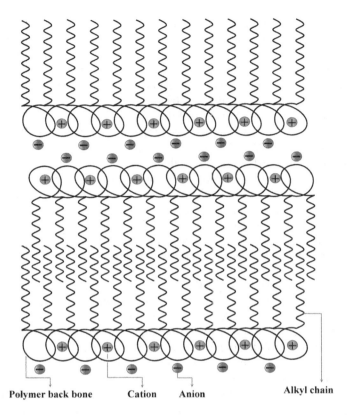

FIGURE 3.4 A diagram of the bulk structure of C1605 Langmuir–Blodgett (LB) films of low-dimensional polymer electrolytes.

electrolytes were formed, inserting alkali metal cations into the helical tubes, whilst the anions occupy the spaces between the helices as shown in Figure 3.4. Different alkali metal salts, such as $LiClO_4$, $LiBF_4$, $LiCF_3SO_3$ or $NaCF_3SO_3$, are used to prepare the LB films of low-dimensional polymer electrolytes and investigate the transport properties.

After doping with a suitable lithium salt, the polymer adopts regular crystalline structures, in which side chains interdigitate in a hexagonal layer between polyether helices, into which cations are encapsulated and anions lie in the inter-lamellar spaces, which are about 40–45 Å, [71, 73, 78, 79] as shown in Figure 3.4.

The melting temperature of hydrophobic side chains is in the range 25–45°C, above the side chain melting point. The salt is complexed with the polyether helical component, which imparts a liquid crystalline character, since the salt-free polymers become isotropic [72–74]. Below the side-chain melting temperature of 42°C, the side chains form a hexagonal crystalline layer and the materials are hard, waxy solids. Above this temperature, the bi-refringence persists, and the system is liquid crystalline (the smectic phase) until *ca.* 150°C [73].

Some of these materials displayed low-temperature dependence of conductivity over 10^{-3} S cm^{-1} at 20–110°C. Sub-ambient measurements to –10°C gave a conductivity of 4×10^{-5} S cm^{-1}, which is the conductivity at *ca.* 25°C as observed for conventional amorphous systems. Prompted by molecular modelling, these spectacular conducting properties are attributed to the specific conduction mechanism in the low-temperature-dependent regime, involving Li$^+$-hopping between decoupled neutral aggregates, such as ion pairs or 'quadrupoles' (double ion pairs) as depicted in Figure 3.5a within the channels of diameter ~1 nm (Figure 3.5b) [74].

It is apparent that, at temperatures above those corresponding to side chain melting, the conductivities are significantly higher than usual for some low-dimensional polymer electrolyte systems, lying in the range 10^{-4}–10^{-2} S cm^{-1}, with a much reduced dependence on temperature. However, with side chain: alkyl-*co*-crystallisation, the conductivity falls to 10^{-7}–10^{-6} S cm^{-1} at ambient temperatures. Some low-dimensional polymer electrolyte systems with the large bis trifluoromethane sulfonimide (CF$_3$SO$_2$)$_2$N anion (TFSI) do not form the extensive smectic layers formed by smaller anions, such as ClO$_4^-$, BF$_4^-$ or CF$_3$SO$_3^-$. This presumably arises due to insufficient area in the *bc* plane (determined by the hexagonal alkane layer) for larger anions. The high conductivity of the low-dimensional polymer electrolyte with the amphiphilic polymer poly[2,5,8,11,14-pentaoxapentadecamethylene-(5-alkyloxy-1,3-phenylene): C$_{16}$H$_{33}$OH: LiTFSI (1:1:1), close to 10^{-3} S cm^{-1} at 40°C, is greater than that of conventional "amorphous PEO" complexes at the same temperature, but, as for the system with poly[2,5,8,11,14-pentaoxapentadecamethylene-(5-alkyloxy-1,3-phenylene) C$_{18}$H$_{38}$:LiClO$_4$ (1:1:1) smectic type B systems, the conductivities decline sharply as the side chains crystallise [73].

Similar mechanisms of ion transport are also suggested for the systems based on other highly polar polymeric matrices, like polyethers [81], polyacrylonitrile [82] or acrylonitrile copolymers [83], which do not dissolve lithium salts but are able to stabilise amorphous ionic clusters. In these systems, the concentration of lithium salts is very high (70–90 wt.%), hence their name as the polymer-in-salt system. However, some of these systems form flexible solid electrolyte membranes, exhibiting glass transition temperatures lower than that of the pure parent organic polymers. Unfortunately, most of these electrolyte systems are thermodynamically unstable, which leads to phase separation (separation of organic polymer and lithium salts) over time. This phase separation results in the colonisation of organic (polymer) and inorganic (lithium salt) domains and adversely affects the ion transport mechanism and kinetics, so that it has no practical application in rechargeable batteries [84]. Recently, these systems were significantly improved by application of di-block

PEO-Based Solid Electrolytes

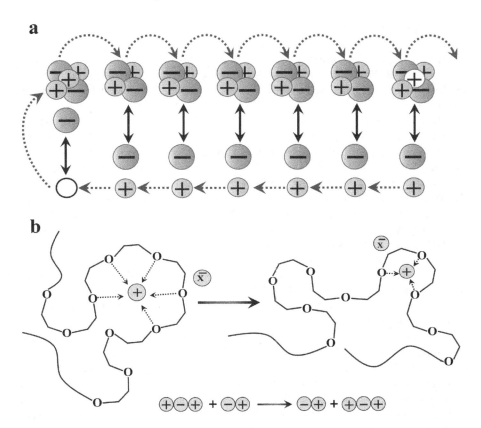

FIGURE 3.5 (a) A plausible mechanism for extended Li+ ion transport along rows of decoupled neutral aggregates, such as ion pairs or 'quadrupoles' (double ion pairs) in a 'blocky' copolymer electrolyte. The parallel row of separated ions provides "charge balancing" by local reorganisation. (b) The ion transport mechanism in the PEO-based solid polymer electrolyte. Adapted and reproduced with permission from Ref. [74]. Copyright © 2005 Elsevier.

copolymers of PEO, with random copolymers of methyl methacrylate with lithium methacrylate (SE-14) [59, 85] showing the ability to solvate alkali metal salts.

$$H-[CH_2-CH_2-O]_{68}-block-[CH_2-\underset{\underset{O-CH_3}{C=O}}{\overset{CH_3}{C}}-ran-CH_2-\underset{\underset{O^-Li^+}{C=O}}{\overset{CH_3}{C}}]_5-H \quad \text{SE-14}$$

As shown in the polymer structure, the second block, which consists of a random copolymer of methyl methacylate MMA and the lithium salt of methacrylic acid,

MAALi⁺, displayed an ability to incorporate lithium ions within the microphase-separated spherical domains of the di-block copolymer PEO-*b*-PMMA-*ran*-PMAALi⁺ (Figure 3.2), creating a secondary lithium source [59, 85]. The di-block copolymer PEO-*b*-PMMA-*ran*-PMAALi, with LiBOB in the molar ratio ethylene oxide: LiBOB = 3:1, was used to form flexible translucent films, which exhibited an average ionic conductivity value of 1.26×10^{-5} S cm^{-1} at room temperature (21°C) which is one order of magnitude higher than the ionic conductivity measured (2.6×10^{-6} S cm^{-1}) for a PEO homopolymer of similar molecular weight (3.5 k) and molar composition of LiBOB. Furthermore, the conductivity value obtained for the di-block copolymer electrolyte is nearly two orders of magnitude greater than that exhibited by traditional high-molecular-weight PEO homopolymer electrolytes, in the absence of ceramic fillers or similar additives [4, 6]. The physical appearance of the di-block copolymer electrolyte PEO-*b*-PMMA-*ran*-PMAALi also differed with varying salt content. Polymer electrolyte films with a high salt loading (EO: LiBOB) of 2:1 or low salt loading (EO: LiBOB) of 10:1 were brittle and opaque, whereas intermediate salt content films were flexible and translucent [85]. The average transference number values of 0.89 and 0.78 were obtained for the salt-optimised polymer electrolyte (EO: LiBOB = 3:1) materials whereas that for the electrolyte having molar ratio of EO: LiBOB= 4:1 was found to be 0.83 and 0.75 at temperatures 21–23°C and 60°C, respectively [59]. Previous work by Appetecchi et al. [86] on LiBOB and PEO homopolymers of higher M_w (1×10^5) reported Li⁺ transference numbers (T_{Li+}) ranging from 0.25 to 0.30 at elevated temperatures 60 to 100°C [86]. A similar rise in T_{Li+} values with increasing salt concentration T_{Li+} = 0.6±0.03 for EO: lithium bis trifluoromethane sulfoneimide (LiTFSI) = 5:1 at 85°C was reported and discussed at length in the Edman et al. [87] study, where solid polymer electrolytes comprising PEO and LiNSO$_2$CF$_3$, the LiTFSI salt, were studied. The increase in transference number was attributed to the reduction in free volume with increasing material density and increasing salt content, resulting in lower anionic mobility [87]. The preliminary tests on lithium battery Li/SPE/Li cells showed that the electrolyte membrane also has good electrochemical stability (>5.0 V) and excellent interfacial stability behaviour with the lithium metal electrode during Li plating/stripping cycles [59].

3.4 CONCLUSIONS

Polymer electrolytes have been a promising alternative for electrolytes in lithium-ion batteries. The initial investigation on polymer electrolytes concentrated on polyethylene oxide. PEO has been widely investigated as a solid polymer electrolyte. It exhibits low glass transition temperature and shows excellent interfacial stability with lithium, which makes the formation of complexes with lithium salts easier. PEO shows low ionic conductivity at room temperature as it exhibits a high degree of crystallinity at room temperature. Several methods have been adopted to overcome these issues. Polymer blending and ceramic filler addition are some of the methods employed for the enhancement of ionic conductivity. Blending of amorphous polymers with highly crystalline PEO helps to enhance the amorphous domains in the system, generating a path for lithium ion conduction. Appropriate tuning of the materials can achieve the commercialisation of the polyethylene oxide-based electrolytic system.

ACKNOWLEDGMENT

Authors Dr Jabeen Fatima M. J. and Dr Prasanth Raghavan would like to acknowledge Kerala State Council for Science, Technology and Environment (KSCSTE), Kerala, India, for financial assistance.

REFERENCES

1. Tarascon JM, Armand M (2001) Issues and challenges facing rechargeable lithium batteries. *Nature* 414:359–367. https://doi.org/10.1038/35104644
2. Nishi Y (2001) Lithium ion secondary batteries; past 10 years and the future. *J Power Sources* 100:101–106. https://doi.org/10.1016/S0378-7753(01)00887-4
3. Armand M, Tarascon JM (2008) Building better batteries. *Nature* 451:652–657. https://doi.org/10.1038/451652aF
4. Scrosati B, Croce F, Persi L (2000) Impedance spectroscopy study of {PEO}-based nanocomposite polymer electrolytes. *J Electrochem Soc* 147: 1718–1721. https://doi.org/10.1149/1.1393423
5. Appetecchi GB, Croce F, Persi L, et al. (2000) Transport and interfacial properties of composite polymer electrolytes. *Electrochim Acta* 45:1481–1490. https://doi.org/10.1016/S0013-4686(99)00363-1
6. Croce F, Persi LL, Scrosati B, et al. (2001) Role of the ceramic fillers in enhancing the transport properties of composite polymer electrolytes. *Electrochim Acta* 46:2457–2461. https://doi.org/10.1016/S0013-4686(01)00458-3
7. Deng F, Wang X, He D, et al. (2015) Microporous polymer electrolyte based on PVdF/PEO star polymer blends for lithium ion batteries. *J Memb Sci* 491:82–89. https://doi.org/10.1016/j.memsci.2015.05.021
8. Polu AR, Rhee HW (2015) Nanocomposite solid polymer electrolytes based on poly(ethylene oxide)/POSS-PEG (n=13.3) hybrid nanoparticles for lithium ion batteries. *J Ind Eng Chem* 31:323–329. https://doi.org/10.1016/j.jiec.2015.07.005
9. Appetecchi GB, Croce F, Hassoun J, et al. (2003) Hot-pressed, dry, composite, PEO-based electrolyte membranes: I. Ionic conductivity characterization. *J Power Sources* 114:105–112. https://doi.org/10.1016/S0378-7753(02)00543-8
10. Shin JH, Lim YT, Kim KW, et al. (2002) Effect of ball milling on structural and electrochemical properties of (PEO)$_n$LiX (LiX = LiCF$_3$SO$_3$ and LiBF$_4$) polymer electrolytes. *J Power Sources* 107:103–109. https://doi.org/10.1016/S0378-7753(01)00990-9
11. Kwon SJ, Kim DG, Shim J, et al. (2014) Preparation of organic/inorganic hybrid semi-interpenetrating network polymer electrolytes based on poly(ethylene oxide-co-ethylene carbonate) for all-solid-state lithium batteries at elevated temperatures. *Polymer (Guildf)* 55:2799–2808. https://doi.org/10.1016/j.polymer.2014.04.051
12. Choi J-W, Cheruvally G, Kim Y-H, et al. (2007) Poly(ethylene oxide)-based polymer electrolyte incorporating room-temperature ionic liquid for lithium batteries. *Solid State Ionics* 178:1235–1241. https://doi.org/10.1016/j.ssi.2007.06.006
13. Rolland J, Brassinne J, Bourgeois J-P, et al. (2014) Chemically anchored liquid-PEO based block copolymer electrolytes for solid-state lithium-ion batteries. *J Mater Chem A* 2:11839–11846. https://doi.org/10.1039/C4TA02327G

14. Tao C, Gao M-H, Yin B-H, et al. (2017) A promising TPU/PEO blend polymer electrolyte for all-solid-state lithium ion batteries. *Electrochim Acta* 257:31–39. https://doi.org/10.1016/j.electacta.2017.10.037
15. Patla SK, Ray R, Asokan K, Karmakar S (2018) Investigation of ionic conduction in PEO–PVDF based blend polymer electrolytes. *J Appl Phys* 123:125102. https://doi.org/10.1063/1.5022050
16. Rupp B, Schmuck M, Balducci A, et al. (2008) Polymer electrolyte for lithium batteries based on photochemically crosslinked poly(ethylene oxide) and ionic liquid. *Eur Polym J* 44:2986–2990. https://doi.org/10.1016/j.eurpolymj.2008.06.022
17. Zhang ZC, Jin JJ, Bautista F, et al. (2004) Ion conductive characteristics of cross-linked network polysiloxane-based solid polymer electrolytes. *Solid State Ionics* 170:233–238. https://doi.org/10.1016/j.ssi.2004.04.007
18. Sharma A, Sharma PK, Sharma AK, Sadiq M (2016) Effect of Nano filler on PEO based polymer electrolytes for energy storage devices applications. *Int Res Adv* 4:1–4
19. Shin JH, Henderson WA, Passerini S (2005) PEO-based polymer electrolytes with ionic liquids and their use in lithium metal-polymer electrolyte batteries. *J Electrochem Soc* 152:978–983. https://doi.org/10.1149/1.1890701
20. Singh PK, Bhattacharya B, Nagarale RK (2010) Effect of nano-TiO$_2$ dispersion on PEO polymer electrolyte property. *J Appl Polym Sci* 118:2976–2980. https://doi.org/10.1002/app.32726
21. Pitawala HMJC, Dissanayake MAKL, Seneviratne VA (2007) Combined effect of Al$_2$O$_3$ nano-fillers and EC plasticizer on ionic conductivity enhancement in the solid polymer electrolyte (PEO)$_9$LiTf. *Solid State Ionics* 178:885–888. https://doi.org/10.1016/j.ssi.2007.04.008
22. Abdullah A, Abdullah SZ, Ali AMM, et al. (2009) Electrical properties of PEO-LiCF$_3$SO$_3$-SiO$_2$ nanocomposite polymer electrolytes. *Mater Res Innov* 13:255–258. https://doi.org/10.1179/143307509X440451
23. Wieczorek W, Stevens JR, Florjańczyk Z (1996) Composite polyether based solid electrolytes. The Lewis acid-base approach. *Solid State Ionics* 85:67–72. https://doi.org/10.1016/0167-2738(96)00042-2
24. Qian X, Gu N, Cheng Z, et al. (2001) Methods to study the ionic conductivity of polymeric electrolytes using a.c. impedance spectroscopy. *J Solid State Electrochem* 6:8–15. https://doi.org/10.1007/s100080000190
25. Mohan VM, Raja V, Sharma AK, Rao VVRN (2005) Ionic conductivity and discharge characteristics of solid-state battery based on novel polymer electrolyte (PEO+NaBiF4). *Mater Chem Phys* 94:177–181. https://doi.org/10.1016/j.matchemphys.2005.05.030
26. Sreekanth T, Jaipal Reddy M, Ramalingaiah S, Subba Rao U V (1999) Ion-conducting polymer electrolyte based on poly (ethylene oxide) complexed with NaNO$_3$ salt-application as an electrochemical cell. *J Power Sources* 79:105–110. https://doi.org/10.1016/S0378-7753(99)00051-8
27. Jonscher AK (1999) Dielectric relaxation in solids. *J Phys D Appl Phys* 32:. https://doi.org/10.1088/0022-3727/32/14/201
28. Sadoway DR (2004) Block and graft copolymer electrolytes for high-performance, solid-state, lithium batteries. *J Power Sources* 129:1–3. https://doi.org/10.1016/j.jpowsour.2003.11.016
29. Nagaoka K, Naruse H, Shinohara I, Watanabe M (1984) High ionic conductivity in poly(dimethyl siloxane-co-ethylene oxide) dissolving lithium perchlorate. *J Polym Sci Polym Lett Ed* 22:659–663. https://doi.org/10.1002/pol.1984.130221205
30. Craven JR, Mobbs RH, Booth C, Giles JRM (1986) Synthesis of oxymethylene-linked poly(oxyethylene) elastomers. *Die Makromol Chemie, Rapid Commun* 7:81–84. https://doi.org/10.1002/marc.1986.030070208

31. Nicholas C V, Wilson DJ, Booth C, Giles JRM (1988) Improved synthesis of oxymethylene-linked poly(oxyethylene). *Br Polym J* 20:289–292. https://doi.org/10.1002/pi.4980200321
32. Blonsky PM, Shriver DF, Austin P, Allcock HR (1984) Polyphosphazene solid electrolytes. *J Am Chem Soc* 106:6854–6855. https://doi.org/10.1021/ja00334a071
33. Fish D, Khan I, Smid J (1986) Conductivity of solid complexes of lithium perchlorate with poly{[ω-methoxyhexa(oxyethylene)ethoxy]methylsiloxane}. *Die Makromol Chemie, Rapid Commun* 7:115–120. https://doi.org/10.1002/marc.1986.030070303
34. Cowie JMG, Martin ACS (1987) Ionic conductivity in poly(di-poly(propylene glycol) itaconate)-salt mixtures. *Polymer (Guildf)* 28:627–632. https://doi.org/10.1016/0032-3861(87)90479-4
35. Al H et (1986) Ion conductivity in polysiloxane comb polymers with ethylene glycol teeth. 27:98–100.
36. Such K, Florjañczyk Z, Wieczorek W PJ (1989) No Title. In: B Scrosati (ed) *Proc. II International Symposium on Polymer Electrolytes*. Elsevier Applied Science, Siena.
37. Florjańczyk Z, Krawiec W, Wieczorek W, Przyłuski J (1991) Polymer solid electrolytes based on ethylene oxide copolymers. *Die Angew Makromol Chemie* 187:19–32. https://doi.org/10.1002/apmc.1991.051870103
38. Florjańczyk Z, Krawiec W, Wieczorek W, Siekierski M (1995) Highly conducting solid electrolytes based on poly (ethylene oxide-co-propylene oxide). *J Polym Sci Part B Polym Phys* 33:629–635. https://doi.org/10.1002/polb.1995.090330413
39. Nishimoto A, Watanabe M, Ikeda Y, Kohjiya S (1998) High ionic conductivity of new polymer electrolytes based on high molecular weight polyether comb polymers. *Electrochim Acta* 43:1177–1184. https://doi.org/10.1016/S0013-4686(97)10017-2
40. Marchese L, Andrei M, Roggero A, et al. (1992) A new class of polymer electrolytes based on chain-extended polyepoxides and LiClO$_4$. *Electrochim Acta* 37:1559–1564. https://doi.org/10.1016/0013-4686(92)80111-X
41. Florjañczyk Z (1991) On the reactivity of sulfur dioxide in chain polymerization reactions. *Prog Polym Sci* 16:509–560. https://doi.org/10.1016/0079-6700(91)90009-A
42. Florjañczyk Z, Raducha D (1993) Copolymerization of ethylene oxide and sulfur dioxide initiated by lewis bases. *Die Makromol Chemie* 194:2605–2613. https://doi.org/10.1002/macp.1993.021940916
43. Florjanczyk Z, Zygadlo-Monikowska E, Raducha D, et al. (1992) Polymeric electrolytes based on sulfur dioxide copolymers. *Electrochim Acta* 37:1555–1558. https://doi.org/10.1016/0013-4686(92)80110-8
44. Fish D, Khan IM, Smid J (1986) Conductivity of solid complexes of lithium perchlorate with poly{[ω-methoxyhexa (oxyethylene) ethoxy] methylsiloxane}. *Makromol Chem Rapid Commun* 7:115.
45. Le Mehaute A, Crepy G, Marcellin G, et al. (1985) Polymer electrolytes. *Polym Bull* 14:233–237. https://doi.org/10.1007/BF00254943
46. Maccallum JR, Smith MJ, Vincent CA (1984) The effects of radiation-induced cross-linking on the conductance of LiClO$_4$·PEO electrolytes. *Solid State Ionics* 11:307–312. https://doi.org/10.1016/0167-2738(84)90022-5
47. Fiona M. Gray (1991) *Solid Polymer Electrolytes: Fundamentals and Technological Applications*. WILEY-VCH Verlag.
48. Bouridah A, Dalard F, Deroo D, et al. (1985) Poly(dimethylsiloxane)-poly(ethylene oxide) based polyurethane networks used as electrolytes in lithium electrochemical solid state batteries. *Solid State Ionics* 15:233–240. https://doi.org/10.1016/0167-2738(85)90008-6
49. MacCallum JR, Vincent CA (1989) *Polymer Electrolyte Reviews*. Springer Netherlands.

50. Gray FM, MacCallum JR, Vincent CA, Giles JRM (1988) Novel polymer electrolytes based on ABA block copolymers. *Macromolecules* 21:392–397. https://doi.org/10.1021/ma00180a018
51. Le Nest JF, Callens S, Gandini A, Armand M (1992) A new polymer network for ionic conduction. *Electrochim Acta* 37:1585–1588. https://doi.org/10.1016/0013-4686(92)80116-4
52. Floriańczyk Z, Such K, Wieczorek W, Wasiucionek M (1991) Highly conductive poly(ethylene oxide)-poly(methyl methacrylate) blends complexed with alkali metal salts. *Polymer (Guildf)* 32:3422–3425. https://doi.org/10.1016/0032-3861(91)90548-W
53. Wieczorek W, Such K, Florjanczyk Z, Stevens JR (1995) Polyacrylamide based composite polymeric electrolytes. *Electrochim Acta* 40:2417–2420. https://doi.org/10.1016/0013-4686(95)00206-T
54. Allcock H (2012) *Phosphorus-Nitrogen Compounds: Cyclic, Linear, and High Polymeric Systems*. Elsevier Science.
55. Blonsky PM, Shriver DF, Austin P, Allcock HR (1986) Complex formation and ionic conductivity of polyphosphazene solid electrolytes. *Solid State Ionics* 18–19:258–264. https://doi.org/10.1016/0167-2738(86)90123-2
56. Bannister DJ, Davies GR, Ward IM, McIntyre JE (1984) Ionic conductivities of poly(methoxy polyethylene glycol monomethacrylate) complexes with $LiSO_3CH_3$. *Polymer (Guildf)* 25:1600–1602. https://doi.org/10.1016/0032-3861(84)90152-6
57. Xia DW, Soltz D, Smid J (1984) Conductivities of solid polymer electrolyte complexes of alkali salts with polymers of methoxypolyethyleneglycol methacrylates. *Solid State Ionics* 14:221–224. https://doi.org/10.1016/0167-2738(84)90102-4
58. Khan I, Fish D, Delaviz Y, Smid J (1989) ABA triblock comb copolymers with oligo(oxyethylene) side chains as matrix for ion transport. *Die Makromol Chemie* 190:1069–1078. https://doi.org/10.1002/macp.1989.021900515
59. Ghosh A, Wang C, Kofinas P (2010) Block copolymer solid battery electrolyte with high Li-Ion transference number. *J Electrochem Soc* 157:A846. https://doi.org/10.1149/1.3428710
60. Niitani T, Shimada M, Kawamura K, Kanamura K (2005) Characteristics of new-type solid polymer electrolyte controlling nano-structure. *J Power Sources* 146:386–390. https://doi.org/10.1016/j.jpowsour.2005.03.102
61. Niitani T, Shimada M, Kawamura K, et al. (2005) Synthesis of Li^+ ion conductive PEO-PSt block copolymer electrolyte with microphase separation structure. *Electrochem Solid-State Lett* 8:385–388. https://doi.org/10.1149/1.1940491
62. Bruce PG (1997) *Solid State Electrochemistry*. Cambridge University Press.
63. Kim B, Chae CG, Satoh Y, et al. (2018) Synthesis of hard-soft-hard triblock copolymers, Poly(2-naphthyl glycidyl ether)- block -poly[2-(2-(2-methoxyethoxy)ethoxy) ethyl glycidyl ether]- block -poly(2-naphthyl glycidyl ether), for solid electrolytes. *Macromolecules* 51:2293–2301. https://doi.org/10.1021/acs.macromol.7b02553
64. Lapienis G, Penczek S (2004) Reaction of oligoalcohols with diepoxides: An easy, one-pot way to star-shaped, multibranched polymers. II. Poly(ethylene oxide) stars—synthesis and analysis by size exclusion chromatography triple-detection method. *J Polym Sci Part A Polym Chem* 42:1576–1598. https://doi.org/10.1002/pola.11099
65. Marzantowicz M, Dygas JR, Krok F, et al. (2009) Star-branched poly(ethylene oxide) $LiN(CF_3SO_2)_2$: A promising polymer electrolyte. *J Power Sources* 194:51–57. https://doi.org/10.1016/j.jpowsour.2009.01.011
66. Higa M, Fujino Y, Koumoto T, et al. (2005) All solid-state polymer electrolytes prepared from a hyper-branched graft polymer using atom transfer radical polymerization. *Electrochim Acta* 50:3832–3837. https://doi.org/10.1016/j.electacta.2005.02.037

67. Niitani T, Amaike M, Nakano H, et al. (2009) Star-Shaped polymer electrolyte with microphase separation structure for all-solid-state lithium batteries. *J Electrochem Soc* 156:A577. https://doi.org/10.1149/1.3129245
68. Hawker CJ, Chu F, Pomery PJ, Hill DJT (1996) Hyperbranched poly(ethylene glycol)s: A new class of ion-conducting materials. *Macromolecules* 29:3831–3838. https://doi.org/10.1021/ma951909i
69. Inoue K, Miyamoto H, Itaya T (1997) Ionic conductivity of complexes of novel multi-armed polymers with phosphazene core and $LiClO_4$. *J Polym Sci Part A Polym Chem* 35:1839–1847. https://doi.org/10.1002/(SICI)1099-0518(19970715)35:9<1839::AID-POLA25>3.0.CO;2-5
70. Pennarun PY, Jannasch P (2005) Electrolytes based on $LiClO_4$ and branched PEG-boronate ester polymers for electrochromics. *Solid State Ionics* 176:1103–1112. https://doi.org/10.1016/j.ssi.2005.01.001
71. Dias FB, Voss JP, Batty SV, et al. (1994) Smectic phases in a novel alkyl-substituted polyether and its complex with lithium tetrafluoroborate. *Macromol Rapid Commun* 15:961–969. https://doi.org/10.1002/marc.1994.030151209
72. Zheng Y, Gibaud A, Cowlam N, et al. (2000) Structure and conductivity in LB films of low-dimensional polymer electrolytes. *J Mater Chem* 10:69–77. https://doi.org/10.1039/a903000j
73. Zheng Y, Chia F, Ungar G, et al. (2001) High ambient dc and ac conductivities in solvent-free, low-dimensional polymer electrolyte blends with lithium salts. *Electrochim Acta* 46:1397–1405. https://doi.org/10.1016/S0013-4686(00)00732-5
74. Zheng Y, Liu J, Liao Y-P, et al. (2005) Low dimensional polymer electrolytes with enhanced Li^+ conductivities. *J Power Sources* 146:418–422. https://doi.org/10.1016/j.jpowsour.2005.03.125
75. Chatani Y, Okamura S (1987) Crystal structure of poly(ethylene oxide)-sodium iodide complex. *Polymer (Guildf)* 28:1815–1820. https://doi.org/10.1016/0032-3861(87)90283-7
76. Lightfoot P, Mehta MA, Bruce PG (1993) Crystal structure of the polymer electrolyte Poly(ethylene oxide)$_3$:$LiCF_3SO_3$. *Science (80-)* 262:883–885. https://doi.org/10.1126/science.262.5135.883
77. Andreev YG, Lightfoot P, Bruce PG (1997) A general monte carlo approach to structure solution from powder-diffraction data: Application to Poly(ethylene oxide)$_3$:$LiN(SO_2CF_3)_2$. *J Appl Crystallogr* 30:294–305. https://doi.org/10.1107/S0021889896013556
78. Dias FB, Batty S V., Ungar G, et al. (1996) Ionic conduction of lithium and magnesium salts within laminar arrays in a smectic liquid-crystal polymer electrolyte. *J Chem Soc - Faraday Trans* 92:2599–2606. https://doi.org/10.1039/ft9969202599
79. Wright P V., Zheng Y, Bhatt D, et al. (1998) Supramolecular order in new polymer electrolytes. *Polym Int* 47:34–42. https://doi.org/10.1002/(SICI)1097-0126(199809)47:1<34::AID-PI2>3.0.CO;2-G
80. Zheng Y, Wright P V., Ungar G (2000) Insertion of ionophobic components into amphiphilic low-dimensional polymer electrolytes. *Electrochim Acta* 45:1161–1165. https://doi.org/10.1016/S0013-4686(99)00376-X
81. Angell CA, Liu C, Sanchez E (1993) Rubbery solid electrolytes with dominant cationic transport and high ambient conductivity. *Nature* 362:137–139. https://doi.org/10.1038/362137a0
82. Forsyth M, Sun J, Macfarlane DR, Hill AJ (2000) Compositional dependence of free volume in PAN/$LiCF_3SO_3$ polymer-in-salt electrolytes and the effect on ionic conductivity. *J Polym Sci Part B Polym Phys* 38:341–350. https://doi.org/10.1002/(SICI)1099-0488(20000115)38:2<341::AID-POLB6>3.0.CO;2-S

83. Florjańczyk Z, Zygadło-Monikowska E, Wieczorek W, et al. (2004) Polymer-in-salt electrolytes based on acrylonitrile/butyl acrylate copolymers and lithium salts. *J Phys Chem B* 108:14907–14914. https://doi.org/10.1021/jp049195d
84. Florjańczyk Z, Zygadło-Monikowska E, Affek A, et al. (2005) Polymer electrolytes based on acrylonitrile–butyl acrylate copolymers and lithium bis(trifluoromethanesulfone) imide. *Solid State Ionics* 176:2123–2128. https://doi.org/10.1016/j.ssi.2004.08.046
85. Ghosh A, Kofinas P (2008) Nanostructured block copolymer dry electrolyte. *J Electrochem Soc* 155:A428. https://doi.org/10.1149/1.2901905
86. Appetecchi GB, Zane D, Scrosati B (2004) PEO-based electrolyte membranes based on LiBC$_4$O$_8$ salt. *J Electrochem Soc* 151:A1369. https://doi.org/10.1149/1.1774488
87. Edman L, Doeff MM, Ferry A, et al. (2000) Transport properties of the solid polymer electrolyte system P(EO)$_n$LiTFSI. *J Phys Chem B* 104:3476–3480. https://doi.org/10.1021/jp993897z

4 Polymer Nanocomposite-Based Solid Electrolytes for Lithium-Ion Batteries

Prasad V. Sarma, Jayesh Cherusseri, and Sreekanth J. Varma

CONTENTS

4.1 Introduction 81
4.2 Active Ceramic Filler-Based PNSEs 84
    4.2.1 Garnet-Type Ceramic Fillers for PNSEs 85
    4.2.2 NASICON-Type Ceramic Fillers for PNSEs 88
    4.2.3 Perovskite-Type Ceramic Fillers for PNSEs 90
    4.2.4 Anti-Perovskite-Type Ceramic Fillers for PNSEs 92
    4.2.5 Sulfide-Type Ceramic Fillers for PNSEs 93
4.3 Inactive Ceramic Oxide-Based PNSEs 95
4.4 Metal-Organic Frameworks (MOFs) as Fillers for PNSEs 96
4.5 Biopolymers as Fillers for PNSEs 98
    4.5.1 Cellulose 99
    4.5.2 Chitosan 100
    4.5.3 Proteins 101
    4.5.4 Starch 102
4.6 Conclusions and Future Perspectives 102
References 102

4.1 INTRODUCTION

Lithium-ion batteries (LIBs) have attracted the energy storage industry due to their high capacity, easy processability, and long cycle life [1, 2]. Although the disposal of LIBs is still a major challenge, one cannot imagine a world without LIBs, due to their importance in our daily life. The market for LIBs is predicted to grow at a compound annual growth rate of 12.31% during the forecast period of 2020–2025. The major factors driving the battery market include declining LIB prices, the rapid growth of the electric vehicle market, the fast-growing renewables sector, and the increased production of consumer electronics [3, 4]. But the unavailability of raw materials may lead to a demand-supply mismatch and that may hinder market growth. The various components of a LIB include anode and cathode electrodes, spacer, electrolyte, sealant, etc. All these components are of equal importance, but the electrolyte has the upper hand in determining the performance of a LIB. In the past, the usage of liquid electrolyte-based LIBs made them less attractive for various applications, such as power supplies for wearable and flexible electronic devices. Since the mobility of the electrolyte ions is very high in the case of liquid electrolytes, it helps in attaining high specific capacity. Among the various end-user segments, the automotive sector has the greatest potential for the rise in LIB demand in the future. The recent developments in the acceptance of electric vehicles, in response to carbon dioxide (CO_2) pollution, is anticipated to accelerate the growth of LIB demand. Petroleum fuel-based vehicles are disappearing from the roads globally, due to the influence of green-energy technologies, ranging from hybrid electric vehicles to electric vehicles.

Nowadays, due to the rapid growth of wearable electronic devices, there is a massive demand for flexible and wearable energy storage to supply power to these devices [5]. The main pre-requisite of an energy storage device to power a wearable electronic device are that it should not leak at any cost [6–9]. Flexible supercapacitors and batteries have achieved great interest in wearable electronic applications [10–14]. The leakage of liquid electrolytes from the LIB cannot be tolerated for these applications. Above all, even the slightest possibility of Li dendrite formation, which might cause an explosion, must be avoided completely in the case of wearable electronic devices [15–17]. Hence, semi-solid-state and solid-state electrolytes have been developed for LIBs recently, with promising power delivery and excellent stability features, including an effective suppression of dendrite formation [18, 19].

Polymers are extensively used for developing solid-state electrolytes for LIBs. Although the lower mobility of the electrolyte ions may be an issue, the requirements for LIBs with non-leaking features necessitates the use of a solid-state electrolyte in a LIB. These challenges are overcome by preparing novel solid-state polymer electrolytes (SPEs) [20–22]. The SPEs are prepared by dissolving Li salts at high concentrations in polymers such as polyethylene oxide (PEO), polyacrylonitrile (PAN), polyvinyl alcohol, etc. In LIBs, a solid-state electrolyte with outstanding flexibility is required to ensure a low interface resistance and excellent mechanical robustness, to overcome any puncture caused by the Li dendrites at the electrolyte/cathode and electrolyte/anode interfaces [23]. SPEs have the advantage of flexibility, ease of processability, low cost and the possibility to blend with a variety of materials, giving the possibilities to engineer the properties suitable for each application. Thin or thick flexible films can be made from these materials, which have the advantage of excellent interface stability with the electrodes.

The overall performance of the battery suffers a setback due to the narrow electrochemical windows, poor ionic conductivities, and low mechanical strength of these polymer electrolytes. The main demerit of using SPEs is the very low ionic conductivity, of the order of 10^{-8}–10^{-5} S cm^{-1}. Furthermore, the highly crystalline nature of SPEs hinders the ion diffusion, ultimately resulting in the delivery of low power. The challenges associated with using SPEs (without any additives) are resolved by preparing polymer nanocomposite-based solid-state electrolytes (PNSEs). The PNSEs are developed by adding nano-/microceramic filler particles

Polymer Nanocomposite Electrolytes

inside a SPE matrix. The addition of secondary filler particles is found to improve the ionic, electrochemical, and mechanical properties of the SPEs. The various types of fillers used to synthesize PNSEs include inorganic material-based fillers, inactive ceramic-based fillers, inactive ceramic oxide-based fillers, metal-organic framework (MOF)-based fillers, and biopolymer-based fillers. This chapter discusses the various types of active ceramics, such as garnet-type, sodium superionic conductor (NASICON)-type, perovskite-type, and sulfide-type fillers. A discussion on inactive ceramic oxide-based fillers for PNSEs is also included. Furthermore, the recent development of MOF filler-based PNSEs is briefly discussed, as are the recent developments in the use of biopolymers, such as cellulose-, chitosan-, protein-, and starch-based PNSEs being discussed. A schematic illustration, showing the categories of various PNSEs prepared using different polymers and fillers, is shown in Figure 4.1.

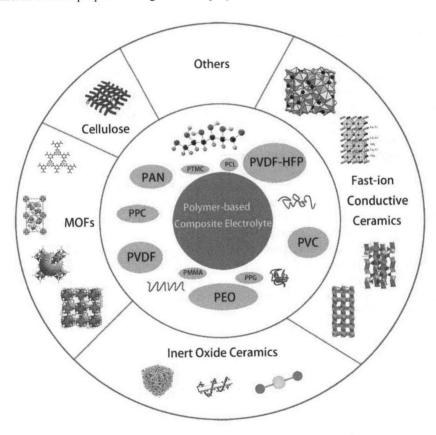

FIGURE 4.1 A schematic diagram showing the various categories of polymer nanocomposite solid-state electrolyte (PNSEs) used in lithium-ion batteries. Adapted and reproduced with permission from Ref. [24]. Copyright © 2019 Yao, Yu, Ding, Liu, Lu, Lavorgna, Wu, and Liu.

4.2 ACTIVE CERAMIC FILLER-BASED PNSEs

Ceramic fillers are generally divided into two categories, inactive and active fillers. Inactive fillers do not get involved in Li$^+$-ion conduction, such as titanium dioxide (TiO$_2$), silica (SiO$_2$), alumina (Al$_2$O$_3$), etc., whereas active fillers, such as Li$_3$N, Li$_{1.3}$Al$_{0.3}$Ti$_{1.7}$(PO$_4$)$_3$ etc., participate in Li$^+$-ion conduction [23]. Inorganic active ceramic electrolytes contain all the inorganic Li$^+$-ion conductors, such as oxides, phosphides, nitrides, and sulfides [24]. These electrolytes exhibit exceptional Li$^+$-ion mobility, even greater than that of their liquid counterparts, and they exhibit high mechanical strength. However, Li dendrite formation and internal short-circuits are not fully suppressed in ceramic electrolytes [25]. Moreover, the electrochemical window of many of the ceramic electrolytes is narrow, which limits their practical applications. The specific energy of the cells is markedly reduced due to the high density of ceramic particles, compared with liquid electrolytes. In addition, it is quite challenging to synthesize large-area, thin, ceramic electrolytes that could meet commercial standards *via* traditional ceramic fabrication routes [25]. Instead of trying to develop solely ceramic electrolytes, it is beneficial to develop PNSEs with ceramic fillers. Uniformly distributed ceramic fillers on a polymer matrix confer all the benefits of ceramic electrolytes, such as increased ionic mobility, increased mechanical strength, etc. PNSEs also have other advantages, such as flexibility, high energy density, low cost, effective suppression of Li dendrite formation, and, most importantly, low fabrication cost. Garnet-type composite, perovskite-type, NASICON-type, and sulfide-type SPE composites are examples of active ceramic electrolytes [26, 27].

There are many inorganic ceramic materials that offer high ionic conductivities and mechanical strength [24, 28]. But these materials, in the absence of a polymer matrix, show poor interfacial contacts with the electrodes, impeding their use as solid-state electrolytes in batteries [28, 29]. A viable solution to overcoming these issues is to incorporate inorganic materials, with excellent ionic conductivity and mechanical strength, into the polymer electrolyte. The resulting polymer composite material will possess all the benefits of the polymer and the inorganic filler. Recently, inorganic fillers, in the form of zero-dimensional (0D) fillers, one-dimensional (1D), i.e. wire- or tube-like fillers, two-dimensional (2D), i.e. sheet-like fillers, and 3-dimensional (3D) filler structures have attracted a lot of scientific interest, most recently due to the possibility of fine-tuning the properties of the composite electrolyte in terms of maximum performance and stability [30]. In a polymer matrix, homogeneously dispersed 0D fillers can effectively aid Li$^+$-ion transport by inhibiting the reorganization of the polymer chains, restricting the crystallization. The most common method to prepare composite electrolyte involves mechanical mixing of the nanofillers in the polymer matrix. But, due to the very high surface energies of the nanomaterials, the fillers will be non-uniformly distributed in the polymer, resulting in local aggregation of nanoparticles. This agglomeration may lead to polymer crystallization within the electrolyte and may reduce the polymer–filler interactions, that can affect the overall performance of the composite electrolyte. As a solution to the agglomeration of nanoparticles, Lin et al. [31] adopted an *in-situ* preparation of the nanoparticles in the polymer electrolyte. It was found to significantly reduce the agglomeration of nanoparticles and restrain crystallization, resulting in an improved ceramic/polymer interaction and increased ionic conductivity. The size of the filler nanoparticles plays a pivotal role in improving the performance of the composite electrolyte. Nano-sized inorganic fillers show enhanced ionic conductivities, when compared with micro-sized analogs, by offering better pathways for Li$^+$-ion diffusion [32, 33]. Reasons for the enhanced ionic conductivity are reported involve the reduction in percolation thresholds, the larger specific surface areas, and increased conductive interfaces, with a decrease in particle size promoting Li$^+$-ion migration [23, 34]. The 1D nanofillers, when compared to their 0D analogs, offer an increased ionic conductivity as a result of the larger area in contact with the polymer matrix [23, 35]. In addition to these lower-dimensional structures, many 2D materials, like graphene [36, 37], transition metal dichalcogenides [38–40], and boron nitride [41, 42], have also been used as fillers in polymer electrolytes, owing to their ultra-thin architecture and a higher degree of chemical functionality [43]. The composite electrolytes with 2D materials have shown more than two orders of magnitude improvements in ionic conductivities and tensile strength, when compared with the pure polymer electrolytes and have inhibited Li dendrite growth [44–46]. This section details the salient features of fast ion-conductors, like garnet-type, perovskite-type, NASICON-type, and sulfide-type ceramic fillers, with different sizes and shapes. These inorganic ceramic fillers can provide impressive ionic conductivities as high as ~10^{-2} S cm^{-1} at room temperature.

4.2.1 Garnet-Type Ceramic Fillers for PNSEs

Garnet-type ceramics are among the most-promising materials for the SPEs in all-solid-state LIBs, owing to the superior Li$^+$-ion conductivity at room temperature and the high shear modulus that protects the battery from Li dendrites [47, 48]. The interest in garnet-type solid-state electrolyte materials started in 2007, when Li$_7$La$_3$Zr$_2$O$_{12}$ (LLZO) was first reported, and the environmental, chemical, and electrochemical stabilities have been extensively studied by many groups. These ceramics generally occur in cubic and tetragonal phases, the latter having higher ionic conductivity. The highest ionic conductivity reported for garnet-type ceramics is ~1.66×10^{-3} S cm^{-1} at room temperature, which is far ahead of other oxide materials used as electrolytes [49]. They are chemically stable, unlike the sulfide electrolytes and many NASICON-type ceramics, and do not undergo reduction in contact with Li [50]. A perspective for the solid-state LIBs, based on garnet-type ceramic fillers, is depicted in Figure 4.2.

Theoretical and experimental research on these types of materials suggested that the structural stability is compromised under ambient conditions and is vulnerable to CO$_2$ and moisture, leading to the formation of Li$_2$CO$_3$, which reduces the ionic conductivity. Although several methods have been shown to overcome these issues [51, 52], the high mass density and brittleness limit the effectiveness of this material for direct use in batteries [48]. PNSEs made of these filler materials have shown superior environmental stability, mechanical robustness, flexibility, and ionic conductivities, which are the combined characteristics of both the filler and the matrix. In a typical

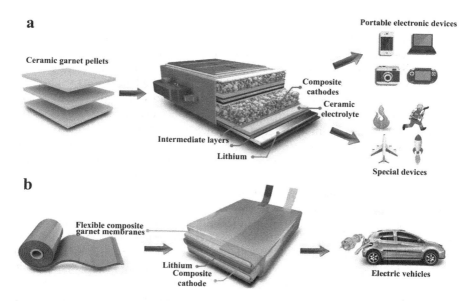

FIGURE 4.2 A schematic illustration displaying the future perspective of lithium-ion batteries, using garnet-type lithium ion-conducting ceramic fillers. Adapted and reproduced with permission from Ref. [50]. Copyright © 2019 Elsevier.

work, Fu et al. [53] have reported the preparation of a flexible, solid-state Li$^+$-ion conducting network membrane electrolyte of PEO and an electrospun garnet-type Li$_{6.4}$La$_3$Zr$_2$Al$_{0.2}$O$_{12}$ nanofiber mat with improved ionic conductivity, electrochemical stability, voltage stability, and mechanical strength. This 3D membrane electrolyte not only promotes long-range Li$^+$-ion diffusion but also effectively blocks the formation of Li dendrites in a symmetrical cell in the Li/electrolyte/Li architecture. With an ionic conductivity of ~2.5×10^{-4} S cm^{-1} for the composite electrolyte, the cell exhibits a current density of 0.5 mA cm^{-2} and 0.2 mA cm^{-2} after more than 300 and 500 hours, respectively, at room temperature. In another study, Xie et al. [48] reported a facile, low-cost preparation of a hybrid electrolyte of PEO with cubic-LLZO nanofibers supported by bacterial cellulose, which exhibits an ionic conductivity of 1.12×10^{-4} S cm^{-1}, comparable with that of a pure LLZO disc.

A technique to prepare a hybrid solid-state electrolyte consisting of well-aligned electrospun Li$_{0.33}$La$_{0.557}$TiO$_3$ (LLTO) nanofibers in PAN has been reported by Liu et al. [35]. They observed an ionic conductivity of 6.05×10^{-5} S cm^{-1} at 30°C, which is an order of magnitude higher than that of the pure polymer electrolyte, and deduced a superior surface conductivity of 1.26×10^{-2} S cm^{-1} at 30°C for the aligned nanowires, which was comparable with the ion conduction in aqueous electrolytes. It was also found that the aligned nanofiber/polymer electrolyte exhibited a ten-times higher ionic conductivity, when compared with the non-aligned nanofiber/polymer electrolyte. A comparison of Li$^+$-ion conduction pathways in nanoparticles, random nanowires, and aligned nanowires is provided in Figure 4.3. It can be seen that, when

Polymer Nanocomposite Electrolytes

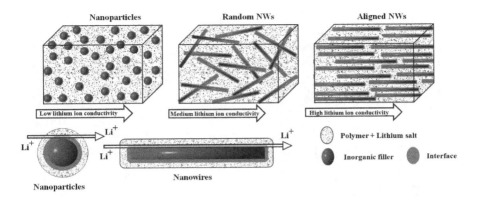

FIGURE 4.3 The lithium-ion conduction pathways in PNSEs with (a) nanoparticles, (b) random nanowires, (c) aligned nanowires, or (d) the surface region of inorganic nanoparticles (NPs) and nanowires (NWs), providing fast lithium-ion conduction.

compared with isolated nanoparticles, random nanowires can provide a continuous fast ion conduction pathway, when compared with isolated nanoparticles.

The role of morphology, doping, and the mechanism of conduction in LLZO nanowire-ceramic fillers was studied by Yang et al. [54]. They noticed an increase in the ionic conductivity of the polymer electrolyte when LLZO nanowires were introduced into the PAN/Li perchlorate (LiClO$_4$) matrix, and found a concentration of 5% LLZO as the optimal mass loading. Nanoparticles of LLZO, prepared by ball-milling, were also used for the study. The ionic conductivity exhibited by PAN/LLZO NP electrolyte was observed to be far inferior to that of the PAN/LLZO nanowire electrolyte, as the nanowires provided long-range conduction pathways for the Li$^+$-ions. It was also found that doping with aluminum (Al) and tantalum (Ta) brought no significant improvement in the conductivity of the hybrid electrolyte films. Unlike that study, there are many reports in which doped garnet-type ceramics show improved characteristics in comparison with the polymer electrolytes. One such study involved a solid flexible electrolyte, containing 20% of Ta-doped garnet-type ceramic filler, Li$_{6.75}$La$_3$Zr$_{1.75}$Ta$_{0.25}$O$_{12}$ (LLZTO), in the PAN-LiClO$_4$ matrix, that displays Li$^+$-ion conductivity in the range 2.2×10^{-4} S cm^{-1} at 40°C [29]. This very high conductivity may be due to the presence of LLZTO that triggers the crystallization and segmentation of PAN chains, improving the thermal and mechanical stabilities of the electrolyte. A symmetrical Li electrolyte/Li cell with LLZTO hybrid electrolyte shows enhanced cycling stability and Li$^+$-ion transference number with an operating voltage of 4.9 V. In another report, Zhao et al. [55] demonstrated an all-solid-state battery with a flexible solid-state electrolyte, prepared using Al-doped LLZTO in a PEO/Li salt matrix that could withstand high temperatures and promote the immobilization of the anions for homogeneous ion distribution and dendrite-free Li accumulation. The LLZTO particles reduce the crystallization of the polymer chains and support ceramic/polymer interactions, leading to an ionic conductivity of ~ 1.12×10^{-5} S cm^{-1} at 25°C, with an operating voltage of 5.5 V. A schematic diagram

88   Polymer Electrolytes

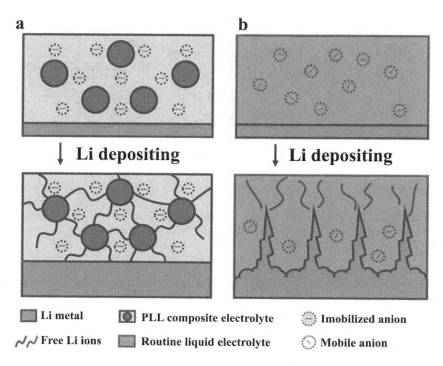

FIGURE 4.4 A schematic illustration of the electrochemical deposition behavior of the lithium anode with the PLL-based solid-state electrolyte containing (a) immobilized anions and (b) the conventional liquid electrolyte with mobile ions. Adapted and reproduced with permission from Ref. [55]. Copyright © 2017 PNAS.

showing the electrochemical deposition behavior of an Li anode with the PEO–lithium bis(trifluoromethylsulfonyl)imide (LiTFSI)-LLZTO (PLL)-based solid-state electrolyte, containing immobilized anions, and the conventional liquid electrolyte with mobile ions are depicted in Figures 4.4a and 4.4b, respectively.

Extensive research has been carried out on hybrid SPEs, containing garnet-type ceramic fillers, their doped forms and extra additives, to study the Li$^+$-ion transport [56, 57]. In all these studies, it was evident that the inclusion of garnet-type ceramic fillers improved the safety and overall properties of the LIB, exploiting the synergistic effects of the polymer/filler combination.

4.2.2 NASICON-Type Ceramic Fillers for PNSEs

Another type of inorganic ceramic fast ion-conductors that has been widely investigated is the NASICON-type, with a general formula of $LiM_2P_3O_{12}$ (M = Sn, Ti, Sr, Ge, Zr, etc.). Li-Al-germanium phosphate (LAGP) is one of the best of its kind, because of its high ionic conductivity and stability with Li metal. The NASICON-type ceramics are hard, brittle, and lack flexibility, which increases the interfacial resistance, leading to poor cell performance [58]. To effectively use these ceramics in LIBs as solid-state electrolytes, hybrid electrolytes are made from polymer electrolytes, like PEO, which offer excellent flexibility and film formation, improving the interfacial contacts with the two electrodes without any short-circuit. Inda et al. [59] reported that a cell with $LiTi_2P_3O_{12}$ in poly(ethylene oxide-*co*-propylene oxide) as the solid-state hybrid electrolyte exhibited a 90% capacity retention after 10 charge/discharge cycles when tested at a current rate of C/12. In a study by Jung et al. [58], improved performance was observed in all-solid-state $Li/LiFePO_4$ cells displaying good cycling stability in the voltage range of 2.6–4 V at a current rate of 0.2 C at 55°C and an initial charge capacity of 137.6 mAh g^{-1}, with a Coulombic efficiency of 96.5%. The cells consisted of a flexible LAGP/PEO hybrid film as the electrolyte, with excellent mechanical properties and ionic conductivity. It was observed that the initial Coulombic efficiency increased and reached 99% after a few cycles, remaining stable thereafter. They also found that a LAGP mass-loading of 60–80% was found to be the ideal composition for this solid composite electrolyte to form a highly conducting, flexible, and mechanically robust film with PEO. Doping of NASICON-type ceramics, like $LiTi_2P_3O_{12}$, with adequate amounts of trivalent ions, like Al^{3+}, Cr^{3+}, and Fe^{3+}, substantially improved the ionic conductivities of the material [60]. Al-doped NASICON ceramic (LATP) of the type $Li_{1+x}Ti_{2-x}P_3O_{12}$, with x=0.3, was found to exhibit the highest ionic conductivity [61]. In a typical study, a composite electrolyte, consisting of a mixture of triboron-based poly(ethylene glycol) (PEG) (BPEG) with 15 wt.% high-molecular-weight PEO and LATP (x=0.3), was used to fabricate a $Li/LiFePO_4$ cell that gave, at 60°C, specific capacities of 158.2 mAh g^{-1} at 0.1 C and 94.2 mAh g^{-1} at 2 C [62]. Here, the BPEG/PEO mixture interconnected different ceramic nanoparticles and performed a leading role in enhancing the ionic conductivity ~2.5×10^{-4} S cm^{-1}. Particles of another NASICON-type ceramic p, Li-tin-zirconium phosphate [$LiSnZr(PO_4)$], were embedded in polyvinylidene difluoride (PVdF)/LLZTO)/PEO matrix and ionic conductivities were monitored [63]. Addition of 15 wt.% ceramic particles offered a high ionic conductivity of 5.76 ×10^{-5} S cm^{-1} at 300 K and a stability window up to 4.73 V. The addition of ceramic fillers was also reported to improve the transference number of the composite electrolyte. A schematic of this ceramic-polymer composite electrolyte-based Li$^+$-ion cell is shown in Figure 4.5.

The synthesis of LATP fast ion-conductors, by spray-drying and sintering processes, was recently reported [64]. In this method, a solid-state hybrid electrolyte of the LATP was prepared with different compositions with PVdF-*co*-hexafluoropropylene, using an inverse-phase method. It was found that the electrolyte with an LATP concentration of 5% with the polymer showed very high thermal stability, displaying a decomposition temperature at 430°C and a stabilized impedance at 472 Ω. An ionic conductivity close to 3.943×10^{-3} S cm^{-1} at 25°C was observed in the hybrid solid-state electrolyte film. A $LiCoO_2$/Li coin-cell, assembled using this membrane, displayed a specific capacity of 145.4 mAh g^{-1} at a rate of 1 C, with a capacity retention of 96.93% even after 25 charge/discharge cycles. Another recent study described the preparation of a quasi-ceramic electrolyte, containing LATP ceramic and PEO, with LiTFSI as the Li salt, at room temperature [65]. The polymer-in-ceramic solid-state

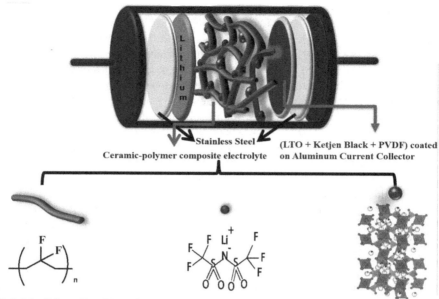

FIGURE 4.5 A schematic illustration of polymer-ceramic composite solid-state electrolyte-based lithium cell. Adapted and reproduced with permission from Ref. [63]. Copyright © 2020 Elsevier.

electrolyte membranes, with 30% PEO/LiTFSI and 70% LATP, displayed ionic conductivity values of 4×10^{-5} S cm^{-1} at 25°C and above 10^{-4} S cm^{-1} at 45°C. All these recent articles point towards the NASICON-type ceramic as a potential filler for the solid polymer hybrid electrolyte system to be used in LIBs, owing to the increased ionic conductivity, mechanical strength, temperature stability, and wide electrochemical window [66]. But its long-term stability in contact with Li is still debatable. Studies are on-going to further cut down the cost of large-scale preparation and improve the ionic conductivities by suitable doping methods and modification of the fabrication processes.

4.2.3 Perovskite-Type Ceramic Fillers for PNSEs

Perovskite-type ceramics are a class of Li ion-conducting materials of the ABO$_3$-type Li lanthanum titanate (LLTO) and its related structures [67, 68], with ionic conductivities that can be enhanced to the order of 10^{-3} S cm^{-1}. It has been observed that the ionic conductivities could be modified by replacing or partially substituting La with rare-earth elements. A decrease in Li$^+$-ion conductivity was observed when La is replaced with rare-earth elements of smaller ionic radii and was found to increase when partially substituted with Sr-like elements, which exhibit a larger value [67, 69–72]. But the addition of Zr at a concentration of up to 5% into the LLTO system by a sintering process was found to increase the ionic conductivities markedly from

Polymer Nanocomposite Electrolytes

~10^{-5} S cm^{-1} to about 1.5×10^{-3} S cm^{-1} at 25°C [73]. Apart from the excellent ionic conductivity values and mechanical strength, the poor thermal stability and the ease of moisture absorption limit the direct use of these ceramics in batteries [68]. Freestanding polypropylene carbonate (PPC)/Li$_{6.75La3}$Zr$_{1.75}$Ta$_{0.25}$O$_{12}$ solid-state hybrid electrolytes with a high ion-transference number of 0.75, a wide electrochemical window of 4.6 V, a mechanical strength of 6.8 MPa, and an ionic conductivity of 5.2×10^{-4} S cm^{-1} at 20°C was prepared by Zhang et al. [69] for a flexible LiFePO$_4$/Li$_4$Ti5O$_{12}$ battery with extended cycling stability and exceptional rate capability. In another study, a significant improvement in the ionic conductivity, mechanical strength and thermal stability was observed in a Li$_{6.75}$La$_3$Zr$_{1.75}$Ta$_{0.25}$O$_{12}$ (LLZTO)/PVdF-based SPE [74]. This was attributed to the LLZTO ceramic particles, which initiated a partial structural reformation in PVdF chains that brought about a conductivity enhancement of 5×10^{-4} S cm^{-1} at 25°C. A mechanism proposed for the PVdF-based systems is shown in Figure 4.6a and the possible complex structures proposed in the PVdF/LLZTO-based composite polymer electrolytes are shown in Figure 4.6b.

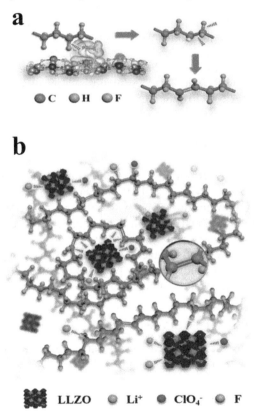

FIGURE 4.6 (a) A proposed mechanism for the PVdF-based polymer nanocomposite solid-state electrolyte systems; (b) proposed possible complex structures in the PVdF/LLZTO-based polymer nanocomposite electrolytes. Adapted and reproduced with permission from Ref. [74]. Copyright © 2017 American Chemical Society.

The main problem seen in the perovskite-type ceramics is the undesirable reaction when it comes into contact with the Li metal at the anode side. This issue was effectively rectified by Liu and co-workers [75] by using PEO layers on either side of the PEO/perovskite-type ceramic composite solid-state electrolyte. This architecture facilitates not only good contact between the electrolyte and the electrodes but also offers low interfacial resistance and appreciable cycling stability. The 3D nanofiber network of LLTO, prepared by electrospinning, endowed the hybrid electrolyte with high ionic conductivity of 0.16×10^{-3} S cm^{-1} at 24°C and superior mechanical strength. They developed a symmetrical Li/solid-state hybrid electrolyte/Li cell using the polymer/3D nanofiber perovskite-type hybrid electrolyte, which delivered a capacity of 135 mAh g^{-1} at a current rate of 2 C. This LIB cell also exhibited a good capacity retention of 79% even after 300 cycles at an operating temperature of 60°C. The device also survived more than 400 cycles without any short-circuit, highlighting the possibility of using sandwiched solid-state hybrid electrolyte architectures in LIBs. Most of the perovskite-type ceramic electrolytes are moisture sensitive, which limits their use in practical applications. $Li_{3/8}Sr_{7/16}Ta_{3/4}Zr_{1/4}O_3$ (LSTZ), which has a bulk ionic conductivity of about 10^{-3} S cm^{-1} at 24°C, has recently emerged as an exception and is now being used for the preparation of moisture-stable hybrid electrolytes. A flexible SPE, consisting of LSTZ, PEO and LiTFSI in Li/LiFePO$_4$ and high-voltage Li/LiNi$_{0.8}$Mn$_{0.1}$Co$_{0.1}$O$_2$ batteries, delivered above 700 hours of operation without Li dendrite growth [76]. The composite electrolyte exhibited ionic conductivities in the range 5.4×10^{-5} and 3.5×10^{-4} S cm^{-1} at 25 and 45°C, respectively, because of the strong interaction between the Ta^{5+} of the perovskite and the F$^-$ of the Li salt at the hybrid electrolyte surface.

Recently, nanowires of LLTO in the PEO/LiClO$_4$ matrix have attracted much attention as hybrid electrolytes due to their capability to mediate Li$^+$-ion transportation, inhibiting any transmission through the PEO or PEO/ceramic interface [77]. A noteworthy ionic conductivity of 4.01×10^{-4} S cm^{-1} was observed at 60°C with an operating voltage of 5.1 V for the SPE. An all-solid-state Li/LiFePO$_4$ battery fabricated using this composite electrolyte exhibited a capacity retention of 92.4% at 1 C at 60°C, even after 100 cycles, and a capacity of 80 mAh g^{-1} at a 2 C rate. With all the advantages, and a few disadvantages, of these types of ceramic fillers and their polymer hybrids, research is advancing to push the properties to their limit and to improve the LIB performance and stability.

4.2.4 Anti-Perovskite-Type Ceramic Fillers for PNSEs

Anti-perovskites or inverse perovskites, as the names indicate, have the same structure as perovskite, with the general formula ABX$_3$, but where the cations take the place of the anions and *vice versa*, generally denoted as A$_3$BX [78]. Unlike the sulfide-type electrolytes, anti-perovskites are stable against contact with Li metal, resulting in low interfacial resistance. Considerable interest in this material took hold with the report on superionic conductivity observed in Li-rich anti-perovskites (LiRAP), which displayed ion conductivities greater than 10^{-3} S cm^{-1} at room temperature and activation energies as low as 0.2–0.3 eV [79]. A superionic conductivity of 10^{-2} S cm^{-1} was attained when the temperature reached the melting point of the material. Zhao et al. have also prepared a range of anti-perovskites, like Li$_3$OCl, using efficient synthesis methods and have modified the structure by replacing Cl$^-$ with Br$^-$ or by using mixed anions to raise the tolerance factor for the formation of the more stable pseudo-cubic phase [79]. These kinds of structural reformations bring about the superionic conduction in anti-perovskites through a mechanism called Frenkel interstitial transport. A similar approach was adopted to increase the ionic conductivity by cationic doping in the anti-perovskite, supporting the ionic mobility ("hopping") through the Schottky channel [79]. In this study, they have obtained ionic conductivities of 0.85×10^{-3} and 1.94×10^{-3} S cm^{-1} for the Li$_3$OCl and Li$_3$OCl$_{0.5}$Br$_{0.5}$ anti-perovskites, respectively, at room temperature, which then increased to 4.82×10^{-3} and 6.85×10^{-3} S cm^{-1}, respectively, as the temperature was increased to 250°C. The synthesis methods described in this article give a clear idea on tailoring the characteristics of anti-perovskites to the desired values to suit specific practical applications. Goodenough and co-workers [80] reported the fabrication of an all-solid-state LiFePO$_4$/Li battery, using a fluorine-doped Li$_2$(OH)X (X=Cl, Br) anti-perovskite-based solid-state electrolyte. The cells were found to be unstable above 3 V due to the hygroscopic nature of the anti-perovskites. To rectify the moisture instability and the high interfacial resistance problems, polymer composites of these materials, which exhibit all the advantages of the filler and the ion-conducting polymer, can be used as the solid-state electrolyte in batteries. Now, there is ample scope for developing efficient and stable flexible solid-state electrolytes, as recent research has now uncovered a range of doped and undoped anti-perovskites [80, 81], double anti-perovskites [82, 83], and those with superionic conductivities [84] in the range of 10^{-2}–10^{-1} S cm^{-1}.

4.2.5 Sulfide-Type Ceramic Fillers for PNSEs

Sulfide-type ceramic fillers are superior to many of the other ceramic fillers, in terms of Li$^+$-ion mobility. As the sulfur anions show very high polarizability, sulfide-type ceramics offer an easy pathway for the Li$^+$-ions and display very high ionic conductivities when compared with their oxide analogs [85]. The sulfide-based PNSEs display excellent Li$^+$-ion transport mobilities in a range around 10^{-2} S cm^{-1}, with a wide potential window (> 5 V vs. Li/Li$^+$) at room temperature. Ionic conductivities of these materials are close to or even better than liquid electrolytes [86]. Sulfide-type ceramics have adequate mechanical properties such as processability, formability, and elastic modulus [87]. Li$_2$S-P$_2$S sulfide glasses exhibit very high shear modulus values around 18–25 GPa [88]. Easy processing techniques, like cold-pressing, are suitable for the synthesis of sulfide-type solid-state electrolytes. Because of the very high ionic conductivities and superior mechanical properties exhibited by the sulfide-type electrolytes, they are considered to be the breakthrough in all-solid-state LIB research. But they are more vulnerable to water vapor, making them less stable under moist conditions, due to the formation of hydrogen sulfide gas [89].

The sulfide-based ceramic materials can be divided into three categories: glass-ceramics, glasses, and ceramics. Thio-Li$^+$-ion superionic conductors (LISICONs) (e.g., $Li_{4-x}M_{1-x}A_xS_4$, where M = Si, Ge, and A = P, Al, Zn, Ga), the LGPS family (e.g., $Li_{10}GeP_2S_{12}$), argyrodites (e.g., Li_6PS_5X, where X=Cl, Br, or I) and other new thio-sulfides, such as Li_4PS_4I, $LiZPS_4$, etc., are those sulfide-based ceramics that act as active fillers. Studies are being performed to incorporate these solid-state electrolytes into a polymer matrix, such as PEO or perfluoropolyether (PFPE), etc., to be used as SPEs for LIB applications [90]. The thio-LISICONs, glass-ceramics (e.g., $Li_7P_3S_{11}$), and glassy materials (e.g., Li_2S-SiS_2-Li_3PO_4), exhibit remarkably high ionic conductivities of the order of 10^{-3} S cm^{-1} at room temperature. But these materials are not suitable electrolytes when used alone, because they cannot stick to the moving boundaries of active particles during the charge/discharge cycle in the electrode. On the other hand, an intimate mixture of organic polymers and sulfide-based glass-ceramics is capable of forming close interfacial contact between electrode and electrolyte in a battery [90]. The polymers, such as silicone polymers, PEO, and styrene-butadiene rubbers, are used as binders for ion-conducting polymer glassy electrolytes. The addition of these polymers is found to enhance the mechanical flexibility at the expense of the ionic conductivity of the electrolyte.

Hayashi et al. [90] synthesized a polymer-ceramic electrolyte of Li_2S-P_2S_5 glass and oligomers (PEG400) *via* a mechano-chemical method. In this study, the incorporation of oligomers into the glass lowered the glass transition temperature that led to a net increase in the ionic conductivity of the electrolyte when compared to pure glass. The addition of salts is another strategy by which to increase the ionic conductivity of the polymeric phase, compromising the single-ion conductor properties of the hybrid electrolyte. Villaluenga et al. [91] reported a single, ion-conducting hybrid polymer synthesized by mechano-chemical mixing of Li_2S P_2S_5, hydroxy-terminated perfluoropolyether (PFPE-diol), and LiTFSI, with the help of a ball-milling machine. In this work, the lithiated phosphorus sulfide group (glass phase) was reacted with hydroxyl groups of PFPE-diol, which resulted in a solid-state hybrid electrolyte with 77% glass-ceramic (75Li_2S: 25P_2S_5) and 23% PFPE-LiTFSI. The ceramic-glass particles were found with dimensions ranging from 1–10 μm. In general, hybrid electrolytes are defined in such a way that there are chemical bonds formed between organic and inorganic entities. It is unlikely that all the particles are attached to the polymer chain, but a significant share of the particles are connected to the polymer, as confirmed through nuclear magnetic resonance spectra in a 75Li_2S: 25P_2S_5-PFPE-LiTFSI-based solid-state electrolyte. This composite polymer electrolyte exhibited a Li$^+$-ion mobility of 10^{-4} S cm^{-1}. The Li$^+$-ion transference number of this hybrid electrolyte was calculated as 0.99, implying that most of the current was carried solely through the Li$^+$-ions [91]. Zhao et al. [72] developed an electrolyte by blending $Li_{10}GeP_2S_{12}$ (LGPS) with the PEO matrix. This novel SPE exhibited an ionic conductivity of 1.18×10^{-3} S cm^{-1} at 80°C and can be operated within an electrochemical potential window of 0–5.7 V. In this SPE, it was found that the crystallinity of PEO increased with increasing LGPS content (5%, 10%) that might be due to the increased number of crosslinking sites, suppressing the plasticizing effect. The best performance was obtained with 1 wt.% LGPS fillers. The interface stability against Li-metal or Li$^+$-ion anode is another key component that defines the cycle life of the solid-state battery. The stability of the PEO/LiTFSI-based SPE did not exhibit good

Polymer Nanocomposite Electrolytes 95

cycling stability. The addition of 1 wt.% LGPS was found to reduce the interfacial resistance markedly. Nam et al. [89] fabricated a flexible sulfide-based solid-state electrolyte on the poly(p-phenylene terephthalamide) (PPTA) non-woven framework. This sulfide electrolyte, composed of a Li_3PS_4 glass-ceramic, exhibited an ionic conductivity of 0.73×10^{-3} S cm^{-1}. Here, PPTA acted as a supporting scaffold, providing mechanical stability and flexibility. The PPTA-Li_3PS_4 assembly displayed an ionic conductivity of around 0.2×10^{-3} S cm^{-1} in a potential window ~2.1 V vs. Li/Li$^+$. A sizeable 3-fold increase in the energy density of the $LiCO_3/Li_4Ti_5O_{12}$ all-solid-state cell was found with the use of non-woven PPTA in the Li_3PS_4 solid-state electrolyte. All these studies point towards a promising solid-state hybrid electrolyte, based on the sulfide-type ceramics, owing to its excellent mechanical properties and Li$^+$-ion conductivities, which surpass the properties of many of the oxide solid electrolytes and conventional liquid electrolytes.

4.3 INACTIVE CERAMIC OXIDE-BASED PNSEs

Incorporation of inert ceramic oxides in the polymer matrix improves the mechanical properties and lowers the polymer crystallinity in such a way as to improve Li$^+$-ion mobility of the polymer SPE. Due to this reason, PNSEs with ceramic fillers have attracted a great deal of attention due to their increased ionic conductivity [26]. It is expected that ceramic fillers can act as cross-linking centers with which to reduce polymer crystallinity. However, agglomeration of ceramic particles in the polymer, weak ceramic-polymer interactions, and high crystallinity of polymers limit further improvements in ionic conductivity of PNSEs. In the early 1980s, Weston et al. [92] showed that a composite of PEO, mixed with Al_2O_3, exhibited improved mechanical properties and ionic conductivity. Later, Capuano et al. [93] further detailed the effects of doping concentration and particle size of $LiAlO_2$ on the conductivity of the solid-state electrolyte. It was observed that conductivity reached its maximum when the doping concentration of $LiAlO_2$ was 10 wt.%, with the ionic conductivity increasing with increasing particle size of ceramic particles, to <10 µm. Al_2O_3 is also expected to exhibit similar properties, such as increasing ionic conductivity and glass transition temperature of PEO in response to doping. By reducing the polymer crystallinity, the free segments in the polymers increase and can effectively stimulate Li$^+$-ion mobility. It is also probable that surface species of ceramic fillers and electrolyte ion species can get into a strong Lewis acid-base interaction. This interaction might stabilize the ions in the electrolyte by enhancing the salt dissociation and also creating additional sites for carrier migration [23]. In a few other studies, various ceramic nanofillers, such as Al_2O_3, SiO_2, TiO_2, barium titanate, etc., were used to decorate a polymer matrix to improve the ionic conductivity at ambient temperature and the mechanical stability of the electrolyte [94, 95]. Most of these elements are considered to be inactive fillers and do not contribute towards the carrier density.

Lin et al. [31] have reported 12-nm SiO_2 nanosphere-decorated PEO that shows significant suppression of polymer crystallization, with an improved dissociation of $LiClO_4$. They synthesized monodispersed SiO_2 nanoparticles decorated uniformly on PEO *via in-situ* hydrolysis of tetraethyl orthosilicate in a PEO solution. It was expected that a strong mechanical/chemical interaction between SiO_2 spheres and PEO takes place during the *in-situ* reaction process. The *in-situ* reaction of SiO_2 nanospheres with PEO might have achieved a particularly strong interaction between the components, leading to the suppression of polymer crystallinity. With a 10 wt.% SiO_2 mass loading, the nanostructured polymer composite electrolyte exhibited an ionic conductivity of 1.2×10^{-3} S cm^{-1} at 60°C and 4.4×10^{-5} S cm^{-1} at 30°C. The elastic modulus of the polymer also plays an important role in improving the performance of the electrolyte. The mobile polymer chains are responsible for softening the polymers, to enhance the ionic conductivity of the polymer electrolyte. Incorporation of materials like aerogels, which have unique properties such as high elastic modulus, large surface-to-volume ratio, and high porosity, in the polymer electrolyte can enhance the mechanical stability of the electrolyte and suppress the Li dendritic growth. SiO_2 aerogel composite-based polymer electrolytes have been reported to improve the elastic modulus, thereby increasing the overall ionic conductivity of the electrolyte [25]. The high porosity of the SiO_2 aerogel was found to incorporate an adequate amount of polymer electrolyte when the electrolyte was infused into SiO_2 aerogel. This well-dispersed arrangement of SiO_2 and polymer could enhance the Lewis acid-base interaction with anions. The SiO_2 aerogel-polymer structure seems to be very promising as it shows high ionic conductivity of 0.6×10^{-3} S cm^{-1} at 30°C with a high modulus of ~0.43 GPa [25]. TiO_2-based inactive fillers are another class of promising materials, suitable for use as additives in hybrid solid-state electrolytes for the PNSEs. In a typical study, poly(methyl methacrylate)-$LiClO_4$-propylene carbonate (PC)-TiO_2 with 1 wt.% TiO_2 free-standing films was prepared using a solution-casting technique, with tetrahydrofuran as solvent. High thermal stability is one of the favorable features exhibited by electrolytes with a high TiO_2 concentration. But electrolytes with as little as 1 wt.% TiO_2 was observed to exhibit an ionic conductivity of 3×10^{-4} S cm^{-1} at 30°C. There were studies reporting the relationship between surface chemistry and its effect on ionic conductivity [96]. The particles with acidic surface groups could increase the ionic conductivity, whereas those with basic or neutral surface chemistry could reduce the ionic conductivity in a PNSE. Al_2O_3 is an example for inactive ceramic filler, which is widely used in SPEs. It has two polymorphs, α-Al_2O_3 and γ-Al_2O_3, the former having Al-atoms (acidic sites) and the latter having Al and oxygen atoms as terminations (acidic and basic sites). A conductivity study performed on SPEs, which are decorated by α-Al_2O_3 and γ-Al_2O_3, confirms the above assumption [96]. A 5 wt.% concentration of neutral Al_2O_3 particles dispersed in the PEO/$LiClO_4$ system exhibited the highest ionic conductivity [96]. The extent of particle dispersion in the polymer matrix and the polymer–particle interactions play key roles in altering the transition temperature of the electrolyte [94].

4.4 METAL-ORGANIC FRAMEWORKS (MOFS) AS FILLERS FOR PNSEs

MOFs are microporous solid materials consisting of a large network of metal atoms that bridge to organic ligands through co-ordination bonds [97]. MOFs are characterized by high surface area, porosity, and polymetallic sites. These materials have found applications in a wide range of areas, like catalysis, gas storage, purification, drug delivery, membrane synthesis, etc. MOFs exhibit properties similar to zeolite fillers, such as high thermal stability, and show the nature of Lewis acidic surface sites. In addition, the organic functional entities of the MOFs are beneficial in improving ionic conductivities and enhancing interfacial compatibilities [98]. Nanocomposites prepared using MOFs and polymer matrices are of great interest in developing all-solid-state polymer electrolytes for LIBs. In most of the reports, nanocomposites of polymers with MOFs are prepared by mechanical blending of the components, leading to the agglomeration of MOF nanoparticles in the polymer matrix. This leads to a significant suppression of ionic conductivity as well as interfacial contact characteristics in MOF-based SPEs. Chemical linkage of MOF nanoparticles with polymer matrices is a promising strategy by which to develop novel materials with a homogeneous distribution of particles in the polymer matrix. Since the MOFs act as inert fillers in a polymer matrix, the Lewis acid surfaces of MOFs interact with the added Li salt and form complexes. This will lead to a structural modification at the filler–polymer interface and a reduction in polymer crystallinity, that aids in increasing the ionic conductivity [98].

Wang et al. [98] synthesized a hybrid covalently-linked M-UiO-66-NH-PEG diacrylate (PEGDA)-based electrolyte through photopolymerization of the vinyl-functionalized MOF, M-UiO-66-NH$_2$, and PEGDA, while using 1-hydroxy-cyclohexyl phenyl ketone as the photo-initiator. A moderate amount of LiTFSI was added to the polymer mixture before photopolymerization. PEGDA exhibited an ionic conductivity of 7.94×10^{-6} S cm^{-1} at 30°C, whereas the nanocomposite MOF-polymer displayed an ionic conductivity of 4.31×10^{-5} S cm^{-1} at 30°C, more than five times that of PEGDA alone. The resulting polymer nanocomposite electrolyte showed an excellent electrochemical window of 5.5 V vs. Li/Li$^+$. Gerbaldi et al. [97] developed Al-based MOF Al-1,3,5-benzenetricarboxylate (Al-BTC) in a PEO matrix nanocomposite with LiTFSI. Al-BTC was synthesized by an electrolytic process and the PNSE was prepared by dispersing adequate amounts of the Al-BTC MOF in the PEO-LiTFSI mixture. For pure PEO/LiTFSI electrolytes, ionic conductivities were obtained in the range of 10^{-4}–10^{-8} S cm^{-1} within a temperature window of 20–80°C. The addition of 2 wt.% Al-BTC MOF to the PEO-LiTFSI matrix led to an increase in ionic conductivity by one order of magnitude. The rate of salt dissociation increased in the polymer matrix due to a decrease in ionic coupling, caused by the Lewis acid-base interactions. The Al-BTC MOF behaves like a Lewis acid which can react with Li$^+$-ions. This, in turn, reduced the crystallinity of the polymer matrix, leading to a marked increase in ionic conductivity. It was also found that the overall interfacial resistance of PNSEs to the Li metal anode was low compared to bare PEO/LiTFSI-based electrolyte. PNSEs with Al-BTC-MOF fillers were found to have greater compatibility with Li electrodes.

A PEO-based electrolyte with Al-1,4-benzenedicarboxylate (MIL-53) [denoted as MIL-53(Al)], where MIL-53 is another MOF that consists of AlO$_4$(OH)$_2$ octahedra, was prepared by Zhu et al. [99]. Here, MIL-53 was the filler material and LiTFSI the Li salt in the PEO matrix. The zeta potential of the MOF particles was measured to be 9. It was observed that, when the pH is lower than 9.1, the surface of the MIL-53(Al) displayed Lewis acid properties. In this study, as the pH of the polymer-MOF-Li salt assembly is only 7, MIL-53(Al) exhibited a strong Lewis acid property. Here, the anions of the LiTFSI salt [N(SO$_2$CF$_3$)$_2$] reside on the MIL-53 surfaces, whereas Li$^+$-ions of the LiTFSI move towards the oxygen atoms in the polymer chains. These interactions in the matrix not only increased the Li salt dissociation but also interrupted the crystallization of PEO chains [99]. The novel PEO/MIL-53(Al)/LiTFSI electrolyte exhibited an ionic conductivity of 3.39×10^{-3} S cm^{-1} at 120°C, which is 3.5 times higher than that of the PEO-LiTFSI electrolyte without MIL-53(Al) fillers. In another study, nano-sized MOF-53 Zn$_4$O(BDC)$_3$, (where BDC is 1,4-benzenedicarboxylate) was synthesized through the solvothermal method and was successfully incorporated onto PEO/LiTFSI-based PNSE through *in-situ* methods [100]. This *in-situ* synthesis method yielded much more uniform MOF-5 distribution in the PEO matrix, with less agglomeration. In this study, the authors varied the PEO: LiTFSI salt ratio at a fixed MOF-5 concentration of 10 wt.%, since this had been experimentally identified to be the ideal loading amount of MOF-5. This electrolyte exhibited an ionic conductivity of 3.16×10^{-5} S cm^{-1} at a ratio of 30:1 (PEO: LiTFSI).

4.5 BIOPOLYMERS AS FILLERS FOR PNSEs

Tremendous efforts are underway to develop synthetic polymer nanocomposites as electrolytes. The primary issue, of the very low ionic conductivity of these compounds (10^{-8}–10^{-4} S cm^{-1}), has not been addressed yet and thus hinders its practical applications. Recently, naturally occurring polymeric materials have attracted significant attention due to their eco-friendliness, low cost, and the versatile functional groups they possess in their complex structure [101]. From a sustainable economic viewpoint, the availability of diverse polymer resources is encouraging the development of natural polymers as promising substitutes for synthetic polymers. These materials will strengthen the trend towards the development of green and flexible LIBs. Most of the battery electrolytes are vulnerable to dendritic growth that may lead to explosions during charge/discharge cycling. Moreover, ionic electrolytes used in batteries are hazardous in nature, which increases the manufacturing cost of the LIBs. The development of SPEs with bio-derivatives will reduce the cost of the batteries, as well as minimizing the environmental impact to a large extent. It was discovered that biopolymers, such as cellulose, starch, protein, chitosan, etc., are potential candidates for SPEs [24, 100]. The biopolymers exhibit very good mechanical properties, such as high moduli, due to their high molecular weights and complex structures. The large functional groups, such as amines, carboxyls, etc., and the active moieties in the biopolymers can be chemically modified in such a way that they could be reactive towards the Li salts or the electrolytes. Natural polymers can dissolve Li salts, and hence these biopolymers can be used as host materials for the Li salts. Also, these natural polymers can be added to the polymer matrix as fillers to enhance the mechanical properties of the composite material.

4.5.1 Cellulose

Cellulose is an earth-abundant material with outstanding properties, such as thermal and chemical stability, mechanical strength, and biodegradability. The empirical structure of cellulose is displayed in Figure 4.7. There are various derivatives of cellulose, such as carboxymethyl cellulose, hydroxyethyl cellulose, etc. Generally, cellulose acts as a porous structure or supporting skeleton, that absorbs electrolytes and provides mechanical stability and flexibility to the polymer electrolyte [102]. Zhao et al. [103] developed a cellulose-supported PPC-based SPE for LiNi$_{0.5}$Mn$_{1.5}$O$_4$-based LIBs. They found that the cellulose skeleton provided improved mechanical strength to the SPE. The polymer electrolyte worked at a high potential window of 5 V vs. Li/Li$^+$ and displayed excellent ionic conductivity of 1.14×10^{-3} S cm^{-1} at room temperature. Zhang et al. [102] synthesized a cellulose non-woven membrane containing PEO: polycyanoacrylate: Li bis(oxalato)borate (LiBOB) in a ratio 10:2:1. Whereas PEO exhibited a mechanical strength of only 2.5 MPa, which is quite low for use in energy storage devices, the solid polymer composed of cellulose displayed a strength of 43 MPa. The synergistic effect between the components was considered to be the reason behind this very high mechanical strength. The addition of cellulose lowers the crystallinity as well as the T_g of PEO. The cellulose-supported PEO: poly(cyano acrylate): LiBOB was stable up to a potential window of 4.6 V vs. Li/Li$^+$, which is higher than the PEO-based SPE (< 4 V vs. Li/Li$^+$) [102]. The ionic conductivity of the electrolyte was found to be 1.4×10^{-3} S cm^{-1} at 160°C and 1.3×10^{-5} S cm^{-1} at 20°C, which is not adequate for practical device uses. But this material is very useful at very high temperatures (160°C) without any internal short-circuit. In further work by the same group, PEO was replaced by PPC and exceptional property enhancement was observed in the SPE [21]. The glass transition temperature of PPC SPE and cellulose-modified PPC SPE were found to be the same, which demonstrates that the addition of cellulose did not enhance the crystallinity of the SPE. They observed that the ionic conductivity of PPC SPE was 3.0×10^{-4} S cm^{-1} at 30°C, which is more than one order of magnitude higher than that of the PEO-based SPEs. In the presence of LiTFSI, three-dimensional cellulose nanocrystals, grafted with polymeric ionic liquid polymer chains, were reported [104]. The SPE displayed an ionic conductivity of 1.9×10^{-4} S cm^{-1} at 30°C, which is promising. In another report, PVdF modified with cellulose acetate butyrate SPE displayed enhanced performance, which was ascribed to the greater compatibility with the electrode and the electrolyte. This SPE showed an ionic conductivity of 2.48×10^{-3} S cm^{-1} at 30°C [105]. These studies identify the

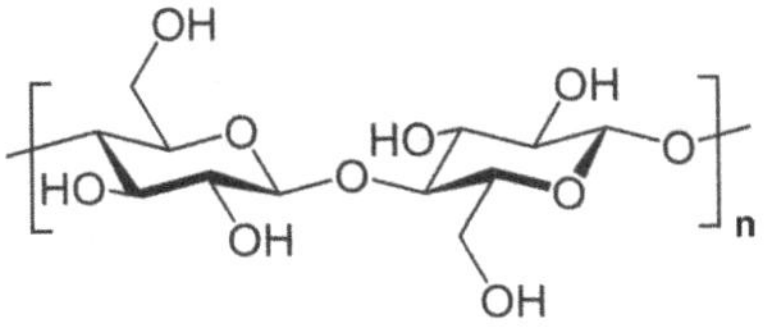

FIGURE 4.7 A schematic representation of the chemical structure of cellulose.

polymer-cellulose hybrid electrolytes as being among the potential candidates for SPEs in LIBs working at high temperatures.

4.5.2 CHITOSAN

Chitosan is another potentially valuable material, having useful properties such as biocompatibility and high activity, etc. Chitosan is a linear polysaccharide composed of randomly distributed β-(1→4)-linked D-glucosamine (deacetylated unit) and N-acetyl-D-glucosamine (acetylated unit), and is commonly prepared by treating the chitin, extracted from the shells of shrimp and other crustaceans, with an alkaline substance, such as sodium hydroxide (Figure 4.8). Chitosan can dissolve Li salts, which makes it a viable candidate for SPEs. The lone pairs of nitrogen and oxygen atoms in the chitosan system might facilitate the coordination of Li$^+$-ions from the added salt [106]. Chitosan can be used as a polymer host in the polymer electrolyte, which is highly crystalline in nature and hence not a good ionic conductor. The low ionic conductivity can be improved by treating chitosan with plasticizing solvents, like diethyl carbonate, ethylene carbonate (EC), and propylene carbonate (PC), or the mixture EC: PC [107–109], which aids the segregation of salts into ions and reduces the glass transition temperature of a polymer. Still, many of the SPEs composed of chitosan display very low ionic conductivities of the order of 10^{-5}–10^{-6} S cm^{-1} and this makes chitosan molecules less attractive for practical applications [110–113].

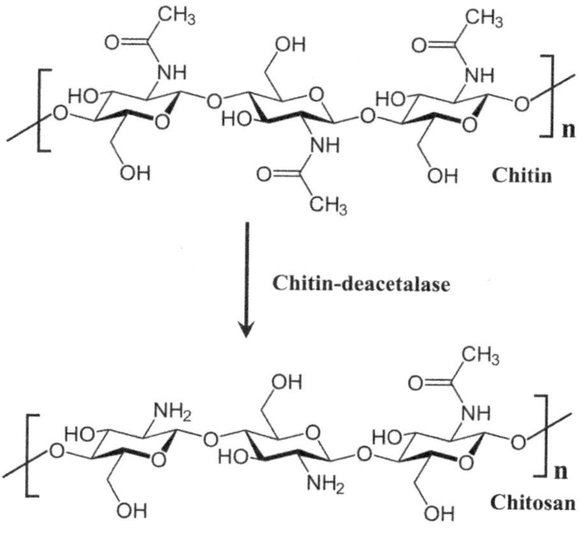

FIGURE 4.8 A schematic illustration of the preparation of chitosan by deacetylation of chitin.

4.5.3 Proteins

Proteins are biopolymers, composed of amino acid chains covalently linked by peptide bonds. Compared with other synthetic polymers, proteins have a highly complex structure (with four levels of structure, namely primary, secondary, tertiary and quaternary structures), that contains various polar and nonpolar functional groups [24, 114]. Hence, use of proteins as a polymer matrix by manipulating its functional groups and structure is not only intriguing scientifically but also an engineering challenge. Li salts also exhibit high solubility in proteins; in turn, proteins exhibit high mechanical moduli, that make them promising material for SPEs. Challenges faced in developing protein-based SPEs are our limited understanding of the Li$^+$-ion–protein chain interactions and of basic transport mechanisms in protein-based SPEs. Gelatin is a widely studied protein in electrolytes, but its main drawback is its low electrochemical window, which is limited to 2 V vs. Li/Li$^+$, due to the decomposition of glycerol that limits its practical applications [114]. Soy protein-based electrolytes have been studied recently for SPE applications [115]. Denatured soy protein, mixed with the Li salt, LiClO$_4$, and PEO (15 wt.%), was used to fabricate SPEs. The ionic conductivity of the as-prepared SPE was found to be 10^{-5} S cm^{-1} at 30 wt.% loading of Li salt. The measured modulus of SPE was ~16 MPa, which is twice that of a conventional SPE composed of PEO alone.

The structure and properties of protein-based electrolytes are sensitive to formation temperature and Li salt loading [114, 115]. Due to the abundant functional groups in proteins, these are not just host materials but also active elements in polymer electrolyte functioning. Protein-based electrolytes display an ionic conduction mechanism that is fundamentally different from traditional PEO-based electrolytes, but similar to that of ceramic materials. Anions of Li salts are locked inside the protein, and ion transport in the matrix is dominated by Li$^+$-ions [115]. A high Li$^+$-ion transference number (0.93) exhibited by protein-based SPEs indicated some decoupled ion transport phenomena, where the anions are locked away. The ionic conductivity of soy protein-based SPEs was observed to be ~10^{-5} S cm^{-1} at room temperature. The soy protein-based SPE exhibited a very high mechanical modulus of 1000 MPa, with good flexibility, where the typical modulus of ceramic conductors is far inferior, being in the range 10–100 GPa. Wang et al. [116] reported the role of gelatin-based protein nanofillers on the adhesion, ionic conductivity, and mechanical properties of polymer electrolytes. They showed that the ionic conductivity can be increased by the addition of protein without compromising the mechanical properties of solid-state electrolytes. This is an effective strategy by which to bypass the critical issue, to achieve a trade-off between ionic conductivity and mechanical strength. The ionic conductivity of protein-based SPEs is still on the lower side, compared with the ionic electrolytes. In the near future, some major improvements can be expected in this area, such as incorporating nanoparticles or manipulating protein structure to achieve better performances.

4.5.4 STARCH

Starch is an abundant polysaccharide biopolymer, which is semi-crystalline in nature, containing branched amylopectin chains and linear amylose chains [117]. Starch offers excellent solubility for Li salts like $LiClO_4$, $LiPF_6$, etc. [114, 118]. The oxygen atoms from hydroxyl groups in the starch carry lone pairs of electrons, that favor solvation of charge carriers in the polymer matrix [118]. Corn starch, loaded with 40 wt.% of $LiClO_4$ SPE, was found to display an excellent ionic conductivity of ~ 1.3×10^{-4} S cm^{-1} at 30°C [119]. Ionic conductivity in starch-ionic salt systems can be further improved by adding plasticizers, like glycerol. Amylopectin-starch, plasticized with 30–35 wt.% glycerol, was reported to display a high ionic conductivity of 1.1×10^{-4} S cm^1 [120]. Plasticizing with ionic liquid is another route by which to remove unstable glycerol out of the equation and to provide stability to the composite matrix. But the starch-based system fails in terms of its mechanical properties. To overcome this issue, ceramic fillers are added, along with the Li salt, in starch-based SPEs, which then display an increase in ionic conductivity and mechanical properties [121]. The mechanical properties of the starch-polymer can be improved by cross-linking polymer chains. KH-560 modified starch-based PNSE exhibited an ionic conductivity ~3.4×10^{-4} S cm^{-1} [122]. In a broader perspective, it can be concluded that the addition of biomaterials, like cellulose, chitosan, proteins, and starch, into electrolytes not only improves the properties of the electrolytes but also reduces the environmental impact caused by non-degradable analogs used in Li$^+$-ion batteries.

4.6 CONCLUSIONS AND FUTURE PERSPECTIVES

SPEs play a crucial role in the fabrication of all-solid-state LIBs. They have advantages, such as non-leakage, reliability, flexibility, thermal stability, and mechanical stability. In this chapter, PNSEs, developed using various filler materials, such as inorganic materials, ceramic oxides, MOFs, and biopolymers, are discussed in detail. The performance of inorganic active fillers, such as garnet-type, NASICON-type, perovskite-type, anti-perovskite-type, and sulfide-type ceramics in LIBs are elaborated. Inorganic filler-based PNSEs exhibit fast ion-conductivities of the order of 10^{-4} S cm^{-1}. The MOFs are found to improve the ionic conductivities, while the enhanced interfacial compatibilities are beneficial for advances in solid-state LIBs. Biopolymers are naturally abundant, environmentally benign, and can act as low-cost fillers in developing PNSEs for LIB applications. The flexible and wearable electronic devices of the future necessitate the use of flexible and bendable power supplies, which therefore calls for the need for flexible LIBs. The evolution and recent developments in PNSE-based LIB fabrication reveal that the PNSEs are indeed a landmark in the history of LIB electrolytes.

REFERENCES

1. Ritchie A, Howard W (2006) Recent developments and likely advances in lithium-ion batteries. *Journal of Power Sources* 162 (2):809–812

2. Yue L, Ma J, Zhang J, Zhao J, Dong S, Liu Z, Cui G, Chen L (2016) All solid state polymer electrolytes for high-performance lithium ion batteries. *Energy Storage Materials* 5:139–164
3. Lu L, Han X, Li J, Hua J, Ouyang M (2013) A review on the key issues for lithium-ion battery management in electric vehicles. *Journal of Power Sources* 226:272–288
4. Kennedy B, Patterson D, Camilleri S (2000) Use of lithium-ion batteries in electric vehicles. *Journal of Power Sources* 90 (2):156–162
5. Varma SJ, Sambath Kumar K, Seal S, Rajaraman S, Thomas J (2018) Fiber-type solar cells, nanogenerators, batteries, and supercapacitors for wearable applications. *Advanced Science* 5(9):1800340. DOI: 10.1002/advs.201800340
6. Zhou G, Li F, Cheng H-M (2014) Progress in flexible lithium batteries and future prospects. *Energy & Environmental Science* 7 (4):1307–1338
7. Chen Y, Au J, Kazlas P, Ritenour A, Gates H, McCreary M (2003) Flexible active-matrix electronic ink display. *Nature* 423 (6936):136–136
8. Gelinck GH, Huitema HEA, Van Veenendaal E, Cantatore E, Schrijnemakers L, Van Der Putten JB, Geuns TC, Beenhakkers M, Giesbers JB, Huisman B-H (2004) Flexible active-matrix displays and shift registers based on solution-processed organic transistors. *Nature Materials* 3 (2):106–110
9. Wang J-Z, Chou S-L, Liu H, Wang GX, Zhong C, Chew SY, Liu HK (2009) Highly flexible and bendable free-standing thin film polymer for battery application. *Materials Letters* 63 (27):2352–2354
10. Cherusseri J, Kar KK (2015) Hierarchically mesoporous carbon nanopetal based electrodes for flexible supercapacitors with super-long cyclic stability. *Journal of Materials Chemistry A* 3 (43):21586–21598
11. Cherusseri J, Kar KK (2016) Ultra-flexible fibrous supercapacitors with carbon nanotube/polypyrrole brush-like electrodes. *Journal of Materials Chemistry A* 4 (25):9910–9922
12. Cherusseri J, Sharma R, Kar KK (2016) Helically coiled carbon nanotube electrodes for flexible supercapacitors. *Carbon* 105:113–125
13. Cherusseri J, Kar KK (2015) Self-standing carbon nanotube forest electrodes for flexible supercapacitors. *Rsc Advances* 5 (43):34335–34341
14. Zeng L, Qiu L, Cheng H-M (2019) Towards the practical use of flexible lithium ion batteries. *Energy Storage Materials* 23:434–438
15. Cheng XB, Hou TZ, Zhang R, Peng HJ, Zhao CZ, Huang JQ, Zhang Q (2016) Dendrite-free lithium deposition induced by uniformly distributed lithium ions for efficient lithium metal batteries. *Advanced Materials* 28 (15):2888–2895
16. Ding F, Xu W, Graff GL, Zhang J, Sushko ML, Chen X, Shao Y, Engelhard MH, Nie Z, Xiao J (2013) Dendrite-free lithium deposition via self-healing electrostatic shield mechanism. *Journal of the American Chemical Society* 135 (11):4450–4456
17. Liu Y, Lin D, Liang Z, Zhao J, Yan K, Cui Y (2016) Lithium-coated polymeric matrix as a minimum volume-change and dendrite-free lithium metal anode. *Nature Communications* 7:10992
18. Chen L, Fan LZ (2018) Dendrite-free Li metal deposition in all-solid state lithium sulfur batteries with polymer-in-salt polysiloxane electrolyte. *Energy Storage Materials* 15:37–45
19. Huo H, Chen Y, Luo J, Yang X, Guo X, Sun X (2019) Rational design of hierarchical "Ceramic-in-Polymer" and "Polymer-in-Ceramic" electrolytes for dendrite-free solid-state batteries. *Advanced Energy Materials* 9 (17):1804004
20. Lu Q, He YB, Yu Q, Li B, Kaneti YV, Yao Y, Kang F, Yang QH (2017) Dendrite-free, high-rate, long-life lithium metal batteries with a 3D cross-linked network polymer electrolyte. *Advanced Materials* 29 (13):1604460

21. Zhang J, Zhao J, Yue L, Wang Q, Chai J, Liu Z, Zhou X, Li H, Guo Y, Cui G (2015) Safety-reinforced poly (propylene carbonate)-based all-solid-state polymer electrolyte for ambient-temperature solid polymer lithium batteries. *Advanced Energy Materials* 5 (24):1501082–1501082. doi:10.1002/aenm.201501082
22. Fan L-Z, Wang X-L, Long F (2009) All-solid state polymer electrolyte with plastic crystal materials for rechargeable lithium-ion battery. *Journal of Power Sources* 189 (1):775–778
23. Liu W, Liu N, Sun J, Hsu PC, Li Y, Lee HW, Cui Y (2015) Ionic conductivity enhancement of polymer electrolytes with ceramic nanowire fillers. *Nano Lett* 15 (4):2740–2745. doi:10.1021/acs.nanolett.5b00600
24. Yao P, Yu H, Ding Z, Liu Y, Lu J, Lavorgna M, Wu J, Liu X (2019) Review on polymer based composite electrolytes for lithium batteries. *Front Chem* 7 (522):17. doi:10.3389/fchem.2019.00522
25. Lin D, Yuen PY, Liu Y, Liu W, Liu N, Dauskardt RH, Cui Y (2018) A silica-aerogel-reinforced composite polymer electrolyte with high ionic conductivity and high modulus. *Advanced Materials* 30 (32):1802661–1802661. doi:10.1002/adma.201802661
26. Wan J, Xie J, Mackanic DG, Burke W, Bao Z, Cui Y (2018) Status, promises, and challenges of nanocomposite solid state electrolytes for safe and high performance lithium batteries. *Materials Today Nano* 4:1–16. doi:10.1016/j.mtnano.2018.12.003
27. Zhao W, Yi J, He P, Zhou H (2019) Solid state electrolytes for lithium-ion batteries: Fundamentals, challenges and perspectives. *Electrochemical Energy Reviews* 2:574–605. doi:10.1007/s41918-019-00048-0
28. Xu RC, Xia XH, Zhang SZ, Xie D, Wang XL, Tu JP (2018) Interfacial challenges and progress for inorganic all-solid state lithium batteries. *Electrochimica Acta* 284:177–187. https://doi.org/10.1016/j.electacta.2018.07.191
29. Zhang X, Xu B-Q, Lin Y-H, Shen Y, Li L, Nan C-W (2018) Effects of Li6.75La3Zr1.75Ta0.25O12 on chemical and electrochemical properties of polyacrylonitrile based solid electrolytes. *Solid State Ionics* 327:32–38. doi:https://doi.org/10.1016/j.ssi.2018.10.023
30. Tan S-J, Zeng X-X, Ma Q, Wu X-W, Guo Y-G (2018) Recent advancements in polymer based composite electrolytes for rechargeable lithium batteries. *Electrochemical Energy Reviews* 1 (2):113–138. doi:10.1007/s41918-018-0011-2
31. Lin D, Liu W, Liu Y, Lee HR, Hsu P-C, Liu K, Cui Y (2016) High ionic conductivity of composite solid polymer electrolyte via in situ synthesis of monodispersed SiO2 nanospheres in poly(ethylene oxide). *Nano Letters* 16 (1):459–465. doi:10.1021/acs.nanolett.5b04117
32. Liu Z, Fu W, Payzant EA, Yu X, Wu Z, Dudney NJ, Kiggans J, Hong K, Rondinone AJ, Liang C (2013) Anomalous high ionic conductivity of nanoporous β-Li3PS4. *Journal of the American Chemical Society* 135 (3):975–978. doi:10.1021/ja3110895
33. Yao X, Liu D, Wang C, Long P, Peng G, Hu Y-S, Li H, Chen L, Xu X (2016) High-energy all-solid state lithium batteries with ultralong cycle life. *Nano Letters* 16 (11):7148–7154. doi:10.1021/acs.nanolett.6b03448
34. Bruce PG, Scrosati B, Tarascon J-M (2008) Nanomaterials for rechargeable lithium batteries. *Angewandte Chemie International Edition* 47 (16):2930–2946. doi:10.1002/anie.200702505
35. Liu W, Lee SW, Lin D, Shi F, Wang S, Sendek AD, Cui Y (2017) Enhancing ionic conductivity in composite polymer electrolytes with well-aligned ceramic nanowires. *Nature Energy* 2 (5):17035. doi:10.1038/nenergy.2017.35
36. Geim AK, Novoselov KS (2007) The rise of graphene. *Nature Materials* 6 (3):183–191. doi:10.1038/nmat1849

37. Rao CNR, Sood AK, Subrahmanyam KS, Govindaraj A (2009) Graphene: The new two-dimensional nanomaterial. *Angewandte Chemie International Edition* 48 (42):7752–7777. doi:10.1002/anie.200901678
38. Varma SJ, Kumar J, Liu Y, Layne K, Wu J, Liang C, Nakanishi Y, Aliyan A, Yang W, Ajayan PM, Thomas J (2017) 2D TiS2 layers: A superior nonlinear optical limiting material. *Advanced Optical Materials* 5 (24):1700713. doi:10.1002/adom.201700713
39. Liu Y, Liang C, Wu J, Varma SJ, Nakanishi Y, Aliyan A, Martí AA, Wang Y, Xie B, Kumar J, Layne K, Chopra N, Odeh I, Vajtai R, Thomas J, Peng X, Yang W, Ajayan PM (2019) Reflux pretreatment-mediated sonication: A new universal route to obtain 2D quantum dots. *Materials Today* 22:17–24. doi:https://doi.org/10.1016/j.mattod.2018.06.007
40. Wang QH, Kalantar-Zadeh K, Kis A, Coleman JN, Strano MS (2012) Electronics and optoelectronics of two-dimensional transition metal dichalcogenides. *Nature Nanotechnology* 7 (11):699–712. doi:10.1038/nnano.2012.193
41. Cheng Q, Li A, Li N, Li S, Zangiabadi A, Li TD, Huang W, Li AC, Jin T, Song Q, Xu W, Ni N, Zhai H, Dontigny M, Zaghib K, Chuan X, Su D, Yan K, Yang Y (2019) Stabilizing solid electrolyte-anode interface in Li-Metal batteries by boron nitride based nanocomposite coating. *Joule* 3 (6):1510–1522. doi:10.1016/j.joule.2019.03.022
42. Dean CR, Young AF, Meric I, Lee C, Wang L, Sorgenfrei S, Watanabe K, Taniguchi T, Kim P, Shepard KL, Hone J (2010) Boron nitride substrates for high-quality graphene electronics. *Nature Nanotechnology* 5 (10):722–726. doi:10.1038/nnano.2010.172
43. Butler SZ, Hollen SM, Cao L, Cui Y, Gupta JA, Gutiérrez HR, Heinz TF, Hong SS, Huang J, Ismach AF, Johnston-Halperin E, Kuno M, Plashnitsa VV, Robinson RD, Ruoff RS, Salahuddin S, Shan J, Shi L, Spencer MG, Terrones M, Windl W, Goldberger JE (2013) Progress, challenges, and opportunities in two-dimensional materials beyond graphene. *ACS Nano* 7 (4):2898–2926. doi:10.1021/nn400280c
44. Shim J, Kim D-G, Kim HJ, Lee JH, Baik J-H, Lee J-C (2014) Novel composite polymer electrolytes containing poly(ethylene glycol)-grafted graphene oxide for all-solid state lithium-ion battery applications. *Journal of Materials Chemistry A* 2 (34):13873–13883. doi:10.1039/C4TA02667E
45. Yuan M, Erdman J, Tang C, Ardebili H (2014) High performance solid polymer electrolyte with graphene oxide nanosheets. *RSC Advances* 4 (103):59637–59642. doi:10.1039/C4RA07919A
46. Shim J, Kim HJ, Kim BG, Kim YS, Kim D-G, Lee J-C (2017) 2D boron nitride nanoflakes as a multifunctional additive in gel polymer electrolytes for safe, long cycle life and high rate lithium metal batteries. *Energy & Environmental Science* 10 (9):1911–1916. doi:10.1039/C7EE01095H
47. Duan H, Zheng H, Zhou Y, Xu B, Liu H (2018) Stability of garnet-type Li ion conductors: An overview. *Solid State Ionics* 318:45–53. doi:https://doi.org/10.1016/j.ssi.2017.09.018
48. Xie H, Yang C, Fu K, Yao Y, Jiang F, Hitz E, Liu B, Wang S, Hu L (2018) Flexible, scalable, and highly conductive garnet-polymer solid electrolyte templated by bacterial cellulose. *Advanced Energy Materials* 8 (18):1703474. doi:10.1002/aenm.201703474
49. Du F, Zhao N, Li Y, Chen C, Liu Z, Guo X (2015) All solid state lithium batteries based on lamellar garnet-type ceramic electrolytes. *Journal of Power Sources* 300:24–28. doi:https://doi.org/10.1016/j.jpowsour.2015.09.061
50. Zhao N, Khokhar W, Bi Z, Shi C, Guo X, Fan L-Z, Nan C-W (2019) Solid garnet batteries. *Joule* 3 (5):1190–1199. doi:https://doi.org/10.1016/j.joule.2019.03.019
51. Li Y, Xu B, Xu H, Duan H, Lü X, Xin S, Zhou W, Xue L, Fu G, Manthiram A, Goodenough JB (2017) Hybrid polymer/garnet electrolyte with a small interfacial resistance for lithium-ion batteries. *Angewandte Chemie* 129 (3):771–774. doi:10.1002/ange.201608924

52. Kazyak E, Chen K-H, Wood KN, Davis AL, Thompson T, Bielinski AR, Sanchez AJ, Wang X, Wang C, Sakamoto J, Dasgupta NP (2017) Atomic layer deposition of the solid electrolyte garnet Li7La3Zr2O12. *Chemistry of Materials* 29 (8):3785–3792. doi:10.1021/acs.chemmater.7b00944
53. Fu K, Gong Y, Dai J, Gong A, Han X, Yao Y, Wang C, Wang Y, Chen Y, Yan C, Li Y, Wachsman ED, Hu L (2016) Flexible, solid state, ion-conducting membrane with 3D garnet nanofiber networks for lithium batteries. *Proceedings of the National Academy of Sciences of the United States of America* 113 (26):7094–7099. doi:10.1073/pnas.1600422113
54. Yang T, Zheng J, Cheng Q, Hu Y-Y, Chan CK (2017) Composite polymer electrolytes with Li7La3Zr2O12 garnet-type nanowires as ceramic fillers: Mechanism of conductivity enhancement and role of doping and morphology. *ACS Applied Materials & Interfaces* 9 (26):21773–21780. doi:10.1021/acsami.7b03806
55. Zhao C-Z, Zhang X-Q, Cheng X-B, Zhang R, Xu R, Chen P-Y, Peng H-J, Huang J-Q, Zhang Q (2017) An anion-immobilized composite electrolyte for dendrite-free lithium metal anodes. *Proceedings of the National Academy of Sciences* 114 (42):11069. doi:10.1073/pnas.1708489114
56. Zheng J, Dang H, Feng X, Chien P-H, Hu Y-Y (2017) Li-ion transport in a representative ceramic–polymer–plasticizer composite electrolyte: Li$_7$La$_3$Zr$_2$O$_{12}$–polyethylene oxide–tetraethylene glycol dimethyl ether. *Journal of Materials Chemistry A* 5 (35):18457–18463. doi:10.1039/C7TA05832B
57. Chen L, Li Y, Li S-P, Fan L-Z, Nan C-W, Goodenough JB (2018) PEO/garnet composite electrolytes for solid state lithium batteries: From "ceramic-in-polymer" to "polymer-in-ceramic". *Nano Energy* 46:176–184. doi:https://doi.org/10.1016/j.nanoen.2017.12.037
58. Jung YC, Lee SM, Choi JH, Jang SS, Kima DW (2015) All solid state lithium batteries assembled with hybrid solid electrolytes. *Journal of the Electrochemical Society* 162 (4):A704–A710. doi:10.1149/2.0731504jes
59. Inda Y, Katoh T, Baba M (2007) Development of all-solid lithium-ion battery using Li-ion conducting glass-ceramics. *Journal of Power Sources* 174 (2):741–744. doi:https://doi.org/10.1016/j.jpowsour.2007.06.234
60. Aono H, Sugimoto E, Sadaoka Y, Imanaka N, Adachi G-y (1990) Ionic conductivity and sinterability of lithium titanium phosphate system. *Solid State Ionics* 40–41:38–42. doi:https://doi.org/10.1016/0167-2738(90)90282-V
61. Pérez-Estébanez M, Isasi-Marín J, Többens DM, Rivera-Calzada A, León C (2014) A systematic study of nasicon-type Li$_{1+x}$M$_x$Ti$_{2-x}$(PO$_4$)$_3$ (M: Cr, Al, Fe) by neutron diffraction and impedance spectroscopy. *Solid State Ionics* 266:1–8. doi:https://doi.org/10.1016/j.ssi.2014.07.018
62. Yang L, Wang Z, Feng Y, Tan R, Zuo Y, Gao R, Zhao Y, Han L, Wang Z, Pan F (2017) Flexible composite solid electrolyte facilitating highly stable "Soft Contacting" Li–Electrolyte interface for solid state lithium-ion batteries. *Advanced Energy Materials* 7 (22):1701437. doi:10.1002/aenm.201701437
63. Pareek T, Dwivedi S, Ahmad SA, Badole M, Kumar S (2020) Effect of NASICON-type LiSnZr(PO$_4$)$_3$ ceramic filler on the ionic conductivity and electrochemical behavior of PVDF based composite electrolyte. *Journal of Alloys and Compounds* 824:153991. doi:https://doi.org/10.1016/j.jallcom.2020.153991
64. Kou Z, Miao C, Wang Z, Mei P, Zhang Y, Yan X, Jiang Y, Xiao W (2019) Enhanced ionic conductivity of novel composite polymer electrolytes with Li$_{1.3}$Al$_{0.3}$Ti$_{1.7}$(PO$_4$)$_3$ NASICON-type fast ion conductor powders. *Solid State Ionics* 338:138–143. doi:https://doi.org/10.1016/j.ssi.2019.05.026
65. Bonizzoni S, Ferrara C, Berbenni V, Anselmi-Tamburini U, Mustarelli P, Tealdi C (2019) NASICON-type polymer-in-ceramic composite electrolytes for lithium batteries. *Physical Chemistry Chemical Physics* 21 (11):6142–6149. doi:10.1039/C9CP00405J

66. Hou M, Liang F, Chen K, Dai Y, Xue D (2020) Challenges and perspectives of NASICON-type solid electrolytes for all-solid state lithium batteries. *Nanotechnology* 31 (13):132003. doi:10.1088/1361-6528/ab5be7
67. Stramare S, Thangadurai V, Weppner W (2003) Lithium lanthanum titanates: A review. *Chem. Mater.* 15(21):3974–3990. doi:10.1021/cm0300516
68. Hou M, Liang F, Chen K, Dai Y, Xue D (2020) Challenges and perspectives of NASICON-type solid electrolytes for all-solid state lithium batteries. *Nanotechnology* 31 (13):132003–132003. doi:10.1088/1361-6528/ab5be7
69. Zhang J, Zang X, Wen H, Dong T, Chai J, Li Y, Chen B, Zhao J, Dong S, Ma J, Yue L, Liu Z, Guo X, Cui G, Chen L (2017) High-voltage and free-standing poly(propylene carbonate)/Li$_{6.75}$La$_3$Zr$_{1.75}$Ta$_{0.25}$O$_{12}$ composite solid electrolyte for wide temperature range and flexible solid lithium ion battery. *Journal of Materials Chemistry A* 5 (10):4940–4948. doi:10.1039/C6TA10066J
70. Zeier WG (2014) Structural limitations for optimizing garnet-type solid electrolytes: A perspective. *Dalton Transactions* 43 (43):16133–16138. doi:10.1039/C4DT02162B
71. Cheng SH-S, He K-Q, Liu Y, Zha J-W, Kamruzzaman M, Ma RL-W, Dang Z-M, Li RKY, Chung CY (2017) Electrochemical performance of all-solid state lithium batteries using inorganic lithium garnets particulate reinforced PEO/LiClO4 electrolyte. *Electrochimica Acta* 253:430–438. https://doi.org/10.1016/j.electacta.2017.08.162
72. Zhao Y, Wu C, Peng G, Chen X, Yao X, Bai Y, Wu F, Chen S, Xu X (2016) A new solid polymer electrolyte incorporating Li$_{10}$GeP$_2$S$_{12}$ into a polyethylene oxide matrix for all-solid state lithium batteries. *Journal of Power Sources* 301:47–53. https://doi.org/10.1016/j.jpowsour.2015.09.111
73. Inaguma Y, Chen L, Itoh M, Nakamura T (1994) Candidate compounds with perovskite structure for high lithium ionic conductivity. *Solid State Ionics* 70–71:196–202. https://doi.org/10.1016/0167-2738(94)90309-3
74. Zhang X, Liu T, Zhang S, Huang X, Xu B, Lin Y, Xu B, Li L, Nan C-W, Shen Y (2017) Synergistic coupling between Li$_{6.75}$La$_3$Zr$_{1.75}$Ta$_{0.25}$O$_{12}$ and Poly(vinylidene fluoride) induces high ionic conductivity, mechanical strength, and thermal stability of solid composite electrolytes. *Journal of the American Chemical Society* 139 (39):13779–13785. doi:10.1021/jacs.7b06364
75. Liu K, Zhang R, Sun J, Wu M, Zhao T (2019) Polyoxyethylene (PEO)|PEO–Perovskite|PEO composite electrolyte for all-solid state lithium metal batteries. *ACS Applied Materials & Interfaces* 11 (50):46930–46937. doi:10.1021/acsami.9b16936
76. Xu H, Chien P-H, Shi J, Li Y, Wu N, Liu Y, Hu Y-Y, Goodenough JB (2019) High-performance all-solid state batteries enabled by salt bonding to perovskite in poly(ethylene oxide). *Proceedings of the National Academy of Sciences* 116 (38):18815. doi:10.1073/pnas.1907507116
77. He K-Q, Zha J-W, Du P, Cheng SH-S, Liu C, Dang Z-M, Li RKY (2019) Tailored high cycling performance in a solid polymer electrolyte with perovskite-type Li0.33La0.557TiO3 nanofibers for all-solid state lithium ion batteries. *Dalton Transactions* 48 (10):3263–3269. doi:10.1039/C9DT00074G
78. Krivovichev Sergey V (2008) Minerals with antiperovskite structure: A review. *Zeitschrift für Kristallographie – Crystalline Materials* 223(1–2):109–113. doi:10.1524/zkri.2008.0008
79. Zhao Y, Daemen LL (2012) Superionic conductivity in lithium-rich anti-perovskites. *Journal of the American Chemical Society* 134 (36):15042–15047. doi:10.1021/ja305709z
80. Li Y, Zhou W, Xin S, Li S, Zhu J, Lü X, Cui Z, Jia Q, Zhou J, Zhao Y, Goodenough JB (2016) Fluorine-doped antiperovskite electrolyte for all-solid state lithium-ion batteries. *Angewandte Chemie International Edition* 55 (34):9965–9968. doi:10.1002/anie.201604554

81. Yin L, Yuan H, Kong L, Lu Z, Zhao Y (2020) Engineering Frenkel defects of anti-perovskite solid state electrolytes and their applications in all-solid state lithium-ion batteries. *Chemical Communications* 56 (8):1251–1254. doi:10.1039/C9CC08382K
82. Wang Z, Xu H, Xuan M, Shao G (2018) From anti-perovskite to double anti-perovskite: Tuning lattice chemistry to achieve super-fast Li+ transport in cubic solid lithium halogen–chalcogenides. *Journal of Materials Chemistry A* 6 (1):73–83. doi:10.1039/C7TA08698A
83. Xu H, Xuan M, Xiao W, Shen Y, Li Z, Wang Z, Hu J, Shao G (2019) Lithium ion conductivity in double antiperovskite Li6.5OS1.5I1.5: Alloying and boundary effects. *ACS Applied Energy Materials* 2 (9):6288–6294. doi:10.1021/acsaem.9b00861
84. Fang H, Jena P (2017) Li-rich antiperovskite superionic conductors based on cluster ions. *Proceedings of the National Academy of Sciences of the United States of America* 114 (42):11046–11051. doi:10.1073/pnas.1704086114
85. Lau J, DeBlock RH, Butts DM, Ashby DS, Choi CS, Dunn BS (2018) Sulfide solid electrolytes for lithium battery applications. *Advanced Energy Materials* 8 (27):1800933. doi:10.1002/aenm.201800933
86. Hayashi A, Sakuda A, Tatsumisago M (2016) Development of sulfide solid electrolytes and interface formation processes for bulk-type all-solid state Li and Na batteries. *Front Energy Res* 4 (25):1–13
87. Kamaya N, Homma K, Yamakawa Y, Hirayama M, Kanno R, Yonemura M, Kamiyama T, Kato Y, Hama S, Kawamoto K, Mitsui A (2011) A lithium superionic conductor. *Nature Materials* 10 (9):682–686. doi:10.1038/nmat3066
88. Hayashi A, Hama S, Morimoto H, Tatsumisago M, Minami T (2001) Preparation of Li2S–P2S5 amorphous solid electrolytes by mechanical milling. *Journal of the American Ceramic Society* 84 (2):477–479. doi:10.1111/j.1151-2916.2001.tb00685.x
89. Nam YJ, Cho S-J, Oh DY, Lim J-M, Kim SY, Song JH, Lee Y-G, Lee S-Y, Jung YS (2015) Bendable and thin sulfide solid electrolyte film: A new electrolyte opportunity for free-standing and stackable high-energy all-solid state lithium-ion batteries. *Nano Letters* 15 (5):3317–3323. doi:10.1021/acs.nanolett.5b00538
90. Hayashi A, Harayama T, Mizuno F, Tatsumisago M (2006) Mechanochemical synthesis of hybrid electrolytes from the Li2S–P2S5 glasses and polyethers. *Journal of Power Sources* 163 (1):289–293. https://doi.org/10.1016/j.jpowsour.2006.06.018
91. Villaluenga I, Wujcik KH, Tong W, Devaux D, Wong DHC, DeSimone JM, Balsara NP (2016) Compliant glass–polymer hybrid single ion-conducting electrolytes for lithium batteries. *Proceedings of the National Academy of Sciences* 113 (1):52. doi:10.1073/pnas.1520394112
92. Weston JE, Steele BCH (1982) Effects of inert fillers on the mechanical and electrochemical properties of lithium salt-poly(ethylene oxide) polymer electrolytes. *Solid State Ionics* 7 (1):75–79. https://doi.org/10.1016/0167-2738(82)90072-8
93. Capuano F, Scrosati B (1991) Composite polymer electrolytes. *J Electrochem Soc* 138 (7):1918–1922
94. Pal P, Ghosh A (2018) Influence of TiO2 nano-particles on charge carrier transport and cell performance of PMMA-LiClO4 based nano-composite electrolytes. *Electrochimica Acta* 260:157–167. https://doi.org/10.1016/j.electacta.2017.11.070
95. Liang B, Tang S, Jiang Q, Chen C, Chen X, Li S, Yan X (2015) Preparation and characterization of PEO-PMMA polymer composite electrolytes doped with nano-Al2O3. *Electrochimica Acta* 169:334–341. https://doi.org/10.1016/j.electacta.2015.04.039
96. Fullerton-Shirey SK, Maranas JK (2010) Structure and mobility of PEO/LiClO4 solid polymer electrolytes filled with Al2O3 nanoparticles. *The Journal of Physical Chemistry C* 114 (20):9196–9206. doi:10.1021/jp906608p

97. Gerbaldi C, Nair JR, Kulandainathan MA, Kumar RS, Ferrara C, Mustarelli P, Stephan AM (2014) Innovative high performing metal organic framework (MOF)-laden nanocomposite polymer electrolytes for all-solid state lithium batteries. *Journal of Materials Chemistry A* 2 (26):9948–9954. doi:10.1039/C4TA01856G
98. Wang Z, Wang S, Wang A, Liu X, Chen J, Zeng Q, Zhang L, Liu W, Zhang L (2018) Covalently linked metal–organic framework (MOF)-polymer all-solid state electrolyte membranes for room temperature high performance lithium batteries. *Journal of Materials Chemistry A* 6 (35):17227–17234. doi:10.1039/C8TA05642K
99. Zhu K, Liu Y, Liu J (2014) A fast charging/discharging all-solid state lithium ion battery based on PEO-MIL-53(Al)-LiTFSI thin film electrolyte. *RSC Advances* 4 (80):42278–42284. doi:10.1039/C4RA06208F
100. Yuan C, Li J, Han P, Lai Y, Zhang Z, Liu J (2013) Enhanced electrochemical performance of poly(ethylene oxide) based composite polymer electrolyte by incorporation of nano-sized metal-organic framework. *Journal of Power Sources* 240:653–658. https://doi.org/10.1016/j.jpowsour.2013.05.030
101. Colò F, Bella F, Nair JR, Destro M, Gerbaldi C (2015) Cellulose based novel hybrid polymer electrolytes for green and efficient Na-ion batteries. *Electrochimica Acta* 174:185–190. https://doi.org/10.1016/j.electacta.2015.05.178
102. Zhang J, Yue L, Hu P, Liu Z, Qin B, Zhang B, Wang Q, Ding G, Zhang C, Zhou X, Yao J, Cui G, Chen L (2014) Taichi-inspired rigid-flexible coupling cellulose-supported solid polymer electrolyte for high-performance lithium batteries. *Scientific Reports* 4 (1):6272–6272. doi:10.1038/srep06272
103. Zhao J, Zhang J, Hu P, Ma J, Wang X, Yue L, Xu G, Qin B, Liu Z, Zhou X, Cui G (2016) A sustainable and rigid-flexible coupling cellulose-supported poly(propylene carbonate) polymer electrolyte towards 5V high voltage lithium batteries. *Electrochimica Acta* 188:23–30. https://doi.org/10.1016/j.electacta.2015.11.088
104. Shi QX, Xia Q, Xiang X, Ye YS, Peng HY, Xue ZG, Xie XL, Mai Y-W (2017) Self-assembled polymeric ionic liquid-functionalized cellulose nano-crystals: Constructing 3D Ion-conducting channels within ionic liquid based composite polymer electrolytes. *Chemistry – A European Journal* 23 (49):11881–11890. doi:10.1002/chem.201702079
105. Liu J, Li W, Zuo X, Liu S, Li Z (2013) Polyethylene-supported polyvinylidene fluoride–cellulose acetate butyrate blended polymer electrolyte for lithium ion battery. *Journal of Power Sources* 226:101–106. doi:https://doi.org/10.1016/j.jpowsour.2012.10.078
106. Navaratnam S, Ramesh K, Ramesh S, Sanusi A, Basirun WJ, Arof AK (2015) Transport mechanism studies of chitosan electrolyte systems. *Electrochimica Acta* 175:68–73. doi:10.1016/j.electacta.2015.01.087
107. Sudhakar YN, Selvakumar M, Bhat DK (2013) LiClO4-doped plasticized chitosan and poly(ethylene glycol) blend as biodegradable polymer electrolyte for supercapacitors. *Ionics* 19 (2):277–285. doi:10.1007/s11581-012-0745-5
108. Winie T, Arof AK (2006) Effect of various plasticizers on the transport properties of hexanoyl chitosan based polymer electrolyte. *Journal of Applied Polymer Science* 101 (6):4474–4479. doi:10.1002/app.24284
109. Xue Z, He D, Xie X (2015) Poly(ethylene oxide) based electrolytes for lithium-ion batteries. *Journal of Materials Chemistry A* 3 (38):19218–19253. doi:10.1039/C5TA03471J
110. Osman Z, Ibrahim ZA, Arof AK (2001) Conductivity enhancement due to ion dissociation in plasticized chitosan based polymer electrolytes. *Carbohydrate Polymers* 44 (2):167–173. https://doi.org/10.1016/S0144-8617(00)00236-8
111. Mobarak NN, Ahmad A, Abdullah MP, Ramli N, Rahman MYA (2013) Conductivity enhancement via chemical modification of chitosan based green polymer electrolyte. *Electrochimica Acta* 92:161–167. https://doi.org/10.1016/j.electacta.2012.12.126

112. Idris NH, Majid SR, Khiar ASA, Hassan MF, Arof AK (2005) Conductivity studies on chitosan/PEO blends with LiTFSI salt. *Ionics* 11 (5):375–377. doi:10.1007/BF02430249
113. Fuentes S, Retuert PJ, González G (2007) Lithium ion conductivity of molecularly compatibilized chitosan–poly(aminopropyltriethoxysilane)–poly(ethylene oxide) nanocomposites. *Electrochimica Acta* 53 (4):1417–1421. https://doi.org/10.1016/j.electacta.2007.05.057
114. Xuewei Fu YWLSWZ (2018) Review: Natural polymer electrolytes for lithium ion batteries. *Journal of Harbin Institute of Technology (New Series)* 25 (1):1–17
115. Fu X, Jewel Y, Wang Y, Liu J, Zhong W-H (2016) Decoupled ion transport in a protein based solid ion conductor. *The Journal of Physical Chemistry Letters* 7 (21):4304–4310. doi:10.1021/acs.jpclett.6b02071
116. Wang X, Fu X, Wang Y, Zhong W (2016) A protein-reinforced adhesive composite electrolyte. *Polymer* 106:43–52. doi:https://doi.org/10.1016/j.polymer.2016.10.052
117. Mattos RI, Tambelli CE, Donoso JP, Pawlicka A (2007) NMR study of starch based polymer gel electrolytes: Humidity effects. *Electrochimica Acta* 53 (4):1461–1465. doi:https://doi.org/10.1016/j.electacta.2007.05.061
118. Liew C-W, Ramesh S (2013) Studies on ionic liquid based corn starch biopolymer electrolytes coupling with high ionic transport number. *Cellulose* 20 (6):3227–3237. doi:10.1007/s10570-013-0079-0
119. Teoh KH, Lim C-S, Ramesh S (2014) Lithium ion conduction in corn starch based solid polymer electrolytes. *Measurement* 48:87–95. doi:https://doi.org/10.1016/j.measurement.2013.10.040
120. Marcondes RFMS, D'Agostini PS, Ferreira J, Girotto EM, Pawlicka A, Dragunski DC (2010) Amylopectin-rich starch plasticized with glycerol for polymer electrolyte application. *Solid State Ionics* 181 (13):586–591. doi:https://doi.org/10.1016/j.ssi.2010.03.016
121. Teoh KHRS, Arof AK (2012) Investigation on the effect of nanosilica towards corn-starch lithium perchlorate based polymer electrolytes. *J Solid State Electrochem* 16 (10):3165–3170
122. Lin Y, Li J, Liu K, Liu Y, Liu J, Wang X (2016) Unique starch polymer electrolyte for high capacity all-solid state lithium sulfur battery. *Green Chemistry* 18 (13):3796–3803. doi:10.1039/C6GC00444J

5 Poly(Vinylidene Fluoride) (PVdF)-Based Polymer Electrolytes for Lithium-Ion Batteries

Jishnu N. S., Neethu T. M. Balakrishnan, Akhila Das, Jarin D. Joyner, Jou-Hyeon Ahn, Jabeen Fatima M. J., and Prasanth Raghavan

CONTENTS

5.1 Introduction 111
5.2 Structure and Ionic Interactions with Lithium Ions 112
5.3 Methods of Preparation of PVdF-Based Electrolytes 114
    5.3.1 Solvent Casting 114
    5.3.2 Phase Inversion 118
    5.3.3 Electrospinning 121
5.4 Conclusion 127
Acknowledgment 127
References 127

5.1 INTRODUCTION

Miniaturized electronic and microelectronic devices are ruling today's market. The growing demands for such small and lightweight portable electronic devices have led to the use of small-sized components and thin polymer films as electrolytes in batteries. The exploration of liquid electrolytes in batteries requires separate containers that will increase the size of the system. Furthermore, the safety hazards cause the system containing liquid components to create serious problems. During continuous use of a battery, the internal reaction will generate heat, which will further heat up the liquid electrolyte, that could eventually lead to explosion of the battery. This scenario brought the polymer electrolytes to front of stage. Gel polymer electrolytes, that consist of a polymer membrane with a lithium salt and a plasticizer, can effectively substitute the liquid electrolytes. The use of gel polymer electrolytes enhances the safety and prevents the leakage of electrolytes inside the battery [1, 2]. Due to their high energy and power density, lithium-ion batteries with polymer electrolytes are extensively used in most of the portable electronic devices; they can deliver without compromising the safety. The developments of polymer electrolytes were made possible by the end of the 1970s [3]. Gel polymer electrolytes, consisting of a polymer matrix, a lithium salt and organic electrolytes, were first described by Feuillade and Perche [4] and further characterized by Abraham and Alamgir [5, 6]. The main obstacle faced while using polymer electrolyte systems is their low ionic conductivity and low mechanical stability. The synthesis of polymer electrolytes with a suitable amorphous phase, with the addition of fillers and blending with different polymers, is examined to enhance mechanical stability and other basic properties, along with the electrochemical performance. Different polymers have been chosen as the matrix for polymer electrolytes in lithium-ion batteries (LIBs), including polyethylene oxides (PEO) [7], polymethyl methacrylate (PMMA) [8], polyvinylidene difluoride-co-hexafluoropropylene (PVdF-co-HFP) [9], polyacrylonitrile (PAN), polyvinylidene difluoride (PVdF) [10], polyvinyl chloride (PVC) [11] and polystyrene (PS) [12]. Some of the physical properties of the polymer matrices are shown in Table 5.1 Each polymer matrix has characteristic properties that could be exploited for the fabrication of the electrolyte matrix. PAN-based polymer systems can be widely used for high-temperature batteries because of their high thermal stability, whereas PMMA and PEO can deliver high values of electrolyte uptake, ionic conductivity and electrochemical properties [13, 14]. PS-based polymer electrolytes can deliver high amorphous content in the electrolyte, increasing the ionic conductivity through the channels, whereas PVdF-co-HFP-based systems have been widely studied because of their semi-crystalline nature and the superior electrochemical properties that they deliver [15].

Polymer electrolytes fabricated using PVdF are best performing, because of their semi-crystalline nature, high anodic stability and the strong electron-withdrawing group present in them. In addition, they also exhibit high dielectric constants, low dissipation factors and high permittivity, that will allow for the dissociation of large number of lithium salts, ultimately resulting in the availability of large numbers of charge carriers. Different methods of preparation, such as electrospinning, phase inversion and solvent casting, have been proposed for PVdF-based polymer electrolytes. Different methods can result in polymer electrolytes with different morphologies, which may be consistent with their effective use as electrolytes in LIBs.

5.2 STRUCTURE AND IONIC INTERACTIONS WITH LITHIUM IONS

The PVdF polymer appears to exhibit five different crystalline polymorphs, namely α, β, γ, δ and ε, which are interconvertible through different methods such as solvent casting [16], phase inversion [17], spin coating [18] and electrospinning [19]. The polymorphs of PVdF (α, β, γ) are depicted in Figure 5.1. The α phase is a readily obtainable phase, whereas the β phase, which exists as a zig-zag structure, provides a high dielectric constant and maximum electrolyte uptake along with good electrical and thermal properties. This β phase is predominant below 70°C, whereas, above

TABLE 5.1
Physical Properties of the Selected Polymer Host Matrix for the Preparation of Polymer Electrolytes Employed in Lithium-Ion Batteries

Polymer host	Repeating unit	Glass transition temperature T_g (°C)	Melting point T_m (°C)	Dielectric constant
Polyethylene oxide	$-(CH_2CH_2O)_n-$	−67	65	2.25
Polypropylene oxide	$-(CH(-CH_3)CH_2O)_n$	−60	-	4
Polyacrylonitrile	$-(CH_2-CH(-CN))_n-$	95	-	5.5
Polymethyl methacrylate	$-(CH_2C(-CH_3)(-COOCH_3))_n$	105	160	4.9
Polyvinyl chloride	$-(CH_2-CHCl)_n-$	85	240	4
Polyvinylidene fluoride	$-(CH_2CF_2)_n-$	−38	171	8.4
Polyvinylidene fluoride-*co*-hexafluoropropylene	$-(CH_2CF_2)_{-x}[-CF_2CF(CF_3)-]_y$	−35	143	8.4

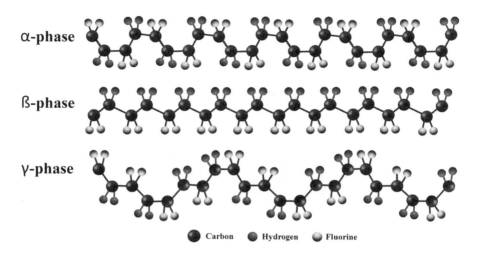

FIGURE 5.1 Schematic representation of different crystal phases (α, β and γ phases) of polyvinylidene difluoride (PVdF).

110°C, the α phase is predominant, with the γ phase only observable at the melting temperature (T_m) of PVdF, which is 160–170°C. The presence of fluorine in PVdF is an important factor that determines its ionic interactions. Due to the presence of this polar group, the polymer itself exhibits a high dielectric constant ($\varepsilon \sim 8.4$) that helps to achieve easier interactions with the lithium salts, which dissociate into ions. It can further enhance the presence of lithium ions within the polymer matrix. For the same reason, PVdF-based electrolytes are promising for use as polymer electrolytes in lithium-ion batteries. Different methods can be used for the fabrication of the PVdF-based polymer electrolytes.

5.3 METHODS OF PREPARATION OF PVDF-BASED ELECTROLYTES

Polymer electrolytes can be fabricated by using different methods, which can yield polymer membranes with different electrochemical performances. Different methods can form membranes with distinct morphologies, that will influence the performance of the resulting electrolyte [20]. Among those different methods, solution casting [21], phase inversion [22] and electrospinning [23] techniques for the fabrication of PVdF-based polymer electrolytes will be discussed in detail in the subsequent sections of this chapter.

5.3.1 Solvent Casting

Solvent casting is a film-forming technique in which the polymer solution is prepared by dissolving the polymer in a suitable solvent, such as an organic solvent, and casting it over glass plates, allowing the solvent to evaporate and leaving a polymer film. It is a long-established technique used for the synthesis of polymer electrolytes.

PVdF-Based Gel Electrolytes

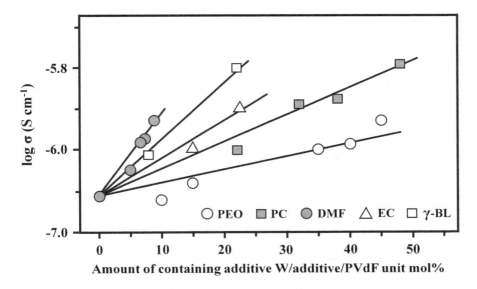

FIGURE 5.2 Effect of polar additives on the lithium ionic conductivity of PVdF/LiClO$_4$ hybrid electrolytes ('w' is the mole ratio of additive and PVdF units). Adapted and reproduced with permission from Ref. [29]. Copyright © 1983 Elsevier.

For PVdF, different solvents can be employed for the dissolution, such as dimethyl fluoride (DMF) [24], dimethyl sulfoxide (DMSO) [25], N-methyl pyrrolidone (NMP) [26] or dimethylacetamide (DMAc) [27]. Tsunemi et al. [28] fabricated the PVdF-based polymer electrolyte by a simple solvent-casting technique. In that study, they chose a hybrid containing PVdF, along with the lithium salt, lithium perchlorate (LiClO$_4$), in the solvent acetone. For enhancement of the ionic conductivity of the resulting polymer electrolyte, propylene carbonate (PC) was selected as the additive. The hybrid film, containing PVdF and LiClO$_4$, exhibits an ionic conductivity of about 10^{-6} S cm^{-1}; with the inclusion of the additives, the ionic conductivity is observed to be about 10^{-5} S cm^{-1}. This increase is conductivity was attributed to the presence of lithium perchlorate and the plasticizing additives, such as PC. Later, this same group investigated the mechanism behind the ionic conductivity of this polymer electrolyte [29]. Different additives, such as DMF, ethylene carbonate (EC), propylene carbonate (PC), polyethylene oxide (PEO) and γ-butyrolactone (γ-BC) were selected to study the dependence of viscosity on the ionic conductivity. From Figure 5.2, it is clear that the viscosity has a remarkable effect on the ionic conductivity of the polymer electrolyte. Addition of PEO, having a comparably high viscosity, exhibits a lower ionic conductivity, although it is higher than for γ-BC [29].

Blending is considered to be the most feasible and effective method available for improving the properties of polymer electrolytes. Blending can be done easily and therefore it is used extensively for the fabrication of polymer electrolytes. Blends of PVdF are achieved with different polymers, such as PVA, PVC and PMMA. Rajendran et al. [30] fabricated the blend of PVdF with PVA by incorporating LiClO$_4$, as the

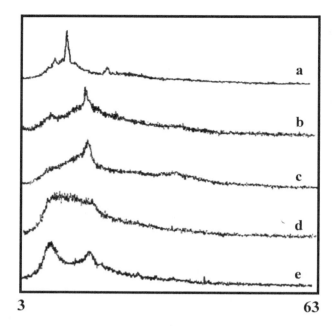

FIGURE 5.3 X-ray diffraction pattern of [(1-x)PMMA–xPVdF]–LiClO$_4$ polymer complex with (a) x= 0, (b) x= 0.25, (c) x= 0.5, (d) x= 0.75 or (e) x= 1. Adapted and reproduced with permission from Ref. [32]. Copyright © 2002 Elsevier.

lithium salt, using the solvent-casting technique. The solid polymer blend was prepared with different compositions of PVA, PVdF and LiClO$_4$. The highest ionic conductivity was observed for the blend containing PVA (65.7)–PVdF (22.5)– LiClO$_4$ (10, wt.%). The highest ionic conductivity observed is about 3.0364×10^{-5} S cm^{-1} at 29°C. Apart from this, the thermal stability of the polymer blend with this composition appears to be reached at 273°C [30]. Compared with this polymer blend, the PVdF blend with PMMA exhibits high ionic conductivity [31]. PMMA (7.5)–PVdF (17.5)–LiClO$_4$ (8)– dimethyl phthalate (DMP) (67, wt.%) exhibits an ionic conductivity of about 4.2×10^{-3} S cm^{-1}. The higher ionic conductivity of this polymer blend electrolyte is attributed to the high amorphous content, caused by the presence of PMMA. Similar polymer blend electrolytes synthesized by Rajendran et al. [32] further revealed the role of PMMA and PVdF in reducing the crystallinity and thereby improving the ionic conductivity of the system. Figure 5.3 illustrates the X-ray diffraction pattern of polymer blend electrolytes containing the [(1-x)PMMA–xPVdF]–LiClO$_4$ polymer complex with (a) x=0, (b) x=0.25, (c) x=0.5, (d) x=0.75 or (e) x=1. With the addition of 75% PVdF, the crystallinity of the sample appears to decrease considerably, and this electrolyte system can deliver an ionic conductivity of about 3.14×10^{-5} S cm^{-1}, which is much lower than that of the PMMA/PVdF blend reported by Mahendran et al. [31].

The PVdF/PVC blend [33] system, along with the EC/PC plasticizer, was fabricated by using the solvent-casting technique. PVC, along with the plasticizers,

PVdF-Based Gel Electrolytes

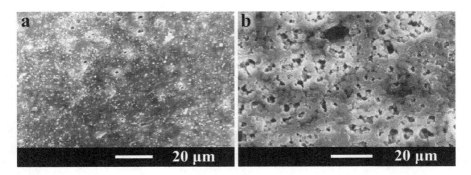

FIGURE 5.4 SEM photographs of (a) PVC (25 wt.%)–(60:40) weight ratio of EC/PC (67 wt.%)–LiClO$_4$ (8 wt.%) and (b) PVdF (20 wt.%)–PVC (5 wt.%)–(60:40) weight ratio of EC/PC (67 wt.%)–LiClO$_4$ (8 wt.%). Adapted and reproduced with permission from Ref. [33]. Copyright © 2008 Elsevier.

forms an immiscible blend (Figure 5.4a), whereas, following the addition of PVdF to the PVC matrix, the pore size increases (Figure 5.4b). The presence of plasticizers helps to enhance the pore structure of the film, resulting in an enhanced ionic conductivity of 3.68×10^{-3} S cm^{-1} [33]. Filler incorporation or the fabrication of a composite are other methods for improving the electrochemical properties. Lithium bisoxalate borate and zirconium dioxide, incorporated into a PVdF/PVC blend, displayed a higher conductivity (about 1.5×10^{-3} S cm^{-1} at 343 K) than other polymer electrolytes prepared by solvent casting [34]. From the scanning electron microscopy (SEM) image (Figure 5.5), it can observe that, with increasing filler concentration, up to a maximum of 2.5 wt.% ZrO$_2$, the porosity and pore structure of the host polymer membrane improves. The variation in the surface morphology and pore structure of the porous polymer membrane, containing different filler loading (0–5 wt.% ZrO$_2$), is shown in Figure 5.5. Compared with the membrane without filler (Figure 5.5a, b, and c) and the membrane with 5 wt.% ZrO$_2$, the membrane containing 2.5 wt.% ZrO$_2$ (Figure 5.5d, e, and f) showed greater porosity and more uniform pore morphology. Hence, the composite membrane with 2.5 wt.% ZrO$_2$ showed high ionic conductivity compared with the rest of the membranes, which suggests that the porosity and pore morphology have pronounced effects on the ionic conductivity of the porous polymer electrolytes. In addition to the porosity and pore morphology, the ceramic fillers act as promoters of ionic conduction by reducing the crystallinity and Lewis acid-base interactions between the filler and the polymer chain [35–37]. However, the electrolyte containing 5 wt.% loading of ZrO$_2$ showed inferior ionic conductivity and transport properties, which could be ascribed to the non-uniform distribution/aggregation of fillers and the excess of ZrO$_2$, which clutches the lithium ions at higher filler loadings [35–39]. For this same reason, the system containing 2.5 wt.% filler exhibited a higher ionic conductivity (1.53×10^{-3} S cm^{-1}) [34].

118   Polymer Electrolytes

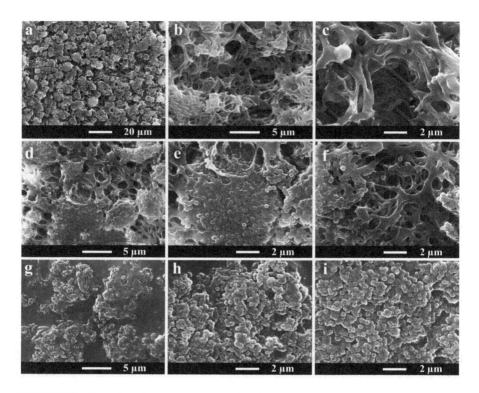

FIGURE 5.5 Scanning electron micrographs (SEM) of CPEs containing different filler loading (wt.%) with different magnifications (a), (b), and (c) 0 wt.%, (d), (e), and (f) 2.5 wt.%, and (g), (h), and (i) 5 wt.%. Adapted and reproduced with permission from Ref. [34]. Copyright © 2003 Elsevier.

5.3.2 Phase Inversion

Phase inversion is a film-forming technique that can result in more porous structures than those of membranes synthesized by solvent casting. Among the different types of phase-inversion techniques, phase inversion by immersion precipitation is normally employed to generate a higher concentration of porous structures [40]. The semi-crystalline porous structure of the PVdF membrane produced by Magistris et al. [41] forms a sponge-like or finger-like structure, depending on the precipitation condition. The membrane is capable of exhibiting porosity of about 75%, with an ionic conductivity of 2 mS cm^{-1}. The sponge- and finger-like structures formed in the membrane by the phase-inversion method are displayed in Figure 5.6 [41].

Hollow-structured PVdF membranes are fabricated by using DMAc as the solvent and LiClO$_4$ as the lithium salt, with water as the coagulating medium. With the addition of lithium perchlorate, the effect of the coagulation bath temperature and the effect of the internal coagulant over the membrane structure was investigated. NMP/water was chosen as the internal coagulating medium. Unlike the

PVdF-Based Gel Electrolytes

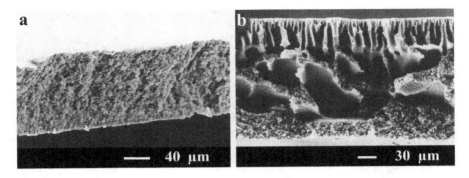

FIGURE 5.6 SEM micrographs of polymer membranes prepared by the phase-inversion method having (a) sponge-like or (b) finger-like structures. Adapted and reproduced with permission from Ref. [41]. Copyright © 2001 Elsevier.

normal phase-inversion method, here, a hollow-fiber membrane was spun by the dry-wet phase-inversion technique, using PVdF/DMAc/LiClO$_4$ spinning dope. While investigating the effect of the internal coagulant on the structure of the membrane (Figure 5.7), with the introduction of 25 vol.% of NMP, finger-like structures that formed from the inner wall became less apparent; when about 75 vol.% of NMP was used, the finger-like structures had disappeared [42].

PVdF gel polymer electrolytes were prepared with a thickness of 350 nm and activated by using the liquid electrolyte, containing a 1 M solution of LiPF$_6$ EC/DEC (1:1). The porosity, and hence the electrolyte uptake of the film, were determined by the preparation conditions of the membrane. The membrane with a high uptake of liquid electrolytes exhibits high ionic conductivity [43]. To exploit the synergistic advantages of individual polymers, blending is selected as the best method. The PVdF/PEO blend [44] polymer electrolyte was fabricated by using the phase-inversion technique. PEO can increase the porosity and the porous structure of the polymer blend. The highest ionic conductivity was observed for the blend containing 50% PEO, attributed to the formation of the honeycomb-like, interconnected, porous structure inside the membrane. Furthermore, the non-solvent, glycerol, was selected to achieve stronger interactions, through hydrogen bonds, that can also result a denser, more porous structure. A polymer blend electrolyte with PVdF- and PEO-based star polymer was fabricated by Deng et al. [45] to form microporous polymer electrolyte. The electrochemical properties of the microporous polymer electrolytes (MPE) depend on the PEO chain length of the star polymer, with the ionic conductivity increasing with increased chain length. The star polymer was synthesized by using triethylamine, tetrahydrofuran (THF) and divinylbenzene. The synthesis procedure for the star polymer is shown in Figure 5.8a and the schematic structure of the resulting MPE is illustrated in Figure 5.8b. This polymer blend can exhibit an initial discharge capacity of 146.8 and 128.8 mAh g^{-1} at 0.1 and 1 C, respectively, using the LiFePO$_4$ cathode. Even after 70 cycles, the cell exhibits 90% of the capacity, with about 94.4% of coulombic efficiency. Blending of PVdF with polydimethylsiloxane (PDMS), synthesized within the ratio (7:3) was reported. This

120   Polymer Electrolytes

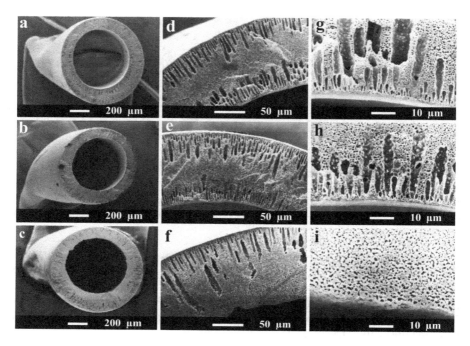

FIGURE 5.7 Morphology of hollow electrospun fibers from spinning dope containing 20 wt.% PVdF, 74 wt.% DMAc, 6 wt.% LiClO$_4$, using (a), (d) and (g) 25 vol.% NMP (b), (e) and (h) and 75 vol.% NMP (c), (f) and (i) as internal coagulant, and using 50°C water as coagulation medium. Adapted and reproduced with permission from Ref. [42]. Copyright © 2005 Elsevier.

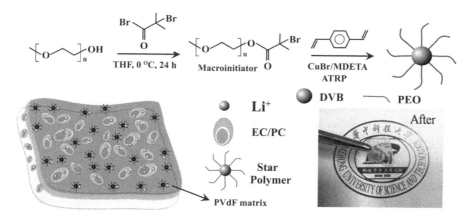

FIGURE 5.8 Synthesis strategy of PEO-based star polymers *via* the 'arm-first' method by atom transfer radical polymerization (ATRP) (top) and schematic illustration of MPE structure (bottom). Adapted and reproduced with permission from Ref. [45]. Copyright © 2015 Elsevier.

PVdF-Based Gel Electrolytes

blend electrolyte exhibits an electrolyte uptake of 250%, with an ionic conductivity of 1.17×10^{-3} S cm^{-1} [46]. Even though PDMS can stabilize the entrapped electrolyte, the performance exhibited by this membrane is still lower than that of a star polymer electrolyte [40], which can exhibit an ionic conductivity of about 3.03×10^{-3} S cm^{-1} [46]. Even though the PVdF/PDMS blend electrolyte, fabricated using glycerol as the co-solvent, with different ratio of PVdF/PDMS, exhibits similar morphology porosity and pore structure (Figure 5.9), the ionic conductivity is superior for the polymer blend electrolyte of PVdF/PEO (star type polymer, PEO). This may be caused by the presence of hydrophilic ethylene oxide (EO) segments, that improve the wettability of the electrolyte, leading to higher electrolyte absorption by the membrane; this can clearly be seen from the contact angle test (Figure 5.10) for PVdF and PVdF blended with the star type PEO [45].

By combining the phase-inversion technique with the chemical reaction method, the PVdF/PAN/SiO$_2$ composite membranes were prepared by Liu et al. [17]. The blend was prepared with different proportions of PVdF and PAN (100/0, 85/15, 80/20, 75/25, 70/30, 65/35, 60/40 or 0/100, w/w). At 70:30 ratio, PVdF/PAN exhibited a high electrolyte uptake of about 246.8% and an ionic conductivity of 3.32×10^{-3} S cm^{-1}. The cell can exhibit an electrochemical stability of approximately 5 V, and the Li/PE/LFP cell delivers a discharge capacity of about 149 mAh g^{-1} at 0.2 C [17].

5.3.3 Electrospinning

Electrospinning is an advanced technique used for the synthesis of fibrous membranes with definite porous structure, that can effectively be used as electrolytes with properties suitable for electrochemical batteries. In the electrospinning technique, a polymer solution is allowed to fall on a collector, employing high voltage. Electrospinning can form fibers with nanometer dimensions [19, 47]. A schematic visualization of the process of electrospinning is displayed in Figure 5.11. Electrospun polymer membranes of PVdF were prepared by Andrew et al. [48] and Choi et al. [49]. Andrew et al. [48] dissolved PVdF in DMF solvent whereas Choi et al. [49] dissolved the PVdF polymer in an acetone: DMAc solvent (17%) for lithium-ion batteries. The effect of electrospinning on the structure of PVdF was investigated by Andrew et al. [48]. The PVdF membranes were dissolved at different weight percentages, varying from 10 to 20 wt.%. This study revealed that, when the polymer jet travelled from the syringe tip to the collector, the crystalline β-phase was formed. On increasing the applied voltage and decreasing the viscosity of the electrospun solution and the amount of the β-phase polymorph formed increased. The electrospun PVdF membrane synthesized by Kim et al. [49] had three-dimensional porous structures (Figure 5.12), with high electrolyte uptake and greater mechanical stability. This polymer membrane exhibits an ionic conductivity of about 1.0×10^{-3} S cm^{-1} at room temperature. In addition, this electrolyte system is able to exhibit a wide electrochemical window of about 0–4.5 V Li/Li$^+$. The high electrolyte uptake is attributed by the spacing between ultrathin fibers.

An electrospun PVdF membrane was fabricated by Woo et al. [50] and hot-pressed, leading to a decrease in porosity and electrolyte uptake. The electrochemical

122   Polymer Electrolytes

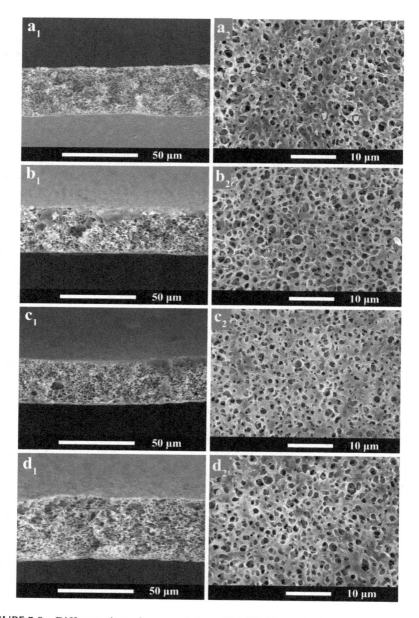

FIGURE 5.9 Difference in surface morphology (FE-SEM images) of porous separators prepared with different PVdF/PDMS ratios in the casting solutions: (a) 10:0; (b) 9:1; (c) 8:2; or (d) 7:3, with 1: cross-section; 2: top surface. Adapted and reproduced with permission from Ref. [45]. Copyright © 2015 Elsevier.

PVdF-Based Gel Electrolytes

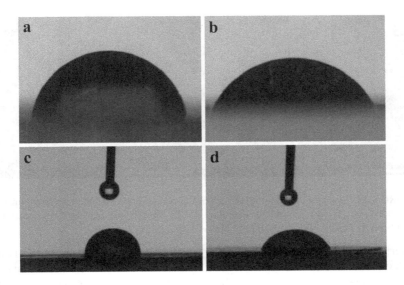

FIGURE 5.10 Hydrophilicity of the PVdF-based membrane (contact angle measurements) prepared with (or without) star polymer: (a) and (c) pure PVdF membrane, and (b) and (d) PVdF blended with 30 wt.% of star polymer. Adapted and reproduced with permission from Ref. [45]. Copyright © 2015 Elsevier.

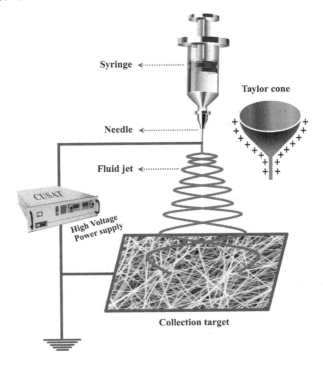

FIGURE 5.11 Schematic visualization of the process and machine set-up of electrospinning.

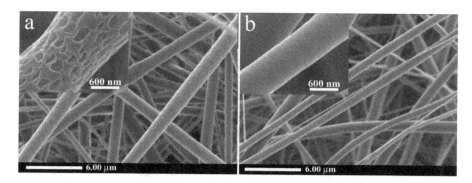

FIGURE 5.12 Surface morphology (FE-SEM images) of PVdF fibers electrospun in different solvents: (a) acetone and (b) acetone: DMAc (50:50). Adapted and reproduced with permission from Ref. [49]. Copyright © 2006 Elsevier.

properties of the hot-pressed electrolyte membranes were inferior to the non-hot pressed membranes at room temperature, even though they have similar ionic conductivity (1.04 mS cm^{-1} at 20°C). Both the pressed and non-pressed electrolyte membranes exhibit a similar initial specific capacity at 0.5 C, which is faded by 14% during the first 110 cycles for the pressed electrolyte membrane, whereas the non-pressed electrolyte membrane almost retained the initial capacity after 100 cycles. In the case of cycle test at 80°C, however, the capacity was only slightly decreased after the initial capacity faded by about 6.5% during the first 10 cycles for the hot-pressed electrolyte membrane. The electrolyte fabricated by using PVdF dissolved in DMAc: acetone delivered an ionic conductivity of 1.0 mS cm^{-1} in 1 M LiPF$_6$ in EC/PC/DMC [51]. The porous membrane exhibited an average fiber diameter (AFD) of 0.45–1.38 μm, achieving a higher electrolyte uptake of 300–350%. To further enhance the properties of lithium nitrate (LiNO$_3$) salt was added to the polymer. As a result, the ionic conductivity was increased to 1.61 mS cm^{-1}. As the percentage of the doping salt increased, increases were observed in the ionic conductivity, the discharge capacity (to 119 mAh g^{-1}) and the voltage stability (to 4.2 V) [52].

Similar to other techniques, a large variety of blends and composites of PVdF could also be synthesized, using the electrospinning technique. Thermoplastic urethane (TPU)-based blends of PVdF, produced by the electrospinning technique, have been widely studied. The first TPU-based PVdF blend was proposed by Wu et al. [53], using DMAc: acetone as the solvent. Later, Santhosh et al. [54] and Peng et al. [55] developed the same blend system. The ionic conductivity exhibited by Santhosh et al. [54] was about ~0.32 mS cm^{-1}, which was greater than that reported by Wu et al. [53]. Peng et al. [55] used PVdF–g–maleic anhydride (PVdF–g–MA) for fabrication of the gel polymer electrolyte, which can exhibit an ionic conductivity of about 4.3 mS cm^{-1} with a high electrochemical window (0–5 V) and a discharge capacity of 166.9 mAh g^{-1} (Li/PE/LFP). Xu et al. [56, 57] improved the mechanical properties of this blend by using PPC and PAN. PPC can act as a good substrate for gel polymer electrolytes (GPE), owing to its good interface effect with electrodes in batteries, and its structure, which is similar to that of the carbonate plasticizer. PVC is another polymer that has been extensively blended with PVdF. Rajendran et al. [58] developed PVdF/PVC blends with different salts of LiClO$_4$ and LiBF$_4$, providing better ionic conductivity than the host polymer. This blend could not achieve a conductivity greater than 10^{-4} S cm^{-1}. Zhong et al. [59] developed this same blend in DMF/acetone, with this polymer matrix achieving higher uptake and porosity than pure PVdF, due to its greater surface area. An AFD of 624 nm and an ionic conductivity of 2.25 mS cm^{-1} at 25°C were achieved with this membrane. The effect of PEO on a PVdF-based PEO polymer blend electrolyte, prepared by electrospinning, was first reported by Prasanth et al. [60], and the effect of different doping salts, such as LiClO$_4$, LiPF$_6$, LiCF$_3$SO$_3$, LiBF$_4$ and LiTFSI was compared. The PVdF/PEO blend electrolytes doped with LiPF$_6$ exhibited a room temperature ionic conductivity of 4.9 mS cm^{-1} and discharge capacity of 168 mAh g^{-1} at 0.1 C. Li et al. [61] reported a similar PVdF/PEO electrospun polymer blend electrolytes with similar ionic conductivity (4.8 mS cm^{-1}) and electrochemical properties (discharge capacity of 171 mAh g^{-1} at 0.1 C) at room temperature. Liang et al. [62] replaced the PEO from PVdF/PEO blend electrolyte [60,61] and reported the effect of PMMA on electrospun PVdF/PMMA blend electrolytes. Compared to PVdF/PEO polymer blend electrolytes, PVdF/PMMA electrolytes exhibited lower ionic conductivity (2.54 mS cm^{-1} at room temperature) and a slightly high-stability electrochemical window (0–5 V). Yvonne et al. [63] reported an electrolyte membrane blend with PVdF/PMMA/cellulose acetate (CA), prepared by the electrospinning technique in the ratio 90:0:10 (PVdF: PMMA: CA). This electrolyte blend system exhibited 99.1% porosity, with a higher melting temperature (161.8°C), higher porosity (324%), lower crystallinity etc., compared with pure PVdF.

The incorporation of fillers into the blended system can further enhance the performance of the polymer electrolyte. Wu et al. [64] reported the effect of fillers on PVdF/thermoplastic polyurethane (TPU) polymer blend gel electrolytes. Studies on the effect of *in-situ*-synthesized SiO$_2$ and TiO$_2$ over this polymer blend revealed that 3 wt.% TiO$_2$ showed a higher electrochemical performance compared with 3 wt.% SiO$_2$. The *in-situ* SiO$_2$ filler, containing a composite electrolyte, exhibited an ionic conductivity of 3.8 mS cm^{-1}, whereas, for TiO$_2$, it was about 4.8 mS cm^{-1}. Ding et al. [65] reported the effect of electrospun composite polymer membranes as separators, in which TiO$_2$ was synthesized by tetrabutyl titanate. The ionic conductivity of PVdF/TiO$_2$ (1.4 mS cm^{-1}) was increased compared with that of pure PVdF membranes (0.89 mS cm^{-1}). Zhang et al. [66] reported a new class of electrospinning, as well as electrospraying, techniques for the fabrication of PVdF/SiO$_2$ composite electrolytes. Compared with polypropylene (PP) (Celgard®) membranes, a slight increase in initial discharge capacity was reported (159 mAh g^{-1} at 0.2 C), with a high cycling performance. Other fillers, like lithium polyvinyl alcohol oxalate borate (LiPVAOB), Al$_2$O$_3$, SiO$_2$, clay, etc., have also been studied with PVdF membranes. Fang et al. [67] reported the addition of the class of clay nanoparticles called 'montmorillonite' to PVdF. An ionic conductivity of 4.2 mS cm^{-1} was observed, which is greater than that achieved with commercial Celgard® separators. Compared with Celgard PP separators, stable cyclic performance was observed in composite membranes, with capacitive retention, even after 50 cycles, at a rate of 0.2 C. The advantages of both alumina

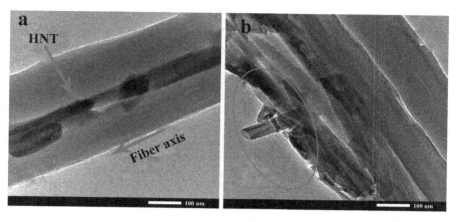

FIGURE 5.13 TEM micrographs of polyvinylidene difluoride/aluminosilicate clay halloysite nanotubes (PVdF/HNTs): (a) and (b) in different positions. Adapted and reproduced with permission from Ref. [70]. Copyright © 2018 Elsevier.

and silica as fillers in the polymer matrix can be exploited by the incorporation of aluminosilicate clay halloysite nanotubes (HNTs), that contain inner cylinder cores similar to those of alumina, whereas the outer surface shows properties similar to that of silica [68, 69]. The combination of HNTs with PVdF enhances the mechanical strength of the latter, as well as increasing its environmentally benign nature and its high cost-effectiveness. Fabricated PVdF/HNT fibers exhibited a tubular structure, with an inner and outer surface similar to those of TiO_2 and Al_2O_3. The transmission electron microscopy (TEM) images (Figure 5.13) [70] showed that the HNTs were aligned along the PVdF polymer. The presence of HNTs can improve the interaction with the lithium salt $LiPF_6$ by the formation of a hydrogen bond with the

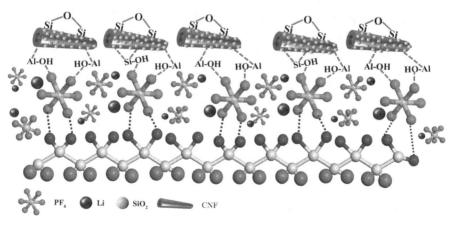

FIGURE 5.14 Schematic illustration of the interactions between nanocomposite nanofibers (NCNFs) and $LiPF_6$. Adapted and reproduced with permission from Ref. [70]. Copyright © 2018 Elsevier.

–OH group of the HNT and the fluorine atom (F) of LiPF$_6$, and the dipole moment existing between the Li$^+$-ions in LiPF$_6$ and the oxygen atom in the HNT, which can enhance the ionic conductivity. The interaction between HNT-incorporated PVdF fibers (NCNF) and LiPF$_6$ is shown in Figure 5.14 [68, 69].

5.4 CONCLUSION

PVdF-based polymer electrolytes have attracted considerable interest for their potential use in lithium-ion batteries, owing to their high dielectric constant, polar nature and their high compatibility with the electrodes in lithium-ion batteries. Since the β-phase of these polymers is electrochemically important to obtain the best battery performance, different fabrication methods have been adopted. Solvent casting, phase inversion and electrospinning are techniques widely used for the fabrication of this polymer membrane. Each method has its own advantages in fabricating the PVdF-based polymer electrolyte. Solvent casting is effectively used to fabricate the electrolyte by a simple casting method, whereas phase inversion is used for improving the porosity of its structure. Further structural improvements can be achieved by using the technique of electrospinning in which a thin fibrous structure can be formed. By using electrospinning, the β-phase formation can be varied, which is significant for achieving appropriate properties of the PVdF-based polymer electrolytes. As with other gel polymer electrolytes, blending and composite formation are also considered to be effective in this membrane as well.

ACKNOWLEDGMENT

Authors Dr. Jabeen Fatima M. J. and Dr. Prasanth Raghavan would like to acknowledge Kerala State Council for Science, Technology and Environment (KSCSTE), Kerala, India, for financial assistance.

REFERENCES

1. Meyer WH (1998) Polymer electrolytes for lithium-ion batteries. *Adv Mater* 10:439–448. https://doi.org/10.1002/(SICI)1521-4095(199804)10:6<439::AID-ADMA439>3.0.CO;2-I
2. Vetter J, Nov P, Wagner MR, Veit C (2005) Ageing mechanisms in lithium-ion batteries. *J Power Sources* 147:269–281. https://doi.org/10.1016/j.jpowsour.2005.01.006
3. Groce F, Gerace F, Dautzemberg G, et al. (1994) Synthesis and characterization of highly conducting gel electrolytes. *Electrochim Acta* 39:2187–2194. https://doi.org/10.1016/0013-4686(94)E0167-X
4. Feuillade G, Perche P (1975) Ion-conductive macromolecular gels and membranes. *J Appl Electrochem* 5:63–69
5. Abraham KM, Alamgir M (1990) Li+-Conductive solid polymer electrolytes with liquid-like conductivity. *J Electrochem Soc* 137:1657–1658. https://doi.org/10.1149/1.2086749

6. Croce F, Scrosati B (1993) Interfacial phenomena in polymer-electrolyte cells: Lithium passivation and cycleability. *J Power Sources* 43:9–19. https://doi.org/10.1016/0378-7753(93)80097-9
7. Kim YT, Smotkin ES (2002) The effect of plasticizers on transport and electrochemical properties of PEO-based electrolytes for lithium rechargeable batteries. *Solid State Ionics* 149:29–37. https://doi.org/10.1016/S0167-2738(02)00130-3
8. Krejza O, Velická J, Sedlaříková M, Vondrák J (2008) The presence of nanostructured Al_2O_3 in PMMA-based gel electrolytes. *J Power Sources* 178:774–778. https://doi.org/10.1016/j.jpowsour.2007.11.018
9. Stephan AM, Nahm KS, Anbu Kulandainathan M, et al. (2006) Poly(vinylidene fluoride-hexafluoropropylene) (PVdF-HFP) based composite electrolytes for lithium batteries. *Eur Polym J* 42:1728–1734. https://doi.org/10.1016/j.eurpolymj.2006.02.006
10. Huang H, Wunder SL (2001) Preparation of microporous PVDF based polymer electrolytes. *J Power Sources* 97–98:649–653. https://doi.org/10.1016/S0378-7753(01)00579-1
11. Ramesh S, Yahaya AH, Arof AK (2002) Dielectric behaviour of PVC-based polymer electrolytes. *Solid State Ionics* 152–153:291–294. https://doi.org/10.1016/S0167-2738(02)00311-9
12. Rohan R, Sun Y, Cai W, et al. (2014) Functionalized polystyrene based single ion conducting gel polymer electrolyte for lithium batteries. *Solid State Ionics* 268:294–299. https://doi.org/10.1016/j.ssi.2014.10.013
13. Appetecchi GB, Croce F, Scrosati B (1995) Kinetics and stability of the lithium electrode in poly(methylmethacrylate)-based gel electrolytes. *Electrochim Acta* 40:991–997. https://doi.org/10.1016/0013-4686(94)00345-2
14. Tan CG, Siew WO, Pang WL, et al. (2007) The effects of ceramic fillers on the PMMA-based polymer electrolyte systems. *Ionics (Kiel)* 13:361–364. https://doi.org/10.1007/s11581-007-0126-7
15. Lehtinen T, Sundholm G, Holmberg S, et al. (1998) Electrochemical characterization of PVDF-based proton conducting membranes for fuel cells. *Electrochim Acta* 43:1881–1890. https://doi.org/10.1016/S0013-4686(97)10005-6
16. Jun MS, Choi YW, Kim JD (2012) Solvent casting effects of sulfonated poly(ether ether ketone) for polymer electrolyte membrane fuel cell. *J Membr Sci* 396:32–37.
17. Liu L, Wang Z, Zhao Z, et al. (2016) PVDF/PAN/SiO_2 polymer electrolyte membrane prepared by combination of phase inversion and chemical reaction method for lithium ion batteries. 699–712. https://doi.org/10.1007/s10008-015-3095-1
18. Park CH, Park M, Yoo SI, Joo SK (2006) A spin-coated solid polymer electrolyte for all-solid-state rechargeable thin-film lithium polymer batteries. *J Power Sources* 158(2):1442–1446.
19. Rao M, Geng X, Liao Y, et al. (2012) Preparation and performance of gel polymer electrolyte based on electrospun polymer membrane and ionic liquid for lithium ion battery. *J Memb Sci* 399–400:37–42. https://doi.org/10.1016/j.memsci.2012.01.021
20. Zhu Y, Xiao S, Shi Y, Yang Y, Hou Y, Wu Y (2014) A composite gel polymer electrolyte with high performance based on poly(vinylidene fluoride) and polyborate for lithium ion batteries. *Adv Energy Mater* 4(1):1300647.
21. Sylla S, Sanchez JY, Armand M (1992) Electrochemical study of linear and crosslinked POE-based polymer electrolytes. *Electrochim Acta* 37:1699–1701. https://doi.org/10.1016/0013-4686(92)80141-8

22. Ma T, Cui Z, Wu Y, et al. (2013) Preparation of PVDF based blend microporous membranes for lithium ion batteries by thermally induced phase separation: I. Effect of PMMA on the membrane formation process and the properties. *J Memb Sci* 444:213–222. https://doi.org/10.1016/j.memsci.2013.05.028
23. Chew SY, Wen J, Yim EKF, Leong KW (2005) Sustained release of proteins from electrospun biodegradable fibers. *Biomacromolecules* 6:2017–2024. https://doi.org/10.1021/bm0501149
24. Jacob MME, Arof AK (2000) FTIR studies of DMF plasticized polyvinyledene fluoride based polymer electrolytes. *Electrochim Acta* 45:1701–1706. https://doi.org/10.1016/S0013-4686(99)00316-3
25. Lin DJ, Chang CL, Lee CK, Cheng LP (2006) Preparation and characterization of microporous PVDF/PMMA composite membranes by phase inversion in water/DMSO solutions. *Eur Polym J* 42:2407–2418. https://doi.org/10.1016/j.eurpolymj.2006.05.008
26. Bottino A, Capannelli G, Comite A (2002) Preparation and characterization of novel porous PVDF-ZrO2 composite membranes. *Desalination* 146:35–40. https://doi.org/10.1016/S0011-9164(02)00469-1
27. Wang D, Li K, Teo WK (1999) Preparation and characterization of polyvinylidene fluoride (PVDF) hollow fiber membranes. *J Memb Sci* 163:211–220. https://doi.org/10.1016/S0376-7388(99)00181-7
28. Tsuchida E, Ohno H, Tsunemi K (1983) Conduction of lithium ions in polyvinylidene fluoride and its derivatives-I. *Electrochim Acta* 28:591–595. https://doi.org/10.1016/0013-4686(83)85049-X
29. Tsunemi K, Ohno H, Tsuchida E (1983) A mechanism of ionic conduction of poly (vinylidene fluoride)-lithium perchlorate hybrid films. *Electrochim Acta* 28:833–837. https://doi.org/10.1016/0013-4686(83)85155-X
30. Rajendran S, Sivakumar M, Subadevi R, Nirmala M (2004) Characterization of PVA-PVdF based solid polymer blend electrolytes. *Phys B Condens Matter* 348:73–78. https://doi.org/10.1016/j.physb.2003.11.073
31. Mahendran O, Rajendran S (2003) Ionic conductivity studies in PMMA/PVdF polymer blend electrolyte with lithium salts. *Ionics (Kiel)* 9:282–288. https://doi.org/10.1007/BF02375980
32. Rajendran S, Mahendran O, Kannan R (2002) Characterisation of [(1 2 x) PMMA ± x PVdF] polymer blend electrolyte with Li 1 ion q. *Fuel* 81:1077–1081
33. Rajendran S, Sivakumar P (2008) An investigation of PVdF/PVC-based blend electrolytes with EC/PC as plasticizers in lithium battery applications. *Phys B Condens Matter* 403:509–516. https://doi.org/10.1016/j.physb.2007.06.012
34. Aravindan V, Vickraman P, Prem Kumar T (2007) ZrO2 nanofiller incorporated PVC/PVdF blend-based composite polymer electrolytes (CPE) complexed with LiBOB. *J Memb Sci* 305:146–151. https://doi.org/10.1016/j.memsci.2007.07.044
35. Raghavan P, Zhao X, Kim J, et al. (2008) Ionic conductivity and electrochemical properties of nanocomposite polymer electrolytes based on electrospun poly(vinylidene fluoride-co-hexafluoropropylene) with nano-sized ceramic fillers. *Electrochem Acta* 54:228–234. https://doi.org/10.1016/j.electacta.2008.08.007
36. Raghavan P, Choi J-W, Ahn J-H, et al. (2008) Novel electrospun poly(vinylidene fluoride-co-hexafluoropropylene)–in situ SiO2 composite membrane-based polymer electrolyte for lithium batteries. *J Power Sources* 184:437–443. https://doi.org/10.1016/j.jpowsour.2008.03.027

37. Raghavan P, Zhao X, Manuel J, et al. (2010) Electrochemical performance of electrospun poly(vinylidene fluoride-co-hexafluoropropylene)-based nanocomposite polymer electrolytes incorporating ceramic fillers and room temperature ionic liquid. *Electrochim Acta* 55:1347–1354. https://doi.org/10.1016/j.electacta.2009.05.025
38. Prasanth R, Shubha N, Hng HH, Srinivasan M (2013) Effect of nano-clay on ionic conductivity and electrochemical properties of poly(vinylidene fluoride) based nanocomposite porous polymer membranes and their application as polymer electrolyte in lithium ion batteries. *Eur Polym J* 49:307–318. https://doi.org/10.1016/j.eurpolymj.2012.10.033
39. Shubha N, Prasanth R, Hoon HH, Srinivasan M (2013) Dual phase polymer gel electrolyte based on non-woven poly(vinylidenefluoride-co-hexafluoropropylene)-layered clay nanocomposite fibrous membranes for lithium ion batteries. *Mater Res Bull* 48:526–537. https://doi.org/10.1016/j.materresbull.2012.11.002
40. Strathmann H, Kock K (1977) The formation mechanism of phase inversion membranes. *Desalination* 21(3):241–255.
41. Magistris A, Mustarelli P, Parazzoli F, et al. (2001) Structure, porosity and conductivity of PVdF films for polymer electrolytes. *J Power Sources* 97:657–660
42. Yeow ML, Liu Y, Li K (2005) Preparation of porous PVDF hollow fibre membrane via a phase inversion method using lithium perchlorate (LiClO 4) as an additive. 258:16–22. https://doi.org/10.1016/j.memsci.2005.01.015
43. Magistris A, Quartarone E, Mustarelli P, et al. (2002) PVDF-based porous polymer electrolytes for lithium batteries. *Solid State Ion* 152:347–354
44. Xi J, Qiu X, Li J, et al. (2006) PVDF-PEO blends based microporous polymer electrolyte: Effect of PEO on pore configurations and ionic conductivity. *J Power Sources* 157:501–506. https://doi.org/10.1016/j.jpowsour.2005.08.009
45. Deng F, Wang X, He D, et al. (2015) Microporous polymer electrolyte based on PVdF/PEO star polymer blends for lithium ion batteries. *J Memb Sci* 491:82–89. https://doi.org/10.1016/j.memsci.2015.05.021
46. Li H, Chen Y, Ma X, et al. (2011) Gel polymer electrolytes based on active PVDF separator for lithium ion battery. I: Preparation and property of PVDF/poly(dimethylsiloxane) blending membrane. *J Memb Sci* 379:397–402. https://doi.org/10.1016/j.memsci.2011.06.008
47. Wu N, Cao Q, Wang X, et al. (2011) A novel high-performance gel polymer electrolyte membrane basing on electrospinning technique for lithium rechargeable batteries. 196:8638–8643. https://doi.org/10.1016/j.jpowsour.2011.04.062
48. Andrew JS, Clarke DR (2008) Effect of electrospinning on the ferroelectric phase content of polyvinylidene difluoride fibers. *Langmuir* 24:670–672. https://doi.org/10.1021/la7035407
49. Kim JR, Choi SW, Jo SM, et al. (2006) Characterization of electrospun PVdF fiber-based polymer electrolytes. *Chem Mater* 19:104–115. https://doi.org/10.1021/cm060223+
50. Woo S, Won S, Mu S, et al. (2006) Electrochemical properties and cycle performance of electrospun poly(vinylidene fluoride)-based fibrous membrane electrolytes for Li-ion polymer battery. 163:41–46. https://doi.org/10.1016/j.jpowsour.2005.11.102
51. Kim JR, Choi SW, Jo SM, et al. (2004) Electrospun PVdF-based fibrous polymer electrolytes for lithium ion polymer batteries. *Electrochim Acta* 50:69–75. https://doi.org/10.1016/j.electacta.2004.07.014
52. Janakiraman S, Surendran A, Ghosh S, Anandhan AV (2018) A new strategy of PVDF based Li-salt polymer electrolyte through electrospinning for lithium battery application. *Mater Res Express* 6(3):035303.

53. Wu N, Jing B, Cao Q, Wang X, Hao Kuang QW (2010) A novel electrospun TPU/PVdF porous fibrous polymer electrolyte for lithium ion batteries. *J Appl Polym Sci* 116:2658–2667. https://doi.org/10.1002/app
54. Santhosh P, Vasudevan T, Gopalan A, Lee KP (2006) Preparation and characterization of polyurethane/poly(vinylidene fluoride) composites and evaluation as polymer electrolytes. *Mater Sci Eng B Solid-State Mater Adv Technol* 135:65–73. https://doi.org/10.1016/j.mseb.2006.08.033
55. Peng S, Cao Q, Yang J, et al. (2015) A novel electrospun poly(vinylidene fluoride)/thermoplastic polyurethane/poly(vinylidene fluoride)-g-(maleic anhydride) porous fibrous polymer electrolyte for lithium-ion batteries. *Solid State Ionics* 282:49–53. https://doi.org/10.1016/j.ssi.2015.09.018
56. Xu J, Liu Y, Cao QI, et al. (2019) A high-performance gel polymer electrolyte based on poly(vinylidene fluoride)/thermoplastic polyurethane/poly(propylene carbonate) for lithium-ion batteries. *J Chem Sci* 131:1–10. https://doi.org/10.1007/s12039-019-1627-4
57. Liu Y, Peng X, Cao Q, et al. (2017) Gel polymer electrolyte based on poly(vinylidene fluoride)/thermoplastic polyurethane/polyacrylonitrile by electrospinning technique gel polymer electrolyte based on poly(vinylidene fluoride)/thermoplastic polyurethane/polyacrylonitrile by E. https://doi.org/10.1021/acs.jpcc.7b03411
58. Rajendran S, Sivakumar P, Shanker R (2007) Studies on the salt concentration of a PVdF – PVC based polymer blend electrolyte. 164:815–821. https://doi.org/10.1016/j.jpowsour.2006.09.011
59. Zhong Z, Cao Q, Jing B, et al. (2012) Electrospun PVdF – PVC nanofibrous polymer electrolytes for polymer lithium-ion batteries. *Mater Sci Eng B* 177:86–91. https://doi.org/10.1016/j.mseb.2011.09.008
60. Prasanth R, Shubha N, Hoon H, Srinivasan M (2014) Effect of poly(ethylene oxide) on ionic conductivity and electrochemical properties of poly(vinylidene fluoride) based polymer gel electrolytes prepared by electrospinning for lithium ion batteries. *J Power Sources* 245:283–291. https://doi.org/10.1016/j.jpowsour.2013.05.178
61. Li W, Wu Y, Wang J, et al. (2015) Hybrid gel polymer electrolyte fabricated by electrospinning technology for polymer lithium-ion battery. *Eur Polym J*. https://doi.org/10.1016/j.eurpolymj.2015.04.014
62. Liang Y, Cheng S, Jianmeng Zhao CZ (2013) Preparation and characterization of electrospun PVDF/PMMA composite fibrous membranes-based separator for lithium-ion batteries. 752:1914–1918. https://doi.org/10.4028/www.scientific.net/AMR.750-752.1914
63. Yvonne T, Zhang C (2014) Properties of electrospun PVDF/PMMA/CA membrane as lithium based battery separator. *Cellulose* 21:2811–2818. https://doi.org/10.1007/s10570-014-0296-1
64. Wu N, Cao Q, Wang X, et al. (2011) In situ ceramic fillers of electrospun thermoplastic polyurethane/poly(vinylidene fluoride) based gel polymer electrolytes for Li-ion batteries. 196:9751–9756. https://doi.org/10.1016/j.jpowsour.2011.07.079
65. Ding Y, Zhang P, Long Z, et al. (2008) Preparation of PVdF-based electrospun membranes and their application as separators. *Sci Technol Adv Mater* 9:. https://doi.org/10.1088/1468-6996/9/1/015005
66. Zhang F, Ma X, Cao C, et al. (2014) Poly(vinylidene fluoride)/SiO$_2$ composite membranes prepared by electrospinning and their excellent properties for nonwoven separators for lithium-ion batteries. *J Power Sources* 251:423–431. https://doi.org/10.1016/j.jpowsour.2013.11.079
67. Fang C, Yang S, Zhao X, et al. (2016) Electrospun montmorillonite modified polyvinylidene fluoride) nanocomposite separators for lithium-ion batteries. 79:1–7. https://doi.org/10.1016/j.materresbull.2016.02.015

68. Tarasova E, Naumenko E, Rozhina E, et al. (2019) Cytocompatibility and uptake of polycations-modified halloysite clay nanotubes. *Appl Clay Sci* 169:21–30. https://doi.org/10.1016/j.clay.2018.12.016
69. Lazzara G, Cavallaro G, Panchal A, et al. (2018) An assembly of organic-inorganic composites using halloysite clay nanotubes. *Curr Opin Colloid Interface Sci* 35:42–50. https://doi.org/10.1016/j.cocis.2018.01.002
70. Khalifa M, Janakiraman S, Ghosh S, et al. (2018) PVDF/halloysite nanocomposite-based non-wovens as gel polymer electrolyte for high safety lithium ion battery. *Polym Compos* 40:2320–2334. https://doi.org/10.1002/pc.25043

6 Poly(Vinylidene Fluoride-co-Hexafluoropropylene) (PVdF-co-HFP)-Based Gel Polymer Electrolyte for Lithium-Ion Batteries

*Akhila Das, Neethu T. M. Balakrishnan,
Jishnu N. S., Jarin D. Joyner, Jou-Hyeon Ahn,
Jabeen Fatima M. J., and Prasanth Raghavan*

CONTENTS

6.1 Introduction 134
    6.1.1 Crystal Phases of PVdF-co-HFP 134
6.2 Preparation of PVdF-co-HFP-Based Polymer Electrolytes 135
    6.2.1 PVdF-co-HFP-Based Electrolytes Prepared by Solution Casting 136
        6.2.1.1 Pure PVdF-co-HFP-Based Polymer Electrolytes 136
        6.2.1.2 PVdF-co-HFP-Based Polymer Blend Electrolytes 136
        6.2.1.3 PVdF-co-HFP-Based Ceramic Filler Composite
                Polymer Electrolytes 138
    6.2.2 Preparation of PVdF-co-HFP-Based Polymer Electrolytes by
        Phase Inversion 139
        6.2.2.1 Pure PVdF-co-HFP-Based Polymer Electrolytes 139
        6.2.2.2 PVdF-co-HFP-Based Polymer Blend Electrolytes 140
        6.2.2.3 PVdF-co-HFP Ceramic Composite Electrolytes 141
    6.2.3 PVdF-co-HFP-Based Polymer Electrolytes Prepared by
        Electrospinning 142
6.3 Conclusion 144
Acknowledgment 145
References 145

6.1 INTRODUCTION

Lithium-ion batteries (LIBs) have played a significant role in next-generation electrochemical energy storage devices, being used for electric/hybrid vehicles, power stations, portable electronics, etc. [1]. LIBs attracted considerable attention because of their superior properties, such as high energy density, compatibility, low self-discharge, light weight, etc. Electrolytes play a vital role in the efficient working of LIBs [2–5]. Conventional electrolytes impede the development of next-generation batteries, because these electrolytes pose a high risk due to their potential for leakage, low working temperature, low electrochemical window, high resistance at solid electrode–electrolyte interfaces, etc. An efficient and safer electrolyte would be desirable to get the most out of this promising battery technology, which is capable of revolutionizing portable devices. Polymer electrolytes are suitable electrolytes for LIBs due to their advantages with respect to being safe, lightweight, more flexible, with wide electrochemical stability windows, etc. Different polymers used as electrolytes include polyvinylidene difluoride (PVdF) [6], polyethylene oxide (PEO) [7], polyvinyl acetate (PVAc) [8] and polyvinylidene fluoride-*co*-hexafluoropropylene (PVdF-*co*-HFP) etc. [9, 10]. PEO is considered to be a dry, solid polymer electrolyte working at ambient temperature, which suffers from low ionic conductivity and poor cycling performance. Polyacrylonitrile (PAN)-based electrolytes possess high ionic conductivity but are expensive [11]. PVAc shows low mechanical stability whereas polyvinyl chloride (PVC) is the least-soluble of the polymers. PVdF-based electrolytes are anodically stable due to the presence of a strong electron-withdrawing group of fluorine atoms, and possess high dielectric constants (8.4 δ) having greater dissolution of salts. The co-polymer PVdF-*co*-HFP shows greater ionic conductivity, compared with PVdF, because of the amorphous nature of HFP, leading to the ease of migration of Li$^+$, along with a high dielectric constant (8.4 δ). Because of these attractive properties of PVdF-*co*-HFP, the potential value of this polymer has been explored in different fields, such as sensors, batteries, supercapacitors, scaffold engineering and textiles.

6.1.1 Crystal Phases of PVdF-*co*-HFP

One of the important co-polymers of PVdF is PVdF-*co*-HFP. The most appropriate ratio of PVdF: HFP, for use as a polymer electrolyte in a battery, is 88:12, which is highly amorphous. The more amorphous the nature, the greater the movement of ions and the higher the ionic conductance. The membrane can easily trap the lithium salts and support the mechanical efficiency of free-standing films. It exhibits both the crystalline phase as well as the amorphous phase. The crystalline phase exhibited by PVdF enhances the mechanical stability of the polymer matrix, which can also act as a separator, and the amorphous nature of HFP improves the ionic conduction of the electrolytes in batteries. The efficiency of this co-polymer matrix depends on the type and amount of doping salts, solvents, etc. PVdF-*co*-HFP exists in different phases, such as Phase I, Phase II, Phase III, etc. Phase I is also known as the α phase, which is the most common phase under normal conditions, and these exist in the TGTG′ (*trans-gauche-trans-gauche′*) nonpolar conformation. When the polymer is

PVdF-HFP-Based Gel Electrolytes

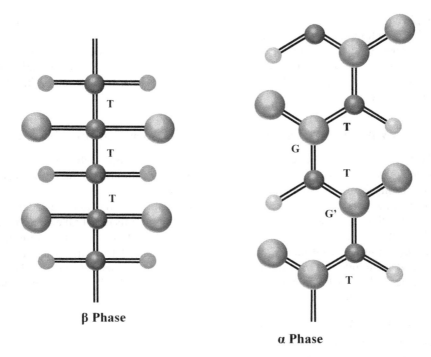

FIGURE 6.1 Schematic illustration of different crystal phases (α phase and β phase) of polyvinylidene difluoride (PVdF). Adapted and reproduced with permission from Ref. [12]. Copyright © 2007 Elsevier.

strained and stretched, a polar conformation, called phase II (or β phase) is formed in a zig-zag fashion [12] (Figure 6.1). Phase III has TTTGTTTG′ conformation, which forms when the polymer is stressed moderately. The last phase, IV, is also known as the δ phase, which exists only under specific conditions and parameters [13, 14]. The addition of plasticizers, such as ethylene carbonate (EC) and propylene carbonate (PC), etc., can transform the crystalline phase from phase I to phase III [15, 16].

6.2 PREPARATION OF PVdF-co-HFP-BASED POLYMER ELECTROLYTES

Polymer electrolytes are electrolytic materials which have a wide range of applications in electrochemical energy storage materials. These materials are synthesized by the dissolution of lithium salts in high-molecular-weight polymers. There are different techniques for the synthesis of polymer electrolytes, including (a) solvent casting [17], (b) phase inversion [18, 19], (c) plasticizer extraction and (d) electrospinning [20]. Plasticizer extraction is a preparation method in which plasticizers are added to improve the mechanical and thermal properties of the polymer matrix. These matrices provide nanoscale pore sizes, having lower porosity (~50%). This technique is a complex process and is expensive, compared with other techniques. Later, advanced synthetic techniques, such as phase-inversion and electrospinning processes, were also introduced, which resulted in greater porosity and conductivity. Detailed explanation of solution- casting, phase-inversion and electrospinning techniques, and their effects on the electrochemical properties of PVdF-*co*-HFP are discussed in the subsequent sections.

6.2.1 PVdF-*co*-HFP-Based Electrolytes Prepared by Solution Casting

Solvent casting is a synthetic procedure by which a polymer is dissolved in a suitable solvent, which is then cast onto a glass plate to produce a film [21]. This simple technique results in uniform porous polymer membrane, having nanoscale-sized pores. However, the handling of large volumes of solvent and the removal of residual solvent is a tedious process. At the same time, the porosity of this membrane is less than 50%, so that modified forms of synthesis techniques, such as phase inversion and electrospinning, were developed. Organic solvents, such as tetrahydro furan (THF), dimethyl acetamide (DMAc) and acetone are used in the synthesis of PVdF-*co*-HFP-based electrolytes. Blending different polymers along with PVdF-*co*-HFP or incorporation of ceramic fillers, etc. enhances the ionic conductivity.

6.2.1.1 Pure PVdF-*co*-HFP-Based Polymer Electrolytes

There are different techniques available for the synthesis of the polymer matrix, and the properties of such matrices are determined by these particular techniques. The solution-casting method is an easier and more conventional technique for the synthesis of most of the polymers. The selected polymer and lithium salt are separately dissolved in a suitable solvent such as dimethylformamide (DMF), acetone or THF, and they are homogeneously mixed, using a magnetic stirrer for a particular period of time. Plasticizers, if necessary, are added, and the resulting homogeneous solution, achieved with continuous mixing, is then poured into a petri dish to allow the solvent to evaporate. The resultant membrane was dried in an oven to remove all traces of the solvent. Earlier polymer membranes were prepared using this solvent-casting technique. Stephan et al. [17] reported the effect of PVdF-*co*-HFP in lithium-ion batteries (LIB), using the solvent-casting technique by incorporating plasticizers, such as ethylene carbonate (EC) and propylene carbonate (PC), and studied the effect of plasticizers on polymer membranes. In the current report, the role of lithium-based anions, such as BF_4^-, ClO_4^-, $CF_3SO_3^-$ etc., was investigated. These different salts, with a fixed salt content (4%), were chosen and underwent electrochemical as well as morphological studies. These three salts, along with PVdF-*co*-HFP, show a good operational temperature range and the maximum conductivity was observed in the salt that contains BF_4^- (greater than 10^{-3} S cm^{-1}). The solvation of Li$^+$ is easier in LiBF$_4^-$ since it possesses low lattice energy, facilitating higher ionic conductivity.

6.2.1.2 PVdF-*co*-HFP-Based Polymer Blend Electrolytes

Blending polymers attracted the interest of researchers, owing to the subsequent improvements in ionic conductivity, ease of preparation, increased thermal stability

PVdF-HFP-Based Gel Electrolytes

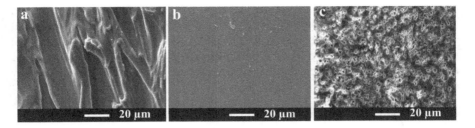

FIGURE 6.2 Scanning electron micrographs depicting the surface morphology of PVdF-co-HFP based polymer electrolytes: (a) pure PVdF-co-HFP membrane, (b) pure PMMA membrane and (c) PVdF-co-HFP: PMMA: LiTFSI gel polymer electrolyte (2500×magnification). Adapted and reproduced with permission from Ref. [21]. Copyright © 2016 Elsevier.

and ability to control the properties by varying the polymer contents. Different polymers have been used for the blending of PVdF-co-HFP. Microporous polymer electrolytes were synthesized by blending a PVdF-co-HFP with PEO and the LiBF$_4$ salt dissolved in PC [22]. Gebreyesus et al. [21] reported the blending of polymer PVdF-co-HFP with polymethyl methacrylate (PMMA), using lithium triflate as the dopant salt. Amorphicity and transparency are the major beneficial effects of PMMA, which is compatible with PVdF-co-HFP (Figure 6.2). Different concentrations of the PVdF-co-HFP/PMMA blend were prepared by the solution casting technique with a fixed doping salt content (25%). The highest conductivity observed was 7.4×10^{-4} S cm^{-1} with a blend of 22.5 wt.% of PMMA content, due to the segmental motion persisting in the polymer chain.

Ulaganathan et al. [23] developed PVdF-co-HFP blended with polyethyl methacrylate (PEMA), with different doping salts such as perchlorate, N[CF$_3$SO$_2$]$_2$, BF$_4^-$ and CF$_3$SO$_3^-$. The PEMA polymer increases ionic conductivity due to the presence of interconnecting pathways by the –COOH groups, as well as by increasing the number of free-volume sites. Among the different salts tested, lithium bis(trifluorosulfonyl) imide (LiN[CF$_3$SO$_2$]$_2$) showed the highest ionic conductivity, to 0.918×10^{-3} S cm^{-1}. Ataollahi et al. [24] reported the same polymer blend combinations, having PVdF-co-HFP as well as the grafted co-polymer of PMMA, called MG49. This polymer possesses oxygen atoms which perform as an electron donor and help in the coordination of the lithium salt. MG49 was dissolved in THF and PVdF-co-HFP was dissolved in acetone as part of a solution-casting technique, with LiBF$_4$ as the doping salt. Different wt.% concentrations of the lithium salt were tested and the maximum ionic conductivity was observed with 30 wt.% of lithium salt (2.32×10^{-4} S cm^{-1}). However, Wu et al. [25], who synthesized a blend with PVdF-co-HFP and the hyperbranched polymer of (poly(3-{2-[2-(2-hydroxyethoxy)ethoxy]ethoxy}methyl-3′-methyloxetane)) (PHEMO), along with the doping salt, lithium bis(trifluorosulfonyl) imide, showed a higher ionic conductivity. The amorphous nature of the polymer matrix results in a greater interaction with the doping salt, as well as providing high mechanical stability to the PVdF-co-HFP polymer. Different wt.% concentrations of PHEMO/PVdF-co-HFP were tested and optimized (2:1), before the doping salt concentration was then optimized. An increase in ionic conductivity was observed (1.75×10^{-3} S cm^{-1}), followed by a decrease in conductivity with further supra-optimal addition of ionic salt concentration. The formation of ion-pairs, which immobilize the polymer chains, and the formation of transient crosslinkers, are the reasons for the subsequent decrease in conductivity. The matrix offered a higher voltage stability of 4.2 V but lacked improved current density.

6.2.1.3 PVdF-co-HFP-Based Ceramic Filler Composite Polymer Electrolytes

Composite polymer electrolytes show increased mechanical stability, improved safety and electrode/electrolyte compatibilities. Additional fillers, such as Al_2O_3, SiO_2, TiO_2, etc., increase electrochemical performance and morphological parameters (Figure 6.3). The ionic conductivity of the polymer electrolytes can increase by up to one order of magnitude, and the conductivity increases with increasing temperature. Increased ionic conductivity was achieved by the addition of ceramic filler up

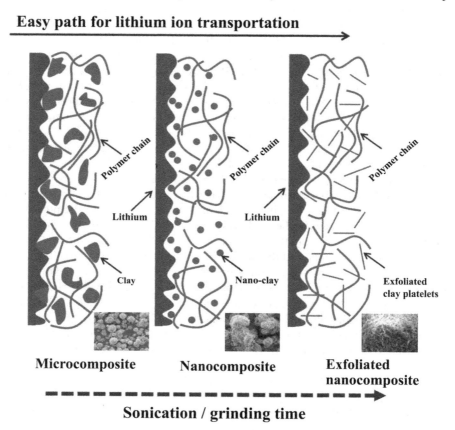

FIGURE 6.3 Schematic representation of the size effect of ceramic fillers (clay is adapted as the representative ceramic filler) on composite morphology and lithium-ion conduction channel formation in composite polymer electrolytes. Adapted and reproduced with permission from Ref. [26]. Copyright © 2006 Elsevier.

to 10 wt.% and then decreased as the filler percentage increased further. Filler additions greater than 10 wt.% led to the dilution effect, decreasing the ionic conductivity [26]. Fillers can stabilize and increase the amorphous nature of PVdF-co-HFP. An "ion-filler" complex is formed when electrolyte species react with Lewis acid-base interactions. Interfacial stability of composite polymer electrolytes is increased by these fillers trapping the remaining organic solvent impurities. Improved cyclability is attributed to the formation of a thin, compact, inorganic layer by the reaction of lithium metal and lithium metal salts in the passivation process.

Additional fillers increase the ionic conductivity of materials as well as enhancing their mechanical properties. Aravindan et al. [12] introduced a gel polymer electrolyte, with the addition of antimony oxide (SbO_3) nanoparticles as a filler in PVdF-co-HFP matrix. Lithium difluoride(oxalate)borate was chosen as the doping salt, with EC and DMC as plasticizers. Lithium difluoro(oxalate)borate (LiDFOB) was synthesized by a solid-state reaction of lithium oxalate and boron trifluoride diethyl etherate in a 1:1 ratio. Different wt.% of filler concentration were added and the electrochemical performances were noted, with 5 wt.% addition of filler showing the best properties.

The addition of fillers leads to the transformation of PVdF-co-HFP from the α phase to the β phase. These fillers serve as crosslinks between polymeric chains and inhibit the formation of the α phase, and thus the domination of β phase offers higher ionic conductivity. With the increase in temperature, the ionic conductivity increases due to the increase in the number of free-volume sites in the polymer matrix. Stephan et al. [26] investigated PVdF-co-HFP polymer hosts with and without the addition of the filler $Al(OH)_n$ into PVdF-co-HFP, with $LiN(C_2F_5SO_2)_2$ as the doping salt dissolved in THF solvent. Variation of ionic conductivity at a different temperature was also noted with optimized concentration (wt.%) of filler.

6.2.2 Preparation of PVdF-co-HFP-Based Polymer Electrolytes by Phase Inversion

Phase inversion is a simple and versatile technique, which consists of participation between a polymer, a solvent and a non-solvent in a ternary system. Porous polymer membranes can readily be obtained by exchanging the solvent as well as the non-solvent during precipitation. First, the polymer is dissolved in a suitable solvent under mechanical stirring until it become homogeneous. The resulting solution is cast on a clean glass plate, using the doctor blade technique, to produce a film with the desired thickness [27, 28]. The cast film is then immersed in a non-solvent bath. The resulting homogeneous film is then precipitated, due to the exchange of solvent and non-solvent, and finger-like as well as sponge-like structures form. PVdF-co-HFP-based polymer membranes prepared by this technique are discussed below.

6.2.2.1 Pure PVdF-co-HFP-Based Polymer Electrolytes

PVdF-co-HFP-based polymer membranes possess a porosity of ~70% of micro- to nanoscale dimensions, with thicknesses ranging from 200 to 250 μm. When the solvent is partially evaporated, the resulting polymer mixture is immersed in a non-solvent bath. Volatile solvents in PVdF-co-HFP are acetone, DMF, THF, etc. and the non-solvent includes butanol, methanol, pentanol or water. The polymer is homogeneously mixed in the presence of a solvent and then induced in a non-solvent in a ternary phase system consisting of a polymer, a non-solvent and a solvent. Water is the best non-solvent for this membrane because of its non-toxicity and its low cost.

6.2.2.2 PVdF-co-HFP-Based Polymer Blend Electrolytes

Microporous PVdF-co-HFP was synthesized by Pu et al. [18], using the phase-inversion method, with acetone as the solvent and water as the non-solvent. PVdF-co-HFP was dissolved in acetone and 5–8% water to obtain a homogeneous solution. The solution was cast using the doctor blade method to produce a film of 250 μm thickness. The amount of non-solvent (water) plays an important role, and 9% water and 5% polymer were adopted, producing a membrane with 85% porosity. The scanning electron microscopic (SEM) images (Figure 6.4) reveal the formation of a nonporous skin layer at the surface of the membrane and the uniform nature of the pores obtained in the sub-layers of the membrane, with small pore size of less than 1 μm in diameter.

PVdF-PEO blends prepared by the phase-inversion technique deliver an ionic conductivity of 2 mS cm^{-1} with a porosity greater than 84%. The addition of PEO in PVdF enhances the pore configuration, including porosity, pore size, pore connectivity, etc., and improving room temperature ionic conductivity. Furthermore, the highly crystalline nature of PVdF membrane diminishes the ionic conductivity and electrochemical properties of the electrolyte. Compared to PVdF, the less crystalline PVdF-co-HFP blended with a block copolymer of polyethylene oxide propylene oxide (P123), which formed amorphous, stable and transparent membranes, having improved pore structure leading to an enhancement in lithium-ion transportation and ionic conductivity (to 4 mS cm^{-1}) [27, 28]. In the same year, Subramania et al. [11] prepared a polymer blend of PVdF-co-HFP with PAN to improve the ionic

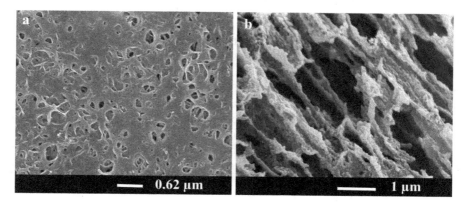

FIGURE 6.4 The scanning electron micrographs visualizing the morphology of polyvinylidene-co-hexafluoropropylene (PVdF-co-HFP)-based porous membrane prepared by the phase-inversion method: (a) lower surface and (b) cross-section. Adapted and reproduced with permission from Ref. [18]. Copyright © 2006 Elsevier.

conductivity of the polymer, and its thermal stability, stability, etc. The PVdF-*co*-HFP/PAN blend was dissolved in DMF and underwent the phase inversion technique, forming large cavities and voids of uniform size leading to higher electrolyte uptake. PVdF-*co*-HFP/PAN showed higher thermal stability (471–571°C) than pure PVdF-*co*-HFP (441°C) and higher ionic conductivity (3.41 mS cm^{-1}) than pure PVdF-*co*-HFP (1.21 mS cm^{-1}) by the phase-inversion method [18]. Incorporation of room temperature ionic liquid (RTIL), along with the polymer blend, enhances the battery performances because ionic liquids offer high thermal and chemical stability, are non-flammable, have high ionic conductivity, etc. The ionic liquid 1-butyl-3-methylimidazolium tetrafluoroborate ([BMIM]BF$_4$) was mixed with the PVdF-*co*-HFP/PMMA blend prepared by Yang et al. [19]. The porosity of the matrix increased up to 56.8% of PMMA content but decreased abruptly at higher PMMA content due to the physical crosslinking in the PMMA/PVdF-*co*-HFP matrix. A microporous gel polymer electrolyte with 40% PVdF-*co*-HFP and 60% PMMA (MIL 60) exhibited a high ionic conductivity (1.43 mS cm^{-1}), high thermal stability (386.6°C), and a large electrochemical stability window (4.5 V).

6.2.2.3 PVdF-*co*-HFP Ceramic Composite Electrolytes

The addition of fillers always increases the ionic conductivity and improves the mechanical properties. An improvement in ionic conductivity was observed for the polymer electrolyte with the addition of various ceramic fillers. Different fillers, such as SiO$_2$, TiO$_2$, Al$_2$O$_3$, etc. [29, 30], were used for the preparation of composite polymer electrolytes prepared by the solution-casting technique. The addition of fillers to the polymer solution synthesized by the phase-inversion method improves porosity, electrolyte uptake and conductivity, etc. [31, 32]. The type of solvent and non-solvent selected in the phase-inversion technique determines the efficiency of the polymer matrix. PVdF-*co*-HFPs with silica filler and having the same solvent (acetone) but different non-solvents (water, isopropyl alcohol or ethanol) behave differently in terms of electrochemical performance; the non-solvent induces a phase separation whereas the solvent, acetone, evaporates. Addition of SiO$_2$ to PVdF-*co*-HFP results in higher mechanical stability [33–35]. Kin et al. [36] compared the effect of the PVdF-*co*-HFP/TiO$_2$ composite polymer membrane synthesized by the conventional solution-casting technique with that synthesized by the phase-inversion technique, in which *N*-methyl pyrrolidone (NMP) is used as the solvent and water as the non-solvent. Surface morphology reveals that addition of 5–30% of the filler TiO$_2$ during the phase-inversion method resulted in a similar honeycomb structure, whereas further addition of filler led to the breakage/elongation of the porous structure. The ionic conductivity of the TiO$_2$/PVdF-*co*-HFP polymer synthesized by the solution-casting method was much lower (~10^{-4} S cm^{-1}) than that achieved by the phase-inversion technique (~10^{-3} S cm^{-1}). Later, Mao et al. [37] reported on the influence of TiO$_2$ as a filler in PVdF-*co*-HFP prepared by the inversion technique. The optimum TiO$_2$ concentration (6.5%) was identified for the highest electrolytic uptake (358%) and ionic conductivity (1.66 mS cm^{-1}). In addition to SiO$_2$ and TiO$_2$, Al$_2$O$_3$-based composite polymer electrolytes were also synthesized, which can act as separators with the same solvent (acetone) and non-solvent (water) as reported before.

PVdF-*co*-HFP/Al$_2$O$_3$-based composite polymer electrolytes exhibit higher thermal stability than pure PVdF-*co*-HFP synthesized by the phase-inversion technique, in which the microporous structure of the composite polymer is controlled by the variation in the amount of the non-solvent used [38].

6.2.3 PVdF-*co*-HFP-Based Polymer Electrolytes Prepared by Electrospinning

Electrospinning is an electro-hydrodynamic phenomenon of generating fibers and particles from the polymer solution. It is a versatile technique used to create a nanofibrous membrane of different dimensions (micro- to nanoscale) with morphological structures. The basic set-up includes a syringe, collector, drum, high-voltage power source, needle and a pump. On applying the high voltage, uniaxial elongation of the viscous polymer comes out from the thin nozzle, forming electrically charged droplets. Charged droplets then form a conically shaped jet (a Taylor cone) and are then ejected toward the collector (Figure 6.5). Collectors may be drums, plates or disks. Non-woven fibers are then collected from the collector. There are several spinning parameters, such as process, system or solution and ambient parameters, that determine the production of desirable nanofibers. Theoretical and computational

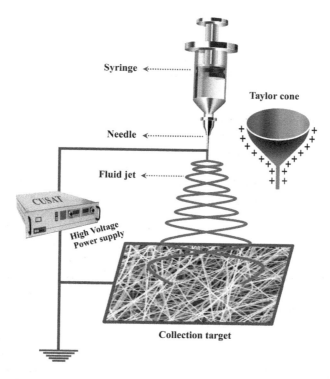

FIGURE 6.5 Schematic diagram of the process and machine set-up of electrospinning.

PVdF-HFP-Based Gel Electrolytes

investigations on a variety of polymer solutions were carried out [39–41]. By controlling different parameters [42–45], PVdF and its copolymer attracted attention because of its high dielectric constant, semi-crystalline nature and high thermal and chemical stability.

Research work on PVdF-*co*-HFPs in LIBs started earlier [46], with fluorinated polymers used to trap liquid electrolytes. Kim et al. [47, 48] prepared electrospun microporous PVdF-*co*-HFP, soluble in a mixture of DMAc and acetone, which showed an average fiber diameter of 0.5 μm. But it also possessed lower ionic conductivity (1 mS cm^{-1}), with an electrochemical stability window of 4.5 V. Later, ionic conductivity was improved by incorporating ionic liquids, such as 1-butyl-3-methylimidazolium tetrafluoroborate (BMIBF$_4$) or 1-butyl-3-methylimidazolium bis(trifluoromethanesulfonyl)imide (BMITFSI) (2.3 mS cm^{-1}) [49]. Blending PVdF-*co*-HFP with another polymer showed a marked effect on the electrochemical performance of batteries. Bansal et al. [50] reported an electrospun PVdF-*co*-HFP/PAN membrane, which is an active material for non-woven separators (ranging up to ~83% porosity), whereas Ding et al. [51] reported that PVdF-*co*-HFP/PMMA gel polymer electrolytes showing a higher electrolytic uptake (377%) and good ionic conductivity (1.99 mS cm^{-1}). Typical pure PVdF-*co*-HFP membrane has an average fiber diameter of 100–250 nm, whereas blending PVdF-*co*-HFP with PMMA or PAN delivers an average fiber diameter of 200–350 nm or 300–650 nm, respectively. Since electrospun PVdF-*co*-HFP/PAN membrane shows a higher average fiber diameter, further electrochemical studies were carried out on this matrix. Raghavan et al. [52] reported electrospun PVdF-*co*-HFP/PAN blend/composite-based gel polymer electrolytes, having average fiber diameter in the range 350–490 nm, with a better ionic conductivity of 10^{-3} S cm^{-1} and anodic stability greater than 4.6 V. The FE-SEM images of electrospun pure PVdF-*co*-HFP membrane and electrospun PVdF-*co*-HFP/PAN-based gel polymer electrolytes are shown in Figure 6.6.

FIGURE 6.6 FE-SEM images on the morphology of electrospun PVdF-*co*-HFP/PAN blend/composite membranes: (a) pure PVdF-*co*-HFP and (b) 50:50 blend of PVdF-*co*-HFP and PAN. Adapted and reproduced with permission from Ref. [52]. Copyright © 2010 Elsevier.

Later, blending PVdF-*co*-HFP with other polymers became prevalent, and blending electrospun PVdF with TPU was reported earlier [53–55], with Zhou et al. [55] reporting that the electrospun PVdF-*co*-HFP/TPU blend delivered a higher ionic conductivity (4.1 mS cm^{-1}). Electrospun PVdF-*co*-HFP can act as a separator-cum-electrolyte membrane [56, 57], having an electrochemical performance superior to commercial Celgard ® separators. PVdF-*co*-HFP can also be synthesized by surface modification, such as plasma modification, as described by Laurita et al. [58]. Using atmospheric pressure in non-equilibrium plasma induction to either pre-treat a PVdF precursor solution before it was electrospun or to post-treat the membrane with the same method led to achieving vast improvements in both electrolyte uptake (1 M LiPF$_6$ in EC: DMC, 1:1 v:v) and mechanical properties. Defect-free nanofibers were obtained by the pretreatment method, whereas, in the post-treatment method, non-woven membranes underwent chemical modification, forming hydroxyl and polar carboxyl groups. At present, room temperature ionic liquids are promising substituents of organic solvents and room temperature ionic liquid (RTIL), along with electrospun PVdF-*co*-HFP showing specific capacities of 143 and 115 mAh g^{-1} at 0.1 and 1 C rate, respectively, with 92% retention at room temperature, but has lower ionic conductivity (1.1×10^{-4} S cm^{-1}). Later, ionic liquid along with grafted PVdF-*co*-HFP was studied and the electrochemica; performance compared [20]. The ionic liquid, 1-butyl-3-methylimidazolium bis(trifluoromethanesulfonyl)imide (BMITFSI), along with grafted poly(polyethylene glycol) methyl ether methacrylate) (PVdF-*co*-HFP-*g*-PPEGMA) or polyvinylidene difluoride-*co*-hexafluoropropylene, were chosen to increase the electrolyte uptake. An improvement in ionic conductivity was observed (to 2.3 mS cm^{-1}) with a wide electrochemical stability window (5.2 V) and specific capacities of 163, 141 and 125 mAh g^{-1} at 0.1, 0.5 and 1 C rates, respectively.

6.3 CONCLUSION

Our energy-dependent society demands portable electronic systems with high energy density, long cycle life, low self-discharge and low memory effect. Compared with other battery systems, rechargeable lithium-ion batteries play a vital role, owing to their higher energy density. However, lithium-ion batteries suffer serious safety problems, which can be rectified by replacing organic liquid electrolytes with gel polymer electrolytes. PVdF-*co*-HFP is a suitable candidate for the synthesis of gel polymer electrolytes for lithium-ion batteries due to its higher chemical and thermal stability, higher mechanical strength andhigh dielectric constant and polarity. Compared with pure PVdF and other co-polymers of PVdF, PVdF-*co*-HFP provides higher ionic conductivity due to the amorphous nature of HFP. Synthesis techniques play a major role in improving the electrochemical performance of batteries, such as cycling stability, anodic stability, ionic conductivity, etc. At present, electrospinning and phase-inversion techniques have become predominant, compared with conventional synthesis methods of solvent casting. Blending PVdF-*co*-HFP with other polymers or the addition of different fillers or plasticizers are the other methods by which to increase the ionic conductivity.

ACKNOWLEDGMENT

Authors Dr. Jabeen Fatima M. J. and Dr. Prasanth Raghavan, would like to acknowledge Kerala State Council for Science, Technology and Environment (KSCSTE), Kerala, India, for financial assistance.

REFERENCES

1. Scrosati B, Garche J (2010) Lithium batteries: Status, prospects and future. *J Power Sources* 195:2419–2430. https://doi.org/10.1016/j.jpowsour.2009.11.048
2. Arya A, Sharma AL (2017) Polymer electrolytes for lithium ion batteries: A critical study. *Ionics (Kiel)* 23:497–540. https://doi.org/10.1007/s11581-016-1908-6
3. Kim JR, Choi SW, Jo SM, et al. (2004) Electrospun PVdF-based fibrous polymer electrolytes for lithium ion polymer batteries. *Electrochim Acta* 50:69–75. https://doi.org/10.1016/j.electacta.2004.07.014
4. Shanthi PM, Hanumantha PJ, Albuquerque T, et al. (2018) Novel composite polymer electrolytes of PVdF-HFP derived by electrospinning with enhanced Li-Ion conductivities for rechargeable lithium-sulfur batteries. *ACS Appl Energy Mater* 1:483–494. https://doi.org/10.1021/acsaem.7b00094
5. Fergus JW (2010) Ceramic and polymeric solid electrolytes for lithium-ion batteries. *J Power Sources* 195:4554–4569. https://doi.org/10.1016/j.jpowsour.2010.01.076
6. Ma T, Cui Z, Wu Y, et al. (2013) Preparation of PVDF based blend microporous membranes for lithium ion batteries by thermally induced phase separation: I. Effect of PMMA on the membrane formation process and the properties. *J Memb Sci* 444:213–222. https://doi.org/10.1016/j.memsci.2013.05.028
7. Guo XZ and Y-G (2013) A PEO-assisted electrospun silicon–graphene composite as an anode material for lithium-ion batteries. *J Mater Chem A Mater Energy Sustain* 1:9019–9023. https://doi.org/10.1039/c3ta11720k
8. Manjuladevi R, Thamilselvan M, Selvasekarapandian S, et al. (2017) Mg-ion conducting blend polymer electrolyte based on poly(vinyl alcohol)-poly (acrylonitrile) with magnesium perchlorate. *Solid State Ionics* 308:90–100. https://doi.org/10.1016/j.ssi.2017.06.002
9. Kanamura K (2005) Electrolytes for lithium batteries. *Fluorinated Mater Energy Convers* 253–266. https://doi.org/10.1016/B978-008044472-7/50039-4
10. Stephan AM (2006) Review on gel polymer electrolytes for lithium batteries. *Eur Polym J* 42:21–42. https://doi.org/10.1016/j.eurpolymj.2005.09.017
11. Subramania A, Sundaram NTK, Kumar GV (2006) Structural and electrochemical properties of micro-porous polymer blend electrolytes based on PVdF-co-HFP-PAN for Li-ion battery applications. *J Polym Sci Part B Polym Phys* 153:177–182. https://doi.org/10.1016/j.jpowsour.2004.12.009
12. Aravindan V, Vickraman P (2007) A novel gel electrolyte with lithium difluoro(oxalato) borate salt and Sb2O3 nanoparticles for lithium ion batteries. *Solid State Sci* 9:1069–1073. https://doi.org/10.1016/j.solidstatesciences.2007.07.011
13. Abbrent S, Plestil J, Hlavata D, et al. (2001) Crystallinity and morphology of PVdF-co-HFP-based gel electrolytes. polymer (Guildf) 42:1407–1416. https://doi.org/10.1016/S0032-3861(00)00517-6
14. Lee S (2011) Crystal structure and thermal properties of poly (vinylidene fluoride-hexafluoropropylene) films prepared by various processing conditions. 12:1030–1036. https://doi.org/10.1007/s12221-011-1030-3
15. Stolarska M, Niedzicki L, Borkowska R, et al. (2007) Structure, transport properties and interfacial stability of PVdF/HFP electrolytes containing modified inorganic filler. *Electrochim Acta* 53:1512–1517. https://doi.org/10.1016/j.electacta.2007.05.079

16. Essalhi M, Khayet M (2010) Effects of PVDF-HFP concentration on membrane distillation performance and structural morphology of hollow fiber membranes. 347:209–219. https://doi.org/10.1016/j.memsci.2009.10.026
17. Stephan AM, Kumar SG, Renganathan NG, Kulandainathan MA (2005) Characterization of poly(vinylidene fluoride-hexafluoropropylene) (PVdF-HFP) electrolytes complexed with different lithium salts. *Eur Polym J* 41:15–21. https://doi.org/10.1016/j.eurpolymj.2004.09.001
18. Pu W, He X, Wang L, et al. (2006) Preparation of PVDF – HFP microporous membrane for Li-ion batteries by phase inversion. 272:11–14. https://doi.org/10.1016/j.memsci.2005.12.038
19. Zhai W, Zhu H, Wang L, et al. (2014) Study of PVDF-HFP/PMMA blended microporous gel polymer electrolyte incorporating ionic liquid [BMIM]BF$_4$ for lithium ion batteries. *Electrochim Acta* 133:623–630. https://doi.org/10.1016/j.electacta.2014.04.076
20. Tong Y, Que M, Su S, Chen L (2016) Design of amphiphilic poly (vinylidene fluoride-co -hexafluoropropylene) -based gel electrolytes for high-performance lithium-ion batteries. https://doi.org/10.1007/s11581-016-1662-9
21. Gebreyesus MA, Purushotham Y, Kumar JS (2016) Preparation and characterization of lithium ion conducting polymer electrolytes based on a blend of poly(vinylidene fluoride-co-hexafluoropropylene) and poly(methyl methacrylate). *Heliyon* 2. https://doi.org/10.1016/j.heliyon.2016.e00134
22. Wang H, Huang H, Wunder SL (2002) Novel microporous poly(vinylidene fluoride) blend electrolytes for lithium-ion batteries. *J Electrochem Soc* 147:2853. https://doi.org/10.1149/1.1393616
23. Ulaganathan M, Mathew CM, Rajendran S (2013) Highly porous lithium-ion conducting solvent-free poly(vinylidene fluoride-co-hexafluoropropylene)/poly(ethyl methacrylate) based polymer blend electrolytes for Li battery applications. *Electrochim Acta* 93:230–235. https://doi.org/10.1016/j.electacta.2013.01.100
24. Ataollahi N, Ahmad A, Hamzah H, et al. (2012) Preparation and characterization of PVDF-HFP/MG49 based polymer blend electrolyte. *Int J Electrochem Sci* 7:6693–6703. https://doi.org/10.1007/s11581-009-0415-4
25. Wu F, Feng T, Bai Y, et al. (2009) Preparation and characterization of solid polymer electrolytes based on PHEMO and PVDF-HFP. *Solid State Ionics* 180:677–680. https://doi.org/10.1016/j.ssi.2009.03.003
26. Stephan AM, Suk K, Kumar TP, et al. (2006) Nanofiller incorporated poly (vinylidene fluoride – hexafluoropropylene) composite electrolytes for lithium batteries. 159:1316–1321. https://doi.org/10.1016/j.jpowsour.2005.11.055
27. Wu CG, Lu MI, Chuang HJ (2005) PVdF-HFP/P123 hybrid with mesopores: A new matrix for high-conducting, low-leakage porous polymer electrolyte. *Polymer (Guildf)* 46:5929–5938. https://doi.org/10.1016/j.polymer.2005.05.077
28. Xi J, Qiu X, Li J, et al. (2006) PVDF-PEO blends based microporous polymer electrolyte: Effect of PEO on pore configurations and ionic conductivity. *J Power Sources* 157:501–506. https://doi.org/10.1016/j.jpowsour.2005.08.009
29. Kim M, Lee L, Jung Y, Kim S (2013) Study on ion conductivity and crystallinity of composite polymer electrolytes based on poly(ethylene oxide)/poly(acrylonitrile) containing nano-sized Al2O3 fillers. *J Nanosci Nanotechnol* 13:7865–7869. https://doi.org/10.1166/jnn.2013.8107
30. Yap YL, You AH, Teo LL, Hanapei H (2013) Inorganic filler sizes effect on ionic conductivity in polyethylene oxide (PEO) composite polymer electrolyte. *Int J Electrochem Sci* 8:2154–2163
31. Xiao W, Li X, Guo H, et al. (2012) Preparation and properties of composite polymer electrolyte modified with nano-size rare earth oxide. *J Cent South Univ* 19:3378–3384. https://doi.org/10.1007/s11771-012-1417-3

32. Liu L, Wang Z, Zhao Z, et al. (2016) PVDF/PAN/SiO 2 polymer electrolyte membrane prepared by combination of phase inversion and chemical reaction method for lithium ion batteries. *J Solid State Electrochem* 20:699–712. https://doi.org/10.1007/s10008-015-3095-1
33. Jeong HS, Noh JH, Hwang CG, et al. (2010) Effect of solvent-nonsolvent miscibility on morphology and electrochemical performance of SiO2/PVdF-HFP-based composite separator membranes for safer lithium-ion batteries. *Macromol Chem Phys* 211:420–425. https://doi.org/10.1002/macp.200900490
34. Wahyudi W, Cao Z, Kumar P, et al. (2018) Phase inversion strategy to flexible freestanding electrode: Critical coupling of binders and electrolytes for high performance Li–S battery. *Adv Funct Mater* 28:1–8. https://doi.org/10.1002/adfm.201802244
35. Huang X (2012) A lithium-ion battery separator prepared using a phase inversion process. *J Power Sources* 216:216–221. https://doi.org/10.1016/j.jpowsour.2012.05.019
36. Kim KM, Park NG, Ryu KS, Chang SH (2006) Characteristics of PVdF-HFP/TiO2 composite membrane electrolytes prepared by phase inversion and conventional casting methods. *Electrochim Acta* 51:5636–5644. https://doi.org/10.1016/j.electacta.2006.02.038
37. Miao R, Liu B, Zhu Z, et al. (2008) PVDF-HFP-based porous polymer electrolyte membranes for lithium-ion batteries. *J Power Sources* 184:420–426. https://doi.org/10.1016/j.jpowsour.2008.03.045
38. Jeong HS, Kim DW, Jeong YU, Lee SY (2010) Effect of phase inversion on microporous structure development of Al2O3/poly(vinylidene fluoride-hexafluoropropylene)-based ceramic composite separators for lithium-ion batteries. *J Power Sources* 195:6116–6121. https://doi.org/10.1016/j.jpowsour.2009.10.085
39. Lim D-H, Haridas AK, Figerez SP, et al. (2018) Tailor-made electrospun multilayer composite polymer electrolytes for high-performance lithium polymer batteries. *J Nanosci Nanotechnol* 18:6499–6505. https://doi.org/10.1166/jnn.2018.15689
40. Chew S, Wen Y, Dzenis Y, Leong K (2006) The role of electrospinning in the emerging field of nanomedicine. *Curr Pharm Des* 12:4751–4770. https://doi.org/10.2174/138161206779026326
41. Dong B, Arnoult O, Smith ME, Wnek GE Electrospinning of collagen nanofiber scaffolds from benign solvents. 539–542. https://doi.org/10.1002/marc.200800634
42. Theron SA, Zussman E, Yarin AL (2017) Experimental investigation of the governing parameters in the electrospinning of polymer solutions. 45:2017–2030. https://doi.org/10.1016/j.polymer.2004.01.024
43. Yo OS, Papila M, Mencelog YZ (2008) Effects of electrospinning parameters on polyacrylonitrile nanofiber diameter : An investigation by response surface methodology. *Mater Des* 29:34–44. https://doi.org/10.1016/j.matdes.2006.12.013
44. Megelski S, Stephens JS, Bruce Chase D, Rabolt JF (2002) Micro- and nanostructured surface morphology on electrospun polymer fibers. *Macromolecules* 35:8456–8466. https://doi.org/10.1021/ma020444a
45. Li X, Cheruvally G, Kim J-K, et al. (2007) Polymer electrolytes based on an electrospun poly(vinylidene fluoride-co-hexafluoropropylene) membrane for lithium batteries. *J Power Sources* 167:491–498. https://doi.org/10.1016/j.jpowsour.2007.02.032
46. Tarascon JM, Gozdz AS, Schmutz C, et al. (1996) Performance of Bellcore's plastic rechargeable Li-ion batteries. *Solid State Ionics* 86–88:49–54. https://doi.org/10.1016/0167-2738(96)00330-X
47. Kim JR, Choi SW, Jo SM, et al. (2005) Characterization and properties of P(VdF–HFP)-based fibrous polymer electrolyte membrane prepared by electrospinning. *J Electrochem Soc* 152:A295. https://doi.org/10.1149/1.1839531
48. Choi SW, Jo SM, Lee WS, Kim YR (2003) An electrospun poly(vinylidene fluoride) nanofibrous membrane and its battery applications. *Adv Mater* 15:2027–2032. https://doi.org/10.1002/adma.200304617

49. Cheruvally G, Kim J-K, Choi J-W, et al. (2007) Electrospun polymer membrane activated with room temperature ionic liquid: Novel polymer electrolytes for lithium batteries. *J Power Sources* 172:863–869. https://doi.org/https://doi.org/10.1016/j.jpowsour.2007.07.057
50. Bansal D, Meyer B, Salomon M (2008) Gelled membranes for Li and Li-ion batteries prepared by electrospinning. *J Power Sources* 178:848–851. https://doi.org/10.1016/j.jpowsour.2007.07.070
51. Ding Y, Zhang P, Long Z, et al. (2009) The ionic conductivity and mechanical property of electrospun P(VdF–HFP)/PMMA membranes for lithium ion batteries. *J Memb Sci* 329:56–59. https://doi.org/10.1016/j.memsci.2008.12.024
52. Raghavan P, Zhao X, Shin C, et al. (2010) Preparation and electrochemical characterization of polymer electrolytes based on electrospun poly(vinylidene fluoride-co-hexafluoropropylene)/polyacrylonitrile blend/composite membranes for lithium batteries. *J Power Sources* 195:6088–6094. https://doi.org/10.1016/j.jpowsour.2009.11.098
53. Na Wu, Bo Jing, Qi Cao, Xianyou Wang, Hao Kuang QW (2010) A novel electrospun TPU/PVdF porous fibrous polymer electrolyte for lithium ion batteries. *J Appl Polym Sci* 116:2658–2667. https://doi.org/10.1002/app
54. Wu N, Cao Q, Wang X, et al. (2011) In situ ceramic fillers of electrospun thermoplastic polyurethane/ poly(vinylidene fluoride) based gel polymer electrolytes for Li-ion batteries. *J Power Sources* 196:9751–9756. https://doi.org/10.1016/j.jpowsour.2011.07.079
55. Zhou L, Cao Q, Jing B, et al. (2014) Study of a novel porous gel polymer electrolyte based on thermoplastic polyurethane/poly(vinylidene fluoride-co-hexafluoropropylene) by electrospinning technique. *J Power Sources* 263:118–124. https://doi.org/10.1016/j.jpowsour.2014.03.140
56. Lee H, Alcoutlabi M, Watson J V, et al. (2013) Electrospun nanofiber-coated separator membranes for lithium-ion rechargeable batteries. 1939–1951. https://doi.org/10.1002/app.38894
57. Zhou X, Yue L, Zhang J, et al. (2013) A core-shell structured polysulfonamide-based composite nonwoven towards high power lithium ion battery separator. *J Electrochem Soc* 160:A1341–A1347. https://doi.org/10.1149/2.003309jes
58. Laurita R, Zaccaria M, Gherardi M, et al. (2016) Plasma processing of electrospun Li-Ion battery separators to improve electrolyte uptake. *Plasma Process Polym* 13:124–133. https://doi.org/10.1002/ppap.201500145

7 Polyacrylonitrile (PAN)-Based Polymer Electrolyte for Lithium-Ion Batteries

Neethu T. M. Balakrishnan, Akhila Das, Jishnu N. S., Jou-Hyeon Ahn, Jabeen Fatima M. J., and Prasanth Raghavan

CONTENTS

7.1 Introduction 149
7.2 Mechanism of Ionic Conductivity in Polyacrylonitrile-Based Polymer Electrolytes 151
7.3 Methods of Preparation of Polyacrylonitrile-Based Polymer Electrolytes 151
    7.3.1 Polyacrylonitrile-Based Gel Polymer Electrolytes Prepared by Solvent Casting 152
    7.3.2 Polyacrylonitrile-Based Polymer Electrolytes Prepared by Phase Inversion 154
    7.3.3 Polyacrylonitrile-Based Polymer Electrolytes Prepared by Electrospinning 157
7.4 Polyacrylonitrile-Based Polymer Blend Electrolytes 158
7.5 Polyacrylonitrile-Based Ceramic Composite Polymer Electrolytes 159
7.6 Conclusion 161
Acknowledgment 162
References 162

7.1 INTRODUCTION

In recent years, marked advances in the continous demand for portable power applications initiated the requirement of lightweight batteries as energy storage devices. To achieve the desired energy storage, lithium-ion batteries (LIBs) are considered to be promising, and are capable of powering from small portable electronic devices up to advanced electric vehicles. Even though LIBs are stalwart in rechargeable energy storage systems, the safety of the LIBs is of paramount importance. The mechanical, electrical and thermal risks associated with LIBs are complicated issues that jeopardize their safety during use. The main reason behind the safety issues associated with the LIBs is the presence of volatile liquid electrolytes inside the battery which are flammable in nature. During high-temperature applications, as well as during continuous use of the battery, the heat that develops within the cell will cause the heating-up of this liquid electrolyte. Whenever this volatile organic material heats up, there will be vaporization of the organic moieties, which will lead to increase of internal pressure, causing the swelling of the battery, which, under extreme conditions, can lead to an explosion. This problem becomes serious in advanced applications, such as electric vehicles [1, 2]. Compared to conventional vehicles, electric vehicles (EVs) consist of a large number of cells connected in series and parallel to operate on a electric motor, instead of an internal-combustion engine that generates power by burning a mix of fuel and gases.

For example, the Tesla model S, with a 99 kWh battery pack, consists of 8256 cylindrical cells, connected in series and parallel [3]. During operation, the EV battery will undergo different kinds of stress situations, like vibrations, extreme temperature, fast charging, etc., that can generate variation within the components, which can adversely affect its performance. So, to improve safety without compromising the performance of LIBs, polymer electrolytes (PEs) were developed, which are the safer route to resolve the safety concerns with LIBs. Different polymer matrices were examined for the fabrication of polymer electrolytes in LIBs, and different methods were used to enhance the mechanical, thermal and electrochemical properties of the polymer electrolyte. The polymer matrices commonly studied for the polymer electrolytes include polyethylene oxides (PEO) [4], polymethyl methacrylate (PMMA) [5], polyvinylidene difluoride-co-hexafluoropropylene (PVdF-co-HFP) [6], polyacrylonitrile (PAN), polyvinylidene fluoride (PVdF) [7], polyvinyl chloride (PVC) [8] and polystyrene (PS) [9]. Compared with LIBs, lithium polymer batteries fabricated by using these electrolytes are used in most of the portable electronic devices, with fewer associated risks. Even though different polymer matrixes have their own advantages as an electrolyte matrix, PAN-based polymer electrolytes have been extensively studied in LIBs due to their ability to form solid as well as gel polymer electrolytes, with excellent electrochemical properties. PAN exhibits high thermal stability, so it is considered to be effective in high-temperature applications of the battery. Unlike other polymer electrolytes, the ionic conductivity of PAN-based solid polymer electrolytes (SPE) depends only on the ionic mobility, not over the segmental motion of the polymer, and they exhibit a decoupled ion transport behavior over a wide range of temperatures [10]. Moreover, they can exhibit a unique mechanism in a system containing polymer–salt combinations, in which the concentration of the salt is higher than in the traditional polymer–salt system. Different methods are employed for the fabrication of PAN-based polymer electrolytes, such as solvent-casting [11, 12], phase-inversion [13, 14] and electrospinning techniques [15, 16]. Each fabrication method involves its own advantages and disadvantages. The limitations of these methods can be resolved up to an extent by modifying the process conditions and composition of the materials. For example, the major concern faced while using the basic solvent-casting technique is the brittleness of the membrane, and this can be avoided to a certain extent by using the flexible polymer matrix, incorporation of plasticizers, blending with other polymers, etc. Likewise, there are the techniques that can provide different membrane properties and electrochemical performances of the PAN-based polymer electrolytes, that will be discussed detail in this chapter.

7.2 MECHANISM OF IONIC CONDUCTIVITY IN POLYACRYLONITRILE-BASED POLYMER ELECTROLYTES

Polyacrylonitrile (PAN), which is also known as polyvinyl cyanide or Creslan 61, is a synthetic semi-crystalline polymer with the chemical formula $(C_3H_3N)_n$. It is a thermoplastic material that exhibits very high thermal stability. Normally, it degrades before it melts and exhibits a degradation temperature above 300°C [17]. The ionic mobility through the PAN-based polymer membranes normally depends on the amorphous content formed in the polymer matrix. Huang et al. [18] reported that the formation of the amorphous phase in the polymer matrix is greatly influenced by the presence and type of lithium salt and plasticizer used in it. The authors examined the ionic conductivity mechanism by using ethylene carbonate (EC) and LiClO$_4$. Li$^+$-ions are found to be distributed in different environments, which were associated with PAN, solvent and amorphous gel-like PAN, that exhibited high segmental mobility [18]. The dissociated Li$^+$-ions became coupled with the C=O group of EC and the –C=N group of PAN. Also, the interaction between PAN and EC was formed through the –C=O and –C=N groups. This forms a co-existed state consisting of three components, which enhances the movement of ions through the polymer matrix [18, The interaction between the Li$^+$-ions and PAN is less dependent on the organic solvent used for the dissolution of lithium salt [19]. Similar observations were made by Watanabe et al. [20], using a system containing PAN/LiClO$_4$ and EC, where they found that the ionic conductivity depends solely upon the ratio of EC/LiClO$_4$ and is independent of PAN [20]. According to Yoon et al. [21], an increase in ionic conductivity was associated with the transition from the composition of the salt in polymer to the polymer in salt. These studies conclude that the mechanism of Li$^+$-ion conductivity through the polymer electrolytes depends purely on the type and composition of salt, solvent and polymer.

7.3 METHODS OF PREPARATION OF POLYACRYLONITRILE-BASED POLYMER ELECTROLYTES

Different methods are proposed for the synthesis of the polymer electrolytes. Normally, PEs are classified as homogeneous or heterogeneous. Homogeneous PEs are synthesized by using normal casting methods and *in-situ* methods, whereas fabrication of heterogeneous PEs involves two steps, such as the fabrication of the porous membrane and activation of the porous membrane, using the liquid organic electrolyte. The PEs fabricated by this method consist of three phases: liquid electrolyte phase, polymer matrix phase and a swelled gel phase [22, 23]. For the fabrication of heterogeneous PEs, the phase-inversion method [14, 24] and an electrospinning method [15, 25] are widely used.

7.3.1 POLYACRYLONITRILE-BASED GEL POLYMER ELECTROLYTES PREPARED BY SOLVENT CASTING

Solvent casting is one of the simplest conventional methods used for the fabrication of PEs. It involves the formation of a homogeneous solution of a polymer in a suitable low-boiling-point solvent, followed by casting it over the substrate to form a porous film. Each of the components and their ratio used to form the porous polymer membrane plays a crucial role in determining the performance of the polymer electrolyte fabricated from those porous membranes [12]. For example, the ionic conductivity of PAN-based PE fabricated by Jyothi et al. [26], using ethylene carbonate (EC) as a plasticizer, dimethyl formamide (DMF) as the solvent and potassium iodide (KI) as the salt, revealing that the ionic conductivity greatly depends on the PAN:KI ratio. The highest conductivity (2.089×10^{-5} S cm^{-1}) observed for the electrolyte had PAN:KI ratio 7:3 [26]. Along with the choice of salts, the correct plasticizer is also important to determine the ionic conductivity of the polymer electrolyte, since it can provide greater plasticity within the polymer. Watanabe et al. [27] used propylene carbonate (PC), along with EC and DMF, for the fabrication of PAN-based PEs, by the solvent-casting method. Comparing the influence of these plasticizers over the ionic conductivity of polymer electrolytes, the order of DMF >EC >PC was observed, which is higher than that of the electrolyte prepared without plasticizer [27]. Since the values for ionic conductivity and mechanical stability are lower for the solvent-cast PEs, most of the studies were conducted using the polymer composites or blend electrolytes. Different polymers were selected for blending with PAN, based on its ability to provide beneficial properties to the finally fabricated PEs. The effect of lithium salt over the blend of PAN/PVC was investigated by Rajendran et al. [28]. The highest ionic conductivity of about 8.35×10^{-5} S cm^{-1} was observed for the PVC/PAN blend electrolyte containing 8% LiClO$_4$-(PVC:PAN: LiClO$_4$:PC::6:24:8:64) [28]. Similar effect of LiClO$_4$ over the PAN/PMMA (75:25) based blend polymer electrolyte was investigated and reported by Flora et al. [29], and the maximum ionic conductivity observed was 1.9×10^{-6} S cm^{-1} at an optimized concentration of LiClO$_4$ (5%), which is lower than that reported by Rajendran et al. [28]. The advantage of blended polymer electrolyte over the individual system can be clearly depicted from the scanning electron microscopy (SEM) images (Figure 7.1) of PAN/LiClO$_4$ (Figure 7.1a), PMMA/LiClO$_4$ (Figure 7.1b) and PAN/PMMA/LiClO$_4$ blend PE membranes (Figure 7.1c). The surface morphology of the solvent-casted polymer blend membrane (Figure 7.1d) consists of numerous microporous structures that can effectively entrap the liquid electrolytes, which is beneficial to increase the ionic conductivity and thereby the electrochemical performance of the resulting electrolyte 29]. Comparing the thermal stability of these two electrolytes, (PVC/PAN and PAN/PMMA) confirms the better performance of PAN/PVC blend system, which can exhibit a higher thermal stability (250°C) [28], than that of PAN/PMMA (230°C) [29]. In order to improve the ionic conductivity, a different approach was proposed by Subramania et al. [30], in which moisture control during the activation of the polymer membrane helped to increase the ionic interactions and thereby resulted in a conductivity in the order of 10^{-3} S cm^{-1}, which

PAN-Based Polymer Electrolytes

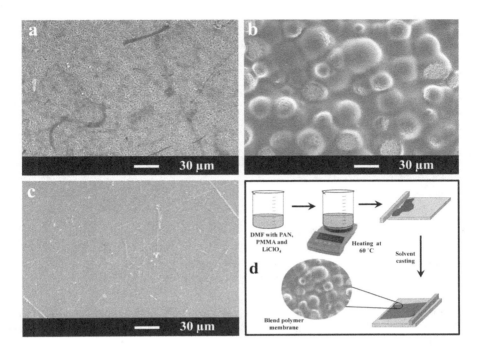

FIGURE 7.1 FE-SEM images on the surface morphology of PAN-based polymer electrolytes of different compositions (in wt.%): (a): PAN (100) –LiClO$_4$ (5), (b) PAN (75)–PMMA (25)– LiClO$_4$ (5), (c) PMMA (10)– LiClO$_4$ (5) and (d) schematic illustration of the preparation of the polymer electrolytes by solvent casting. Adapted and reproduced with permission from Ref. [29]. Copyright © 2012 Elsevier.

is much higher than that reported for other solvent-cast polymer blend electrolytes. The blend electrolyte, of PAN along with PVA and LiClO$_4$, was prepared by this method and exhibited an ionic conductivity of 2.53×10^{-3} S cm^{-1} with an electrochemical stability of 4.7 V [30].

Similar to blending, preparation of composite polymer electrolytes can also enhance their ionic transport properties and electrochemical performance. The fillers incorporated within the polymer matrix will provide a reinforcing effect, which can enhance the mechanical strength and the reduction in crystallinity due to the filler-polymer interaction aid to upright the ionic conduction through the membrane. Rajendran et al. [31] reported an increase in ionic conductivity of the composite polymer electrolyte formed by the incorporation of TiO$_2$ into PAN/PVC blends. The ionic conductivity observed was about 4.46×10^{-3} S cm^{-1}, which was higher than the blend electrolyte (PAN: PVC: LiClO$_4$: EC (19.2:4.8:8:68) without TiO$_2$ (7.57×10^{-5} S cm^{-1}) [31]. The PAN/PEO polymer blend electrolyte, in the ratio 2:8 with 15 wt.% of alumina (Al$_2$O$_3$), was prepared by Kim et al. [32] (Figure 7.2). Compared to the electrolyte fabricated with TiO$_2$ by Rajendran et al. [31], the ionic conductivity observed was lower with the addition of Al$_2$O$_3$ as the filler. Without the filler, the PE is only

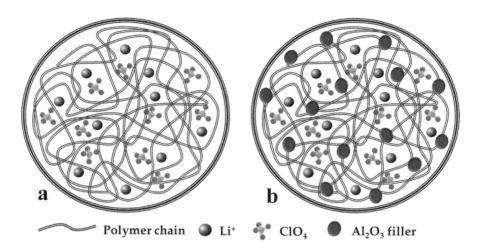

FIGURE 7.2 Schematic illustration of the proposed microstructure of composite polymer electrolytes: (a) without filler and (b) with ceramic filler (Al_2O_3).

capable of exhibiting an ionic conductivity of 3.5×10^{-5} S cm^{-1}, which gets raised to 7.20×10^{-5} S cm^{-1} by the presence of Al_2O_3. The improved ionic conductivity was attributed to the presence of surface ions and charges from Al_2O_3, even though the ionic conductivity observed here was much lower than the value reported by Rajendran et al. [31]. This might be due to the strong influence of TiO_2 on the greater flexibility of the amorphous phase or the less-ordered regions formed in the polymer electrolyte [31] than that offered by the Al_2O_3 fillers. But for both of the systems after the optimum concentration of the filler, a decline in ionic conductivity and electrochemical performance was observed. This, caused by the agglomeration or nonuniform dispersion of the fillers in the polymer matrix, leads to the formation of continuous and non-conducting phase in the composite. The fillers can act as electrically inert components and will block the movement of lithium ions, resulting in an increase in the total resistance within PEs. So, the concentration of filler is always important in determining the performance of a composite polymer electrolyte.

7.3.2 Polyacrylonitrile-Based Polymer Electrolytes Prepared by Phase Inversion

The phase-inversion method involves the controlled transformation of a polymer from the liquid phase to a solid phase. Phase inversion by immersion precipitation is widely employed for the fabrication of the polymer electrolyte membrane. Similar to other polymer electrolyte fabrication techniques, the electrochemical properties of phase-inversion membranes also depend on the salt and the solvent/non-solvent used. The influence of salts over the electrochemical performance of the phase-inversion membrane was studied by using three different salts, along with the EC/DMC combination of organic solvents. Among the three different systems, $LiPF_6$ in EC/DMC,

PAN-Based Polymer Electrolytes

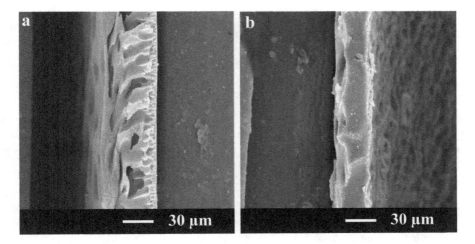

FIGURE 7.3 Scanning electron micrographs of the cross-sectional morphology of: (a) as-prepared PAN membrane and (b) the PAN membrane after gelling with 1 M LiPF$_6$ in EC/DMC (1:1 v/v). Adapted and reproduced with permission from Ref. [33]. Copyright © 2003 Electrochemical Society.

LiClO$_4$ in EC/DMC and LiBF$_4$ in EC/DMC, gel polymer electrolyte (GPE) prepared by LiPF$_6$–EC/DMC forms a 68% porous membrane (Figure 7.3), with an ionic conductivity of 2.8×10^{-3} S cm^{-1}. Additionally, it can exhibit an electrochemical stability of 5 V [33]. The high ionic conductivity of the LiClO$_4$ is attributed to its low dissociation energy, which leads to the dissociation of the salt into Li$^+$-ions at a fast rate [34]. At low temperature, the freezing of solvent occurs, and the EC/DMC domain exists as a crystalline solvent. Even though the GPE prepared with LiClO$_4$-EC/DMC can exhibit a high ionic conductivity, even at −20°C (6×10^{-4} S cm^{-1}), due to the strong interaction that occurs in polymer–solvent–salt mixtures, which hinders the ordering of the solvent molecules [33].

The major problem related to the phase-inversion membrane is its brittleness, which will retard its flexibility, ionic transport properties and electrochemical performance. To improve the flexibility of the phase-inversion membrane, the copolymer of hydroxyethyl acrylate (HEA) and acrylonitrile (AN) is used as the matrix for the preparation of GPE. Wu et al. [35] reported a high ionic conductivity of 3.66×10^{-3} S cm^{-1}, using polyhydroxy ethyl acrylate-*co*-acrylonitrile (PHEA-*co*-AN) based GPE. The dependence on solution concentration of the membrane morphology was studied. The porosity of the membrane prepared with 40 wt.% of (PHEA-*co*-AN) in DMF was highest, exhibiting a large number of honeycomb-like structures beneath the upper layer (skin layer), with small pores inside the large pores. The membrane exhibits a large number of finger-like cavities beneath the upper layer (Figure 7.4a, b). While investigating the effect of polymer concentration over the membrane morphology, the pore size appears to decrease from 10 to 2 µm (Figure 7.4c, d) in response to increasing polymer concentration. The small pore

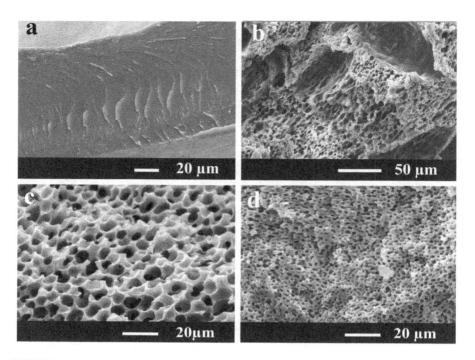

FIGURE 7.4 FE-SEM images visualizing: (a) and (b) the cross-sectional morphology of membrane for (PHEA-*co*-AN) porous membrane prepared from 40 wt.% solution concentration, and (c, d) the inner structure of porous membrane prepared with different solution concentration (wt.%) (c) 35 or (d) 40 solution concentration. Adapted and reproduced with permission from Ref. [35]. Copyright © 2007 Elsevier.

size, high porosity and the honeycomb-like porestructure is beneficial for absorbing and retaining large amounts of electrolyte solution [35].

As a consequence of the highly flexible and porous structure, the electrochemical performance of this membrane is better than that reported by Min et al. [33]. The membrane is capable of exhibiting an electrochemical stability of 4.6 V, which is higher than that of the PAN-based phase-inversion membrane reported by Min et al. [33]. The ion transport properties and electrochemical performances reported for the blends and composites of PAN-based PEs are comparable. The transport property studies on blend polymer electrolyte of PAN and polyvinylidene-*co*-hexafluoropropylene (PVdF-*co*-HFP) revealed that ionic conductivity is increasing with PAN content or $LiClO_4$ concentrations. The highest ionic conductivity (3.41×10^{-3} S cm^{-1} at 25°C) was observed for PAN/PVdF-*co*-HFP (50:50) containing 4 M $LiClO_4$. The Li/LiSr$_{0.25}$Mn$_{1.75}$O$_4$ cell fabricated with this blend electrolytes exhibits a discharge capacity of 135 mAh g^{-1} at 0.1 C which is similar to the capacity of the cell with liquid electrolyte [36]. A PAN/PVdF/SiO$_2$ composite membrane, fabricated using the phase-inversion method, exhibits a similar room temperture ionic conductivity of 3.32×10^{-3} S cm^{-1} (vs. Li/Li$^+$) [24]. In addition the incorporation of filler increases

PAN-Based Polymer Electrolytes

the electrochemical stability of the PE from 4.6 [36] to 5 V [24], which shows the advantageous effect of fillers over the performance of electrolyte.

7.3.3 POLYACRYLONITRILE-BASED POLYMER ELECTROLYTES PREPARED BY ELECTROSPINNING

Electrospinning is a vital technique for the fabrication of fibrous PAN-based polymer electrolytes. The conventional methods, like solvent-casting and phase-inversion methods, produce brittle PAN-based membranes due to the fact that the interaction of adjacent cyanide groups (–C=N) increases the resistance of interior rotation of the main chain and thus decreases the flexibility of the main chain. This issue can be resolved by using the electrospinning technique, which produces flexible and highly porous (>85%) membranes with good mechanical strength. When electrospinning is used, the highly polar nitrile group present in the PAN will hinder the alignment of macromolecules and results in the formation of flexible porous polymer electrolytes [37]. The schematic illustration of the electrospinning process is shown in Figure 7.5. There are several studies reported on the PAN-based polymer electrolytes prepared by the electrospinning method. The electrospinning parameters

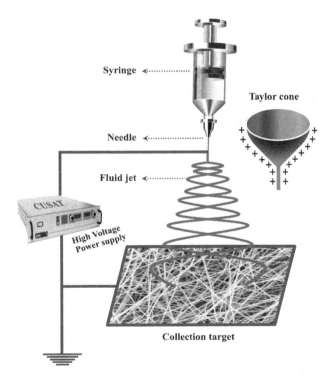

FIGURE 7.5 Schematic diagram on the visualization of the process and machine set-up of electrospinning.

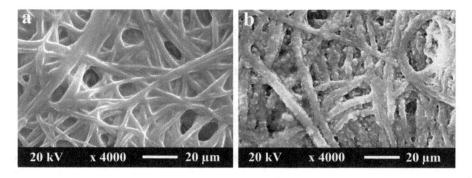

FIGURE 7.6 FE-SEM images of the surface morphology of electrospun polyacrylonitrile fibrous membrane soaked in 1 M LiPF$_6$ in EC/DMC (1:1 v/v) (a) after 1 h and (b) after 750 h. Adapted and reproduced with permission from Ref. [38]. Copyright © 2011 Elsevier.

have greater influence over the morphology of the resulting fiber structure. For the PAN-based PE membranes, the dependence on the electrospinning parameters was studied by Raghavan et al. [38] and Carol et al. [39]. In the Raghavan et al. [38] study, a solution concentration of approximately 16 wt.% of PAN in DMF and an applied voltage of 20 kV were found to be optimal for obtaining the membrane with most uniform fiber morphology. This membrane exhibits high electrolyte uptake, which is attributed to the highly porous nature (>85%) of the membrane. The high gelation of the membrane exhibits its affinity towards the liquid electrolytes, associated with the presence of polar functional groups, as is evident from Figure 7.6. Moreover, this PE is capable of exhibiting an ionic conductivity of >2×10^{-3} S cm^{-1} as well as an oxidative stability >4.7 V. On evaluating the battery performance using a lithium iron phosphate cathode (LiFePO$_4$, or simply LFP), the cell exhibited a capacity 150 mAh g^{-1}, which is 88% of the theoretical capacity. In a similar study carried out by Carol et al. [39], a lower ionic conductivity was reported of about 1.7×10^{-5} S cm^{-1} at 20°C, with the membrane prepared by using 10 wt.% PAN in DMF.

7.4 POLYACRYLONITRILE-BASED POLYMER BLEND ELECTROLYTES

Blends of PAN with PMMA [40], PS [41], PVdF [42] or PVdF-co-HFP [43] have been widely studied for the preparation of PEs in LIBs, which are useful to combine the synergistic advantages of each of the components. The first multi-component polymer blends and composites were proposed by Prasanth et al. [43], based on PVdF-co-HFP and PAN [43]. This group also demonstrated tailor made gel polymer electrolyte of PAN/PMMA/PS [44], PAN/PEO/PAN, PAN/PMMA/PAN and PAN/PVAc/PAN having layer by layer structure. PMMA provides swelling or gelation rather than dissolution, whereas PAN helps to impart mechanical stability and PS is introduced to provide the amorphous path for providing toughness as well as for the ease of ion transportation PAN/PMMA/PS, in the ratios 80:10:10, 90:05:05, 90:10:00 or 90:00:10, were used to fabricate individual electrolyte membranes, With increasing PMMA/PS concentration, the resulting membrane shows irregularity and non-uniformity. The lithium-ion cell assembled in half cell configuration with polymer membrane of PAN/PMMA/PS (80:10:10) exhibited a discharge capacity of 122 mAh g^{-1}, which is close to the theoretical maximum value of LiMn$_2$O$_4$, and is superior to the other combination of blends [44]. By exploring continuous electrospinning technique, Prasanth et al. [45] developed a tailor-made electrospun multicomponent gel polymer electrolyte by sandwiching electrospun PEO or PVAc or PMMA layers in between the electrospun PAN membrane. Among the three, PAN/PVAc/PAN layer by layer membrane activated with 1 M LiPF$_6$ in EC/DMC (1:1 v/v) exhibits high ionic conductivity of 4.72 × 10^{-3} S cm^{-1} at 25°C, which is attributed to the combination effect of high electrolyte content at the pores and good affinity of PVAc to the liquid electrolyte. The same electrolyte, PAN/PVAc/PAN, exhibits high electrochemical stability of 4.6 V (Li/Li$^+$) and an initial discharge capacity of 145 mAh g^{-1} at 0.1 C in Li/LFP cell, which corresponds to 85% of the theoretical capacity of the cathode. A good capacity retention was observed for cell, that persists about 140 mAh g^{-1} of discharge capacity after 50 cycles. Following Prasanth et al. [44], Tan et al. [41] developed a tertiary blend based on PAN, along with thermoplastic polyurethane (TPU) and polystyrene (PS). Electrospinning is carried out with a uniform solution of PAN:TPU:PS (5:5:1) in DMF. The presence of TPU enhances the ionic conductivity and PS provides increased mechanical stability. The blended membranes show smooth and completely stretched fiber morphology (Figure 7.7), with good fatigue resistance, mechanical strength and spatial stability.

Similarly, binary systems of PAN with PMMA [46, 47] or PVC [48] have also been reported. In the binary blend system, PAN/PMMA blends have been extensively studied. It has been reported that PAN/PMMA GPEs with an acrylonitrile to methyl methacrylate ratio of 4:1 exhibit high ionic conductivity and good electrochemical properties [46, 47]. A polymer blend electrolyte of PAN/PMMA was prepared by the incorporation of room temperature ionic liquid (RTIL) N-methyl N-butyl pyrrolidinium bis(trifluoromethanesulfonyl)imide (PYR$_{14}$TFSI), and exhibited a room temperature ionic conductivity of 3.55×10^{-3} S cm^{-1} and an electrochemical stability greater than 5 V. Discharge capacities of 139, 134, 120 and 101 mAh g^{-1} were observed for this system tested in Li/GPE/LFP cells at a discharge rate of 0.1, 0.2, 0.5 and 1 C, respectively [25]. Zhong et al. [48] developed PAN/PVC (8:2 w/w)-based electrolytes, which exhibited an ionic conductivity of 1.05×10^{-3} S cm^{-1} at 25°C, lower than that of the PAN/PMMA blend [46, 47].

7.5 POLYACRYLONITRILE-BASED CERAMIC COMPOSITE POLYMER ELECTROLYTES

Different inorganic fillers, such as aluminum oxide, silica, lanthanum titanate oxide and lithium aluminum titanium phosphate, were incorporated into the PAN and PAN-based blend electrolytes to enhance their Li$^+$-ion transport properties and electrochemical performance. The PAN/SiO$_2$ system exhibited good dimensional stability up to 150°C (Figure 7.8), ionic conductivity and electrochemical properties. The study revealed electrolyte uptake and ionic conductivity of the composite polymer electrolyte were markedly increased with increasing SiO$_2$ content [49]. Similar

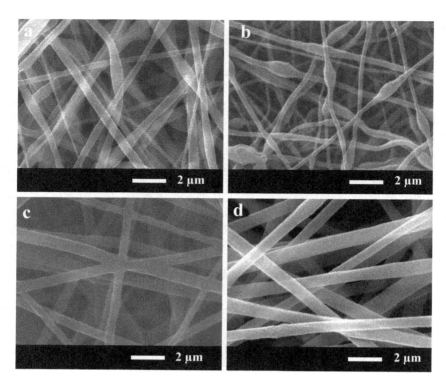

FIGURE 7.7 SEM images of the surface morphology of the PAN-based polymer blend electrospun membranes with different weight ratios (wt.%): (a) PAN/PS (5:1), (b) PAN/TPU (5:5), (c) TPU/PS (5:1) and (d) PAN/TPU/PS (5:5:1). Adapted and reproduced with permission from Ref. [41]. Copyright © 2019 Taylor and Francis.

results can also be seen in the aluminum oxide-containing composite polymer electrolyte system. An electrolyte uptake of more than 400% was exhibited by all the membranes, with a maximum uptake of 561% being shown by the membrane with 40 wt.% alumina content [50]. A composite polymer electrolyte (PAN/Al$_2$O$_3$), which contained 20 wt.% alumina and triethylene glycol diacetate-2-propenoic acid butyl ester (TEGDA-BA) as the liquid organic electrolyte, exhibited an ionic conductivity of 2.35×10^{-3} S cm^{-1} at 25°C and an electrochemical stability greater than 4.5 V. It also showed a discharge capacity of 240.4 mAh g^{-1}, when the cell is cycled with half-cell-configuration containing electrodes such as Li [Li$_{1/6}$Ni$_{1/4}$Mn$_{7/12}$]O$_{7/4}$F$_{1/4}$. Lithium lanthanum titanate (LLTO), which is an ionically conducting material, can provide high bulk Li$^+$-ion conductivity (10^{-3} S cm^{-1}). The PAN-based composite polymer electrolyte system, containing 15 wt.% of LLTO and having particle size of about 20 nm, exhibits very low interfacial resistance, which arises because of the ability to trap the impurities present in the liquid electrolyte. The nanocomposite polymer electrolyte prepared by incorporation of LLTO into PAN exhibited a discharge capacity of about 160 mAh g^{-1}, in a Li/LFP cell, and it retains this capacity for up to 50 cycles.

PAN-Based Polymer Electrolytes

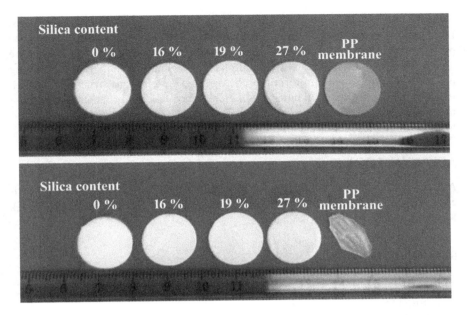

FIGURE 7.8 Photographs of SiO$_2$/PAN hybrid nanofiber membranes with different SiO$_2$ contents and microporous PP membrane: (a) before and (b) after thermal exposure at 150°C for 30 min. Adapted and reproduced with permission from Ref. [48]. Copyright © 2016 Elsevier.

Similar electrochemical characteristics were observed in PAN-based nanocomposite polymer electrolytes incorporated with glass ceramics such as lithium aluminum titanium phosphate (LATP) and lithium aluminum germanium phosphate (LAGP). The Li/PE/LFP cell with PAN/LATP composite electrolyte exhibits an initial discharge capacity of between 145 and 165 mAh g^{-1} with a coulombic efficiency of 93%, which higher than that achieved with Celgard® separators [51].

7.6 CONCLUSION

Electrolytes based on PAN have been extensively studied for their use in lithium-ion batteries. Conventional and other methods have been used for the fabrication of these electrolytes. The ionic conductivity exhibited by the solvent-cast membranes appears to be lower than that of the liquid electrolytes. Blending with other polymers, such as PVA, PVC PEO, and PMMA, are used to enhance the electrochemical performance of the solvent-cast polymer electrolytes; in addition, fillers are used for the same purpose. Fillers can enhance the amorphous content of the polymer and thereby enhance the free movement of ions through the polymer. Phase-inversion methods provide a much more porous structure than membranes prepared by the solvent-casted method. The porous structures help to entrap the liquid electrolyte. Blending of polymer and composite formation are effective methods to improve the ionic conductivity and electrochemical properties of the polymer electrolhytes. Electrospinning is one of the advanced techniques that can be used to unravel the problems caused by conventional methods used for the preparation of electrolyte membranes, such as solvent-casting and phase-inversion methods. Electrospinning produces thin fibers with uniform morphology and large porous structures, providing flexible electrolytes with much better electrochemical performance. Using these methods, PAN-based polymer membranes with enhanced electrochemical performance can be prepared to use in LIBs.

ACKNOWLEDGMENT

Authors Dr. Jabeen Fatima M. J. and Dr. Prasanth Raghavan, would like to acknowledge Kerala State Council for Science, Technology and Environment (KSCSTE), Kerala, India, for financial assistance.

REFERENCES

1. Wen J, Yu Y, Chen C (2012) A review on lithium-ion batteries safety issues: Existing problems and possible solutions. *Mater Express* 2:197–212. https://doi.org/10.1166/mex.2012.1075
2. Liu K, Liu Y, Lin D, et al. (2018) Materials for lithium-ion battery safety. *Sci Adv* 4:. https://doi.org/10.1126/sciadv.aas9820
3. Howey DA, Alavi SMM (2015) Rechargeable battery energy storage system design. *Handb Clean Energy Syst* 1–18. https://doi.org/10.1002/9781118991978.hces212
4. Kim YT, Smotkin ES (2002) The effect of plasticizers on transport and electrochemical properties of PEO-based electrolytes for lithium rechargeable batteries. *Solid State Ionics* 149:29–37. https://doi.org/10.1016/S0167-2738(02)00130-3
5. Krejza O, Velická J, Sedlaříková M, Vondrák J (2008) The presence of nanostructured Al_2O_3 in PMMA-based gel electrolytes. *J Power Sources* 178:774–778. https://doi.org/10.1016/j.jpowsour.2007.11.018
6. Stephan AM, Nahm KS, Anbu Kulandainathan M, et al. (2006) Poly(vinylidene fluoride-hexafluoropropylene) (PVdF-HFP) based composite electrolytes for lithium batteries. *Eur Polym J* 42:1728–1734. https://doi.org/10.1016/j.eurpolymj.2006.02.006
7. Huang H, Wunder SL (2001) Preparation of microporous PVDF based polymer electrolytes. *J Power Sources* 97–98:649–653. https://doi.org/10.1016/S0378-7753(01)00579-1
8. Ramesh S, Yahaya AH, Arof AK (2002) Dielectric behaviour of PVC-based polymer electrolytes. *Solid State Ionics* 152–153:291–294. https://doi.org/10.1016/S0167-2738(02)00311-9
9. Rohan R, Sun Y, Cai W, et al. (2014) Functionalized polystyrene based single ion conducting gel polymer electrolyte for lithium batteries. *Solid State Ionics* 268:294–299. https://doi.org/10.1016/j.ssi.2014.10.013
10. Forsyth M, MacFarlane DR, Hill AJ (2000) Glass transition and free volume behaviour of poly(acrylonitrile)/$LiCF_3SO_3$ polymer-in-salt electrolytes compared to poly(ether urethane)/$LiClO_4$ solid polymer electrolytes. *Electrochim Acta* 45:1243–1247. https://doi.org/10.1016/S0013-4686(99)00387-4
11. Sylla S, Sanchez JY, Armand M (1992) Electrochemical study of linear and crosslinked POE-based polymer electrolytes. *Electrochim Acta* 37:1699–1701. https://doi.org/10.1016/0013-4686(92)80141-8

12. Jun MS, Choi YW, Kim JD (2012) Solvent casting effects of sulfonated poly (ether ether ketone) for Polymer electrolyte membrane fuel cell. *Journal of Membrane Science* 396:32–37.
13. Magistris A, Mustarelli P, Parazzoli F, et al. (2001) Structure, porosity and conductivity of PVdF films for polymer electrolytes. In: *Journal of Power Sources*. pp 657–660
14. Magistris A, Quartarone E, Mustarelli P, et al. (2002) PVDF-based porous polymer electrolytes for lithium batteries. 153:347–354
15. Wu N, Cao Q, Wang X, et al. (2011) A novel high-performance gel polymer electrolyte membrane basing on electrospinning technique for lithium rechargeable batteries. 196:8638–8643. https://doi.org/10.1016/j.jpowsour.2011.04.062
16. Lee SW, Choi SW, Jo SM, et al. (2006) Electrochemical properties and cycle performance of electrospun poly(vinylidene fluoride)-based fibrous membrane electrolytes for Li-ion polymer battery. *J Power Sources* 163:41–46. https://doi.org/10.1016/j.jpowsour.2005.11.102
17. Gupta AK, Paliwal DK, Bajaj P (1998) Melting behavior of acrylonitrile polymers. *J Appl Polym Sci* 70:2703–2709. https://doi.org/10.1002/(sici)1097-4628(19981226)70:13<2703::aid-app15>3.3.co;2-u
18. Huang B, Wang Z, Chen L, et al. (1996) The mechanism of lithium ion transport in polyacrylonitrile-based polymer electrolytes. *Solid State Ionics* 91:279–284. https://doi.org/10.1016/s0167-2738(96)83030-x
19. Long L, Wang S, Xiao M, Meng Y (2016) Polymer electrolytes for lithium polymer batteries. *J Mater Chem A* 4:10038–10039. https://doi.org/10.1039/c6ta02621d
20. Watanabe M, Kanba M, Nagaoka K, Shinohara I (1983) Ionic conductivity of hybrid films composed of polyacrylonitrile, ethylene carbonate and LiClO$_4$. *J Polym Sci Part A-2, Polym Phys* 21:939–948. https://doi.org/10.1002/pol.1983.180210610
21. Yoon HK, Chung WS, Jo NJ (2004) Study on ionic transport mechanism and interactions between salt and polymer chain in PAN based solid polymer electrolytes containing LiCF 3so3. In: *Electrochimica Acta*. pp 289–293
22. Vincent CA (1989) Polymer electrolytes. *Chem Br* 25:. https://doi.org/10.1146/annurev-matsci-071312-121705
23. Wu CG, Lu MI, Chuang HJ (2005) PVdF-HFP/P123 hybrid with mesopores: A new matrix for high-conducting, low-leakage porous polymer electrolyte. *Polymer (Guildf)* 46:5929–5938. https://doi.org/10.1016/j.polymer.2005.05.077
24. Liu L, Wang Z, Zhao Z, et al. (2016) PVDF/PAN/SiO 2 polymer electrolyte membrane prepared by combination of phase inversion and chemical reaction method for lithium ion batteries. *J Solid State Electrochem* 20:699–712. https://doi.org/10.1007/s10008-015-3095-1
25. Rao M, Geng X, Liao Y, et al. (2012) Preparation and performance of gel polymer electrolyte based on electrospun polymer membrane and ionic liquid for lithium ion battery. *J Memb Sci* 399–400:37–42. https://doi.org/10.1016/j.memsci.2012.01.021
26. Krishna Jyothi N, Venkataratnam KK, Narayana Murty P, Vijaya Kumar K (2016) Preparation and characterization of PAN-KI complexed gel polymer electrolytes for solid-state battery applications. *Bull Mater Sci* 39:1047–1055. https://doi.org/10.1007/s12034-016-1241-8
27. Watanabe M, Kanba M, Nagaoka K, Shinohara I (1982) Ionic conductivity of hybrid films based on polyacrylonitrile and their battery application. *J Appl Polym Sci* 27:4191–4198. https://doi.org/10.1002/app.1982.070271110
28. Rajendran S, Babu RS, Sivakumar P (2007) Effect of salt concentration on poly (vinyl chloride)/poly (acrylonitrile) based hybrid polymer electrolytes. *J Power Sources* 170:460–464. https://doi.org/10.1016/j.jpowsour.2007.04.041

29. Flora XH, Ulaganathan M, Babu RS, Rajendran S (2012) Evaluation of lithium ion conduction in PAN/PMMA-based polymer blend electrolytes for Li-ion battery applications. *Ionics (Kiel)* 18:731–736. https://doi.org/10.1007/s11581-012-0690-3
30. Subramania A, Kalyana Sundaram NT, Vijaya Kumar G, Vasudevan T (2006) New polymer electrolyte based on (PVA-PAN) blend for Li-ion battery applications. *Ionics (Kiel)* 12:175–178. https://doi.org/10.1007/s11581-006-0018-2
31. Rajendran S, Babu RS, Sivakumar P (2008) Investigations on PVC/PAN composite polymer electrolytes. *J Memb Sci* 315:67–73. https://doi.org/10.1016/j.memsci.2008.02.007
32. Kim M, Lee L, Jung Y, Kim S (2013) Study on ion conductivity and crystallinity of composite polymer electrolytes based on poly(ethylene oxide)/poly(acrylonitrile) containing nano-sized Al2O3 Fillers. *J Nanosci Nanotechnol* 13:7865–7869. https://doi.org/10.1166/jnn.2013.8107
33. Min HS, Ko JM, Kim DW (2003) Preparation and characterization of porous polyacrylonitrile membranes for lithium-ion polymer batteries. *J Power Sources* 119–121:469–472. https://doi.org/10.1016/S0378-7753(03)00206-4
34. Gurusiddappa J, Madhuri W, Padma SR, Priya Dasan K (2016) Studies on the morphology and conductivity of PEO/LiClO$_4$. *Mater Today Proc* 3:1451–1459. https://doi.org/10.1016/j.matpr.2016.04.028
35. Wu G, Yang HY, Chen HZ, et al. (2007) Novel porous polymer electrolyte based on polyacrylonitrile. *Mater Chem Phys* 104:284–287. https://doi.org/10.1016/j.matchemphys.2007.03.013
36. Subramania A, Sundaram NTK, Kumar GV (2006) Structural and electrochemical properties of micro-porous polymer blend electrolytes based on PVdF-co-HFP-PAN for Li-ion battery applications. *J Polym Sci Part B Polym Phys* 153:177–182. https://doi.org/10.1016/j.jpowsour.2004.12.009
37. Chen R, Hu Y, Shen Z, et al. (2017) Facile fabrication of foldable electrospun polyacrylonitrile-based carbon nanofibers for flexible lithium-ion batteries. *J Mater Chem A* 5:12914–12921. https://doi.org/10.1039/c7ta02528a
38. Raghavan P, Manuel J, Zhao X, et al. (2011) Preparation and electrochemical characterization of gel polymer electrolyte based on electrospun polyacrylonitrile nonwoven membranes for lithium batteries. *J Power Sources* 196:6742–6749. https://doi.org/10.1016/j.jpowsour.2010.10.089
39. Carol P, Ramakrishnan P, John B, Cheruvally G (2011) Preparation and characterization of electrospun poly(acrylonitrile) fibrous membrane based gel polymer electrolytes for lithium-ion batteries. *J Power Sources* 196:10156–10162. https://doi.org/10.1016/j.jpowsour.2011.08.037
40. Wang SH, Kuo PL, Hsieh C Te, Teng H (2014) Design of poly(acrylonitrile)-based gel electrolytes for high-performance lithium ion batteries. *ACS Appl Mater Interfaces* 6:19360–19370. https://doi.org/10.1021/am505448a
41. Tan L, Deng Y, Cao Q, et al. (2019) Gel electrolytes based on polyacrylonitrile/thermoplastic polyurethane/polystyrene for lithium-ion batteries. *Ionics (Kiel)* 25:3673–3682. https://doi.org/10.1007/s11581-019-02940-7
42. Gopalan AI, Santhosh P, Manesh KM, et al. (2008) Development of electrospun PVdF-PAN membrane-based polymer electrolytes for lithium batteries. *J Memb Sci* 325:683–690. https://doi.org/10.1016/j.memsci.2008.08.047
43. Raghavan P, Zhao X, Shin C, et al. (2010) Preparation and electrochemical characterization of polymer electrolytes based on electrospun poly(vinylidene fluoride-co-hexafluoropropylene)/polyacrylonitrile blend/composite membranes for lithium batteries. *J Power Sources* 195:6088–6094. https://doi.org/10.1016/j.jpowsour.2009.11.098

44. Lim DH, Haridas AK, Figerez SP, Raghavan P, Matic A, Ahn JH (2018) Tailor-made electrospun multilayer composite polymer electrolytes for high-performance lithium polymer batteries. *Journal of Nanoscience and Nanotechnology* 18(9):6499–6505.
45. Prasanth R, Aravindan V, Srinivasan M (2012) Novel polymer electrolyte based on cob-web electrospun multi component polymer blend of polyacrylonitrile/poly(methyl methacrylate)/polystyrene for lithium ion batteries – preparation and electrochemical characterization. *J Power Sources* 202:299–307. https://doi.org/10.1016/j.jpowsour.2011.11.057
46. Rao MM, Liu JS, Li WS, et al. (2008) Preparation and performance analysis of PE-supported P(AN-co-MMA) gel polymer electrolyte for lithium ion battery application. *J Memb Sci* 322:314–319. https://doi.org/10.1016/j.memsci.2008.06.004
47. Pu W, He X, Wang L, et al. (2006) Preparation of P(AN-MMA) microporous membrane for Li-ion batteries by phase inversion. *J Memb Sci* 280:6–9. https://doi.org/10.1016/j.memsci.2006.05.028
48. Zhong Z, Cao Q, Jing B, et al. (2012) Novel electrospun PAN – PVC composite fibrous membranes as polymer electrolytes for polymer lithium-ion batteries. 853–859. https://doi.org/10.1007/s11581-012-0682-3
49. Yanilmaz M, Lu Y, Zhu J, Zhang X (2016) Silica/polyacrylonitrile hybrid nanofiber membrane separators via sol-gel and electrospinning techniques for lithium-ion batteries. *J Power Sources* 313:205–212. https://doi.org/10.1016/j.jpowsour.2016.02.089
50. Wang Q, Song W, Fan L, Song Y (2015) Facile fabrication of polyacrylonitrile / alumina composite membranes based on triethylene glycol diacetate-2-propenoic acid butyl ester gel polymer electrolytes for high-voltage lithium-ion batteries. *J Memb Sci* 486:21–28. https://doi.org/10.1016/j.memsci.2015.03.022
51. Liang Y, Lin Z, Qiu Y, Zhang X (2011) Fabrication and characterization of LATP/PAN composite fiber-based lithium-ion battery separators. *Electrochim Acta* 56:6474–6480. https://doi.org/10.1016/j.electacta.2011.05.007

8 Polymer Blend Electrolytes for High-Performance Lithium-Ion Batteries

Jishnu N. S., Neethu T.M. Balakrishnan, Anjumole P. Thomas, Akhila Das, Jou-Hyeon Ahn, Jabeen Fatima M. J., and Prasanth Raghavan

CONTENTS

8.1 Introduction 168
8.2 Polymer Blend Electrolytes 169
    8.2.1 PVdF and PVdF-*co*-HFP-Based Polymer Blend Electrolytes 169
    8.2.2 Polymethyl Methacrylate (PMMA Based) Polymer Blend Electrolytes 169
        8.2.2.1 Polymethyl Methacrylate (PMMA Based) Polymer Blend Electrolytes by Solvent Casting 170
        8.2.2.2 Polymethyl Methacrylate (PMMA Based) Polymer Blend Electrolytes by Phase Inversion 171
        8.2.2.3 PMMA Based Polymer Blend Electrolytes by Electrospinning 171
    8.2.3 Polyethylene Oxide (PEO Based) Polymer Blend Electrolytes 172
        8.2.3.1 Polyethylene Oxide (PEO)-Based Polymer Blend Electrolytes by Phase Inversion 173
        8.2.3.2 Polyethylene Oxide (PEO Based) Polymer Blend Electrolytes by Electrospinning 174
    8.2.4 Polyvinyl Chloride (PVC Based) Polymer Blend Electrolytes 176
        8.2.4.1 Polyvinyl Chloride (PVC Based) Polymer Blend Electrolyte by Solvent Casting 177
        8.2.4.2 Polyvinyl Chloride (PVC Based) Polymer Blend Electrolyte by Electrospinning 179
8.3 Conclusion 181
Acknowledgment 181
References 181

8.1 INTRODUCTION

Polymer electrolytes are the most promising materials for meeting the never-ending demand for safer lithium-ion batteries (LIBs). Such non-inflammable and thermally stable polymer electrolytes can replace their liquid counterparts, which make up a vital part inside the battery. Different polymers used as polymer electrolytes (PEs) in LIBs include polyethylene oxide (PEO) [1], polymethyl methacrylate (PMMA) [2], polyvinylidene difluoride-co-hexafluoropropylene (PVdF-co-HFP) [3], polyacrylonitrile (PAN) [4], polyvinylidene fluoride (PVdF) [5], polyvinyl chloride (PVC) [6] and polystyrene (PS) [7], with more being studied. Each of the polymers has its own benefits as a matrix for polymer electrolytes. PMMA [2] is a thermoplastic material that exhibits a high interfacial stability with the lithium metal, and possesses high electrolyte uptake, high ionic conductivity and a good electrochemical stability [8]. PVAc [9] is another polymer matrix, that exhibits a lower-than-usual interfacial resistance and crystallinity. PEO [1] is a matrix that is extensively studied as a solid polymer electrolyte for LIBs [10]. PAN [4]-based electrolytes are extensively studied for use in LIBs due to their high thermal stability and flame-retardant behavior. PAN-based electrolytes are capable of providing an ionic conductivity between 10^{-5} and 10^{-3} S cm^{-1}. PVdF is a semi-crystalline polymer that exhibits a high dielectric constant. The presence of high electron-withdrawing functional groups further enhances the electrochemical performance of PVdF by enhancing the dissociation of lithium salts. Hexafluoropropylene (HFP) is an amorphous polymer that can incorporate with the VdF building block and modify the properties of the homopolymer, PVdF, the resulting copolymer being used for a wider range of applications. The presence of HFP can entrap more liquid electrolyte, thus improving the performance of the battery, while VdF provides the mechanical support. The copolymer formation achieves a reduction in the crystallinity of the PVdF-co-HFP, that will enhance ionic mobility through the system. Polyvinylpyrrolidone (PVP) is an amorphous polymer containing a (C–O) bond; moreover, this polymer readily dissolves in most organic solvents.

Even though each polymer matrix provides valuable properties for use as an electrolyte in LIBs, however, their ionic conductivity and electrochemical properties are inferior to those of liquid electrolytes. So, different modifications can be adopted to the polymer matrix, in which blending is considered to be one of the simplest methods that can be used for the improvement of ionic conductivity and electrochemical properties of the polymer electrolyte by making use of the synergy between the characteristics of each of the individual components. Different combinations of fillers are examined for use in LIBs that include binary and tertiary blend systems. Compared to the individual polymer matrix, augment in the electrochemical properties is observed for the electrolytes prepared by polymer blends. This chapter summarizes the different polymer blend-based electrolytes that have been studied for use in LIBs and summarizes the improvements made in terms of electrochemical performances compared with the use of PEs based on individual polymers in LIBs.

8.2 POLYMER BLEND ELECTROLYTES

In order to achieve good electrochemical performance in polymer electrolytes, different methods have been proposed. Among these various approaches, polymer blending and polymer nanocomposite preparation have been found to be the most feasible [11]. For the fabrication of polymer blend electrolytes, polymers having complementary properties are selected, such as one having good electrochemical properties and the other with good mechanical performance [12–14]. The ionic conductivity and electrochemical/physical properties exhibited by the polymer blend electrolytes are observed to be superior to those of the individual constituents. Two or more different polymers can form a uniform mixture to form polymer electrolytes with improved properties [15–17]. This chapter explains how the gel polymer electrolytes comprise two or more polymers, which can be advantageously combined to achieve beneficial electrochemical and physical properties from the individual polymers to enhance the battery performance of LIBs

8.2.1 PVdF AND PVdF-co-HFP-Based Polymer Blend Electrolytes

Polyvinylidene difluoride (PVdF) and its co-polymer polyvinylidene difluoride-co-hexafluoropropylene (PVdF-co-HFP) are thermoplastic fluoropolymers. PVdF is produced by the polymerization of vinylidene difluoride ($C_2H_2F_2$). PVdF possesses a greatest purity, as well as resistance to solvents, acids and hydrocarbons. It is widely employed as binder material in both anodes and cathodes in LIBs. PVdF exhibits good electrochemical stability in contact with electrolyte mixtures and with high affinity for liquid electrolytes. For high-voltage operations, NMP (N-methyl-2-pyrrolidone) is used as a solvent, along with PVdF. Since it has good stability in the electrolyte, it is well studied as a polymer matrix and is used for the fabrication of polymer electrolytes in LIBs. Additionally, different copolymers of PVdF, such as polyvinylidene difluoride-co-trifluoroethylene (PVdF-co-TrFE), polyvinylidene difluoride-co-hexafluoropropylene (PVdF-co-HFP) [18] and polyvinylidene fluoride-co-chlorotrifluoroethylene (PVdF-co-CTFE) [19], also exhibit better electrochemical performance as separators in the battery [15, 20–22]. They are capable of exhibiting excellent properties, such as high polarity, excellent thermal and mechanical properties, wettability by organic solvents and chemical inertness and stability in the cathodic environment, as well as possessing tailorable porosity through binary and ternary solvent/non-solvent systems [23, 24]. This polymer is partially fluorinated and semi-crystallized, with its amorphous phase located in between the crystalline lamellae arranged in spherulites [25, 26]. It involves different phases, in which the β-phase is important, since it exhibits ferroelectric, piezoelectric, and pyroelectric properties [25]. The strong electron-withdrawing fluoride group in PVdF provides a high dielectric constant (ε~8.4) [27] and thereby promotes the dissociation of lithium salts, which will be helpful in achieving the availability of a large number of charge carriers [27, 28].

8.2.2 Polymethyl Methacrylate (PMMA Based) Polymer Blend Electrolytes

Blend electrolytes can be fabricated by using solvent-casting, plasticizer extraction, phase-inversion and electrospinning methods. PMMA-based polymer blends have attracted much attention due to the high ionic conductivity and the good mechanical properties that they can deliver. Polymethyl methacrylate (PMMA) is a synthetic polymer resin formed by the polymerization of methyl methacrylate, $-(C_5O_2H_8)-$. It is a transparent and rigid thermoplastic material commonly used in shatterproof windows, skylights, illuminated signs and aircraft canopies. PMMA as a host for polymer electrolytes was first proposed by Iijima et al. [29] and, more recently, by Bohnke et al. [30] and Zhang et al. [4, 31]. The amorphous structure of PMMA is useful in providing high ionic conductivity, and it exhibits excellent interfacial stability with lithium metal. The pendant, $-COOCH_3$, is not likely to crystallize around Li$^+$ as PEO does. PMMA-based electrolytes exhibit high electrolyte uptake, high ionic conductivity and good electrochemical stability [32]. Additionally, these electrolytes have the ability to achieve chemical cross-linking, that will improve mechanical stability, along with electrolyte retention ability [33, 34]. Moreover, PMMA-based electrolytes can avoid lithium dendrite formation [33]. PMMA-based gel polymer electrolytes are insufficient to free-stand when coming into contact with a plasticizer and the PMMA fibers appear to be brittle. As a result, blending with other polymers is an alternative strategy to make them suitable for commercial applications. Blends of PMMA with PVdF [35, 36] or PVdF-co-HFP [37] appear to be the best choices to improve the ionic conductivity and electrochemical properties of PMMA and PVdF based polymer electrolytes.

8.2.2.1 Polymethyl Methacrylate (PMMA Based) Polymer Blend Electrolytes by Solvent Casting

Solvent casting is one of the widely used methods adapted for the preparation of porous polymer membranes. The membranes prepared by this simple method offer a porosity of about 40% and are used as the host membranes for the fabrication of polymer electrolytes in lithium-ion batteries. Rajendran et al. [38] prepared a PMMA/PVdF polymer blend membrane by the solvent-casting method, using dimethyl phthalate (DMP) and $LiCF_2SO_3$ as plasticizer and lithium salt, respectively. This blend is capable of exhibiting an ionic conductivity of 0.914×10^{-3} S cm^{-1} at 30°C [38], which is higher than that achieved by the PVdF-based polymer electrolytes containing the same lithium salt [39]. Using ethylene carbonate (EC) and propylene carbonate (PC) solvents, an enhanced ionic conductivity is observed between 10^{-3} to 10^{-2} S cm^{-1} with the same solvent-casting technique. In addition, the mechanical stability of this polymer blend is observed to be satisfactory over a wide range of temperatures [40]. While investigating the structural morphology of the blend, it is clear that the porous structures have a great influence over the electrochemical characterization. By combining the blending of PVdF-co-HFP along with the polymer dissolution method, an ionic conductivity of 10.23×10^{-3} S cm^{-1} at room temperature could be attributed by the enhanced porosity within the membrane [41]. The influence of the lithium salt concentration on battery performance was studied by Gohel et al. [42]. A concentration of 7.5 wt.% of $LiClO_4$ was observed to exhibit an ionic conductivity of about 2.83×10^{-4} S cm^{-1}, which is lower than that observed using the dissolution process [42].

8.2.2.2 Polymethyl Methacrylate (PMMA Based) Polymer Blend Electrolytes by Phase Inversion

Compared with membrane preparation by the solvent-casting technique, phase-inversion membranes are observed to exhibit more porous structures but only limited studies have been reported on PVdF/PMMA phase-inversion membranes. Different phase-inversion techniques are adopted for the fabrication of polymer electrolyte membranes based on PVdF- or PVdF-*co*-HFP-based PMMA blends. The concentration of PMMA has a great influence on the morphology of the membrane prepared. It is observed that, in the thermally induced phase-inversion method, the membrane morphology is changed from a cellular structure to a network structure with increasing PMMA content [43]. In addition, a maximum ionic conductivity of about 3.38×10^{-3} S cm^{-1} at 25°C was observed with a voltage stability window of up to 4.7 V, which is better than that achieved by solvent-cast membranes [42].

8.2.2.3 PMMA Based Polymer Blend Electrolytes by Electrospinning

Electrospinning is an advanced technique used for the synthesis of fibrous membranes with definite porous structures, that can effectively be used as electrolytes with better electrochemical properties. In the electrospinning technique, a polymer solution is forced through a spinneret under a high voltage to form fibers having nanometer diameters. A schematic visualization of the process of electrospinning is shown in Figure 8.1. An electrospun PVdF/PMMA (80:20 wt./wt.) membrane with fiber diameters of 183 nm [36] or 325 nm [35] was observed with a high spinning voltage, exhibiting an electrolyte uptake of 275% (1 M LiPF$_6$ in EC: DMC) [36] or 285% (1 M LiClO$_4$-PC) [35], even though the two membranes had similar porosity values of about 85%. The membrane activated with 1 M LiClO$_4$-PC exhibited a lower ionic conductivity, 2.95×10^{-3} S cm^{-1} [35], caused by the difference in affinity of the membrane for the different electrolytes. The average fiber diameters of PVdF and PVdF-PMMA (50:50) were found to be 647 nm and 179 nm, respectively (Figure 8.2). An electrolyte uptake of 290% (porosity 87%) and ionic conductivity of 0.15 S cm^{-1} was observed for this membrane at room temperature, which is higher than that of pure PVdF (porosity 78%, electrolyte uptake 260% and ionic conductivity 0.1 S cm^{-1}). The discharge capacity exhibited by this polymer electrolyte is about 144.7 and 150.3 mAh g^{-1} with PVdF/PMMA ratios of 80:20 and 50:50, respectively, so that the blend with 50:50 polymers was found to be a potential separator for LIBs. Relatively high cathode utilization was observed for the membrane, which was 3.5 times higher than that of the blend with 80:20 PVdF: PMMA [36].

The electrochemical performance of PVdF-*co*-HFP-based blends is observed to be higher than that of PVdF, due to the lower crystallinity of PVdF-*co*-HFP resulting from the presence of HFP in the copolymer. The PVdF-*co*-HFP/PMMA blend containing 66% PVdF-*co*-HFP exhibits an ionic conductivity of 2.95×10^{-3} S cm^{-1} [37], whereas the blend having only 50% PVdF shows an ionic conductivity of 15×10^{-2} S cm^{-1} [36]. Compared with the pure PVdF-*co*-HFP membrane, the PVdF-*co*-HFP/PMMA polymer blend membrane showed an uptake and leakage for electrolyte solutions of 377% and 87%, respectively, which is about 75% and 11% higher,

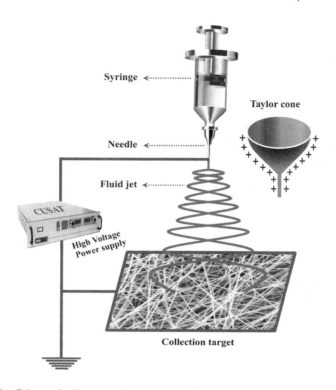

FIGURE 8.1 Schematic diagram of the process and machine set-up of electrospinning.

respectively, than the corresponding values of the pure PVdF-*co*-HFP membrane. The discharge capacity observed with the Li/PE/LiFePO$_4$ cell is about ~145 mAh g^{-1} and retains a capacity of 133.5 mAh g^{-1} (92% retention) after 150 cycles. It is observed to be superior to that of Celgard 2400, which appears to exhibit a fading discharge capacity, of about 115 mAh g^{-1} (79% retention) after 150 cycles. This improved performance of the PVdF-*co*-HFP/PMMA membrane is attributed to the sandwich structure of the cell and its higher porosity [37]

8.2.3 Polyethylene Oxide (PEO Based) Polymer Blend Electrolytes

Polyethylene oxide (PEO), also known as polyethylene glycol (PEG), is a single-chain single crystal, having the chemical formula H–(O–CH$_2$–CH$_2$)$_n$–OH, C$_{2n}$H$_{4n+2}$O$_{n+1}$, with a low glass transition temperature (Tg), good chain flexibility, good thermal properties and mechanical properties, superior electrochemical stability to lithium metal and excellent solubility with conductive lithium salts, is poised to be an enabler for solid-state lithium batteries, but its application is restricted by low ionic conductivity at room temperature and poor mechanical strength at elevated temperatures [44]. PEO exhibits a very low ionic conductivity 10^{-8}–10^{-6} S cm^{-1} [45, 46] due to its highly crystalline structure.

Polymer Blend Electrolytes for LIBs

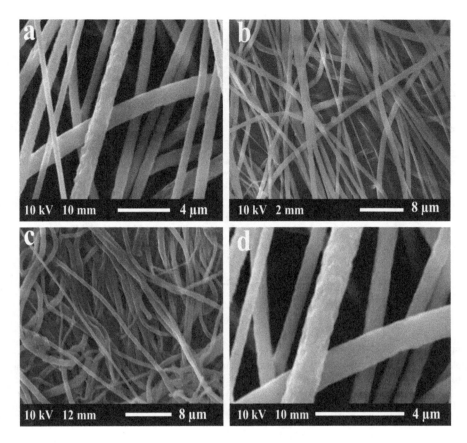

FIGURE 8.2 FE-SEM images on the surface morphology of the electrospun polymer blend nanofibrous membranes of: (a) and (d) pure PVdF, (b) PVdF/PMMA (80:20) and (c) PVdF/PMMA (50:50). Adapted and reproduced with permission from Ref. [36]. Copyright © 2018 Elsevier.

8.2.3.1 Polyethylene Oxide (PEO)-Based Polymer Blend Electrolytes by Phase Inversion

Phase inversion was successfully proposed for the fabrication of the PVdF/PEO polymer electrolyte membrane. The presence of PEO enhances the porosity, porous structure and pore connectivity of the PVdF microporous membranes. The modification of the porosity will help to enhance the room temperature ionic conductivity of the polymer electrolyte [47]. With an increase in the PEO content in the blend, a marked increase in the ionic conductivity was observed (Figure 8.3). But with the increase in PEO content, the mechanical strength of the membrane decreased from ~85 MPa to ~30 MPa [47]. The increase in ionic conductivity was a direct effect of the porous structure formed by the polymer membrane. The surface and cross-sectional morphology of the PVdF/PEO polymer blend membrane as a function of

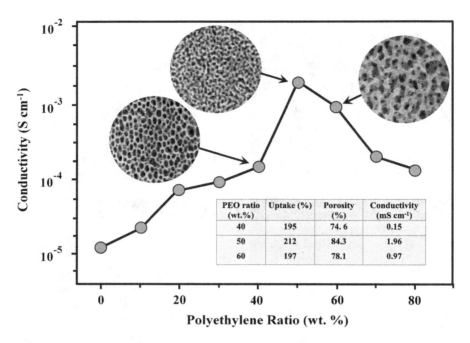

FIGURE 8.3 Room temperature (25°C) ionic conductivity of PVdF/PEO polymer blend electrolytes prepared by activating the polymer membrane with 1 mol L^{-1} LiClO$_4$ in PC as a function of PEO content. Inset shows cross-sectional morphology (FE-SEM images) of the PEO/PVdF polymer blend electrolytes. Adapted and reproduced with permission from Ref. [47]. Copyright © 2006 Elsevier.

PEO content is shown in FE-SEM images in Figure 8.4 and 8.5, respectively. When the PEO contents are low, the pores are not interconnected, as shown in Figure 8.4a-c (surface morphology) and Figure 8.5a and b (cross-sectional morphology), but, when the weight percentage of PEO exceeded 50%, well-interconnected porous structures were observed, as shown in Figure 8.4d-f (surface morphology) and Figure 8.5c-f (cross-sectional morphology) [47]. Moreover, the cell fabricated using this polymer electrolyte exhibited an initial discharge capacity of 150 mAh g^{-1} at 0.1 C.

8.2.3.2 Polyethylene Oxide (PEO Based) Polymer Blend Electrolytes by Electrospinning

Prasanth et al. [48] reported the polymer blend based on PVdF/PEO with and without ceramic fillers. The presence of PVdF enhanced the mechanical stability, while the ether linkage in PEO improved the segmental mobility. The surface morphology of the membrane is shown in Figure 8.6. The PVdF/PEO (5–20% PEO) was synthesized with [49] and without lithium aluminum germanium phosphate (LAGP) as the filler [49]. The blend containing the ceramic filler was washed with boiling water to preferentially remove PEO [49]. The surface morphology of the electrospun membrane before and after *in-situ* porosity generation is shown in Figure 8.7. Figure 8.8 shows

Polymer Blend Electrolytes for LIBs 175

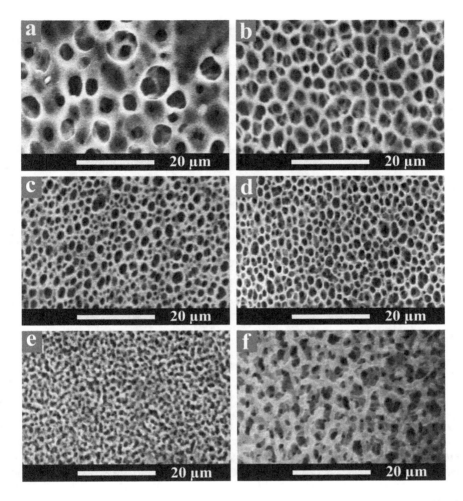

FIGURE 8.4 FE-SEM images of the surface morphology of PVdF/x% PEO polymer blend membranes: (a) x=0 (pure PVdF), (b) x=10, (c) x=20, (d) x=40, (e) x=50 and (f) x=60. Adapted and reproduced with permission from Ref. [47]. Copyright © 2006 Elsevier.

the surface morphology of the PVdF/PEO blend membranes before and after soaking in hot water.

By immersing the membrane in hot water (70°C) for 48 h, PEO was preferentially removed by forming nano-sized pores, that improved the electrochemical performance by increasing electrolyte retention. Using LiTFSI, the initial discharge capacity found was about 130 to 150 mAh g^{-1}. Even after 50 cycles, discharge capacities of about 126, 137 and 150 mAh g^{-1} were observed for the PVdF/PEO blend electrolyte prepared with 5, 10 or 20 wt.% PEO, respectively, after soaking in hot water for the preferential leaching treatment [49].

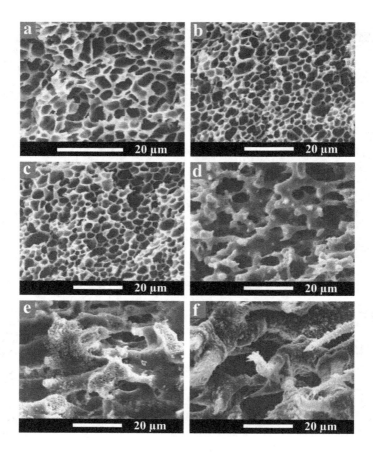

FIGURE 8.5 FE-SEM images of the cross-sectional morphology of PVdF/x% PEO polymer blend membranes: (a) x=0 (pure PVdF), (b) x=20, (c) x=40, (d) x=50, (e) x=60 and (f) x=80. Adapted and reproduced with permission from Ref. [47]. Copyright © 2006 Elsevier.

8.2.4 POLYVINYL CHLORIDE (PVC BASED) POLYMER BLEND ELECTROLYTES

Polyvinyl chloride is a widely used synthetic thermoplastic polymeric material, with the empirical formula $(C_2H_3Cl)_n$. After polyethylene and polypropylene, PVC is the polymer most widely used for the preparation of the separator Celgard®. PVC is an inexpensive material that has a low dielectric constant ($\varepsilon \approx 3$), good processability and which can act as a good mechanical stiffener [32, 50, 51]. The blending of PVC with PVdF will enhance the ionic conductivity by reducing the crystallinity of the polymer but the thermal stability of the blend decreases due to the lower decomposition temperature of PVC (285°C) than that of PVdF (~450°C) [51].

Polymer Blend Electrolytes for LIBs   177

FIGURE 8.6 FE-SEM image of surface morphology of electrospun membrane: (a) pure PVdF, (b) PVdF/PEO (90:10) blend; magnified image of fiber; (c) pure PVdF and (d) PVdF/PEO (90:10) polymer blend fiber. Adapted and reproduced with permission from Ref. [48]. Copyright © 2014 Elsevier.

8.2.4.1 Polyvinyl Chloride (PVC Based) Polymer Blend Electrolyte by Solvent Casting

The PVdF-based polymer electrolytes prepared by the solvent-casting techniques are found to exhibit an ionic conductivity in the range 10^{-4}–10^{-3} S cm^{-1} by the incorporation of plasticizers such as EC, PC and the lithium salt LiBF$_4$ [52]. When the polymer matrix is blended with PVC, the ionic conductivity is found to be increased to 3.68×10^{-3} S cm^{-1} [53]. By varying the concentration of the lithium salts, the electrochemical performance of the polymer electrolyte also varies. A concentration of approximately 8 wt.% lithium salts is found to be optimal to impart better electrochemical performance [52]. For the blend electrolyte, formed by the addition of PVdF to plasticized PVC-LiBF$_4$, the lithium salt is found to be unsuitable for

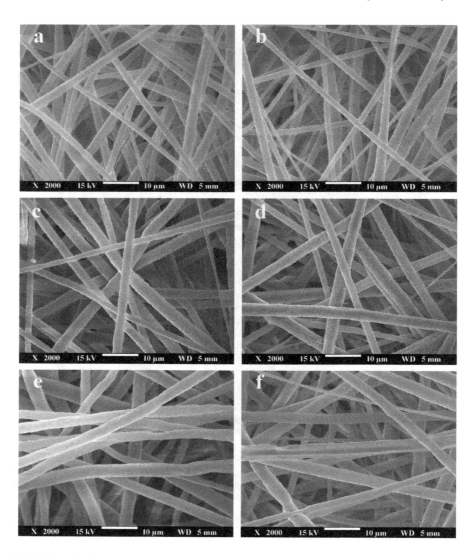

FIGURE 8.7 Surface morphology of electrospun PVdF/PEO polymer blend membranes with lithium aluminum germanium phosphate (LAGP) glass ceramic lithium-ion-conducting filler (PVdF: LAGP, 94:6 wt.%) having x wt.% PEO on the total weight of PVdF and LAGP: before (a) ESM-01 (x=5), (c) ESM-02 (x=10), (e) ESM-03 (x=20); and after (b) IPG-01 (x=5), (d) IPG-02 (x=10) and (f) IPG-03 (x=20) soaking in hot water. ESM: as-spun membrane; IPG: *in-situ* porosity generated membrane). Adapted and reproduced with permission from Ref. [49]. Copyright © 2014 Elsevier.

Polymer Blend Electrolytes for LIBs 179

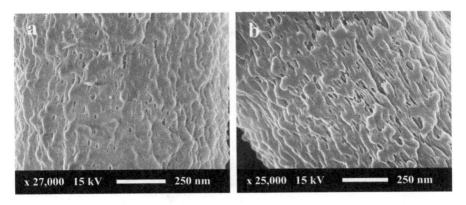

FIGURE 8.8 Magnified images of the surface morphology of electrospun PVdF/PEO polymer blend fibers with lithium aluminum germanium phosphate (LAGP) glass ceramic lithium-ion-conducting filler (PVdF: LAGP, 94:6 wt.%) having x wt.% PEO on the total weight of PVdF and LAGP: (a) IPG-02 (x=10) and (b) IPG-03 (x=20). IPG: *in-situ* porosity generated membrane). Adapted and reproduced with permission from Ref. [49]. Copyright © 2014 Elsevier.

achieving better ionic mobility as a result of the enhanced crystallinity [54]. As with PVdF, PVdF-HFP has also been investigated, in combination with PVC, in LIBs. The PVdF-*co*-HFP/PVC blend exhibits an ionic conductivity of approximately 10^{-11} S cm^{-1}. With the addition of 5 wt.% of the lithium salt, the conductivity increases to 10^{-7} S cm^{-1} and it is found that the conductivity responds positively to increasing salt concentration in a linear manner. With the addition of 35 wt.% of the lithium salt (LiClO$_4$), the ionic conductivity is approximately 1.05×10^{-4} S cm^{-1} at room temperature [55]. At concentrations greater than 35%, the aggregation of ions results in a decrease in ionic mobility and thereby in ionic conductivity. A similar blend electrolyte, with the same composition of LiClO$_4$, exhibits an ionic conductivity of about 2.10×10^{-4} S cm^{-1} [56], which is higher than that reported by Mohamed et al. [55].

8.2.4.2 Polyvinyl Chloride (PVC Based) Polymer Blend Electrolyte by Electrospinning

PVdF/PVC polymer blend electrolytes, fabricated by using the solvent-casting technique, had low ionic conductivity and poor cycling behavior, which could be improved by using the electrospinning technique. Nanofibrous membranes based on PVdF/PVC (8:2, w/w) were prepared by electrospinning and then activated with 1 M LiClO$_4$ in PC/EC (1:1 v/v) to transform the gel polymer electrolytes. With an increase in the PVC content, the electrolyte uptake and ionic conductivity of the system are found to be increased, which are the direct effects of the reduced crystallinity of the blend formed. The blending of PVdF with 20% PVC leads to a reduction in the crystallinity of PVdF, resulting in the formation of the gelled phase in the blend electrolyte. Both membranes fabricated exhibit ultrafine straight fibers

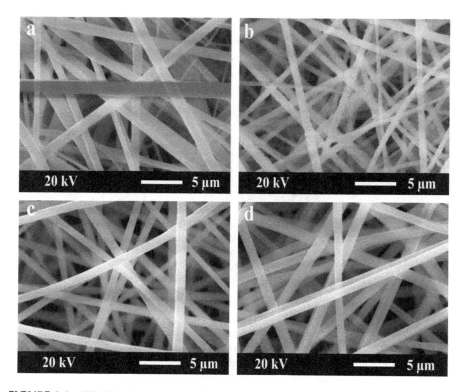

FIGURE 8.9 FE-SEM images of the surface morphology of: (a) pure PVdF and (b) PVdF/PVC (8:2) polymer blend membrane. Adapted and reproduced with permission from Ref. [60]. Copyright © 2012 Springer. PAN/PVC polymer blend membrane: (c) pure PAN and (d) PAN/PVC (8:2) polymer blend membrane. Adapted and reproduced with permission from Ref. [57]. Copyright © 2012 Elsevier.

with varying average fiber diameter (AFD). The pure PVdF membrane exhibits a diameter range of 257–1380 nm (Figure 8.9 a), whereas that for the PVdF/PVC blend is about 385–875 nm (Figure 8.9 b) [57]. The low AFD leads to high surface areas that will promote high electrolyte uptake. The low crystallinity of the blend promotes the mobility of the Li^+-ions [58]. The ionic conductivity exhibited by the blend electrolyte is 2.25 mS cm^{-1}, which is about 35% higher than that of pure PVdF membranes at 25°C. The Li/PE/LiFePO$_4$ charge-discharge studies show that about 85.4% cathode unitization are observed (145 mAh g^{-1}) in the first cycle at a current density of 0.1 C, which shows a Coulombic efficiency of 98.6%. Even after 50 cycles, the cell retained a discharge capacity of 130.8 mAh g^{-1}, representing 90.1% of the initial discharge capacity, which is an indication of good cycling stability and good interfacial stability between the electrode and the PVdF/PVC polymer blend membranes [59]. The three-dimensional network structure exhibited by the PAN/PVC blend (Figure 8.9 c, d) can result in an AFD of 1100 nm. The blend exhibits an ionic conductivity of 1.05×10^{-3} S cm^{-1}, with an improved mechanical strength [57].

8.3 CONCLUSION

PVdF and PVdF-*co*-HFP-based polymer blend electrolytes are prepared between different polymers. With the blending of different polymers, the crystallinity of the polymers decreases, which will enhance the ionic conductivity of the blend. PVdF and PVdF-*co*-HFP are the well studied polymer matrices for the preparation of blend polymer electrolytes. Compared with PVdF, PVdF-*co*-HFP exhibits low crystallinity as a result of the presence of low cryslalline hexafluoropropylene. In addition to this, the presence of the highly electronegative fluorine atom provides high polarity, that can result in the dissociation of more lithium ions, that will help to provide good ionic conductivity and thereby better electrochemical performance. Blending with PMMA, PEO and PVC help to further increase in the electrochemical performance of the polymer. These polymer matrices can provide highly amorphous content for the electrolyte matrix that can facilitate ionic movement. Additionally, these polymers can improve the mechanical strength of the resulting polymer electrolyte, which is important to ensure better battery performance.

ACKNOWLEDGMENT

Authors Anjumole P. Thomas, Dr. Jabeen Fatima M. J. and Dr. Prasanth Raghavan would like to acknowledge the Department of Science and Technology (DST), India and Kerala State Council for Science, Technology and Environment (KSCSTE), Kerala, India, for financial assistance.

REFERENCES

1. Kim YT, Smotkin ES (2002) The effect of plasticizers on transport and electrochemical properties of PEO-based electrolytes for lithium rechargeable batteries. *Solid State Ionics* 149:29–37. https://doi.org/10.1016/S0167-2738(02)00130-3
2. Krejza O, Velická J, Sedlaříková M, Vondrák J (2008) The presence of nanostructured Al$_2$O$_3$ in PMMA-based gel electrolytes. *J Power Sources* 178:774–778. https://doi.org/10.1016/j.jpowsour.2007.11.018
3. Stephan AM, Nahm KS, Anbu Kulandainathan M, et al. (2006) Poly(vinylidene fluoride-hexafluoropropylene) (PVdF-HFP) based composite electrolytes for lithium batteries. *Eur Polym J* 42:1728–1734. https://doi.org/10.1016/j.eurpolymj.2006.02.006
4. Lim D-H, Haridas AK, Figerez SP, et al. (2018) Tailor-made electrospun multilayer composite polymer electrolytes for high-performance lithium polymer batteries. *J Nanosci Nanotechnol* 18:6499–6505. https://doi.org/10.1166/jnn.2018.15689
5. Huang H, Wunder SL (2001) Preparation of microporous PVDF based polymer electrolytes. *J Power Sources* 97–98:649–653. https://doi.org/10.1016/S0378-7753(01)00579-1
6. Ramesh S, Yahaya AH, Arof AK (2002) Dielectric behaviour of PVC-based polymer electrolytes. *Solid State Ionics* 152–153:291–294. https://doi.org/10.1016/S0167-2738(02)00311-9
7. Rohan R, Sun Y, Cai W, et al. (2014) Functionalized polystyrene based single ion conducting gel polymer electrolyte for lithium batteries. *Solid State Ionics* 268:294–299. https://doi.org/10.1016/j.ssi.2014.10.013
8. Vondrák J, Reiter J, Velická J, et al. (2005) Ion-conductive polymethylmethacrylate gel electrolytes for lithium batteries. *J Power Sources* 146:436–440. https://doi.org/10.1016/j.jpowsour.2005.03.048

9. Baskaran R, Selvasekarapandian S, Kuwata N, et al. (2006) Conductivity and thermal studies of polymer blend electrolytes based on PVAc – PMMA. 177:2679–2682. https://doi.org/10.1016/j.ssi.2006.04.013
10. Qian X, Gu N, Cheng Z, et al. (2002) Plasticizer effect on the ionic conductivity of PEO-based polymer electrolyte. *Mater Chem Phys* 74:98–103. https://doi.org/10.1016/S0254-0584(01)00408-4
11. Rajendran S, Mahendran O, Kannan R (2002) Characterisation of [(1 2 x) PMMA ± x PVdF] polymer blend electrolyte with Li 1 ion q. *Fuel* 81:1077–1081
12. Liu Y, Peng X, Cao Q, et al. (2017) Gel polymer electrolyte based on poly(vinylidene fluoride)/thermoplastic polyurethane/polyacrylonitrile by the electrospinning technique. *J Phys Chem C* 121:19140–19146. https://doi.org/10.1021/acs.jpcc.7b03411
13. Liang YZ, Cheng SC, Zhao JM, et al. (2013) Preparation and characterization of electrospun PVDF/PMMA composite fibrous membranes-based separator for lithium-ion batteries. *Adv Mater Res* 750–752:1914–1918. https://doi.org/10.4028/www.scientific.net/amr.750-752.1914
14. Manjuladevi R, Thamilselvan M, Selvasekarapandian S, et al. (2017) Mg-ion conducting polymer blend electrolyte based on poly(vinyl alcohol)-poly (acrylonitrile) with magnesium perchlorate. *Solid State Ionics* 308:90–100. https://doi.org/10.1016/j.ssi.2017.06.002
15. Sousa RE, Nunes-Pereira J, Ferreira JCC, et al. (2014) Microstructural variations of poly(vinylidene fluoride co-hexafluoropropylene) and their influence on the thermal, dielectric and piezoelectric properties. *Polym Test* 40:245–255. https://doi.org/10.1016/j.polymertesting.2014.09.012
16. Ahmad Z, Al-Awadi NA, Al-Sagheer F (2007) Morphology, thermal stability and visco-elastic properties of polystyrene-poly(vinyl chloride) blends. *Polym Degrad Stab* 92:1025–1033. https://doi.org/10.1016/j.polymdegradstab.2007.02.016
17. Nunes-Pereira J, Costa CM, Lanceros-Méndez S (2015) Polymer composites and blends for battery separators: State of the art, challenges and future trends. *J Power Sources* 281:378–398
18. Sohn JY, Im JS, Gwon SJ, et al. (2009) Preparation and characterization of a PVDF-HFP/PEGDMA-coated PE separator for lithium-ion polymer battery by electron beam irradiation. *Radiat Phys Chem* 78:505–508. https://doi.org/10.1016/j.radphyschem.2009.03.035
19. Sousa RE, Kundu M, Gören A, et al. (2015) Poly(vinylidene fluoride-co-chlorotrifluoroethylene) (PVDF-CTFE) lithium-ion battery separator membranes prepared by phase inversion. *RSC Adv* 5:90428–90436. https://doi.org/10.1039/c5ra19335d
20. Costa CM, Silva MM, Lanceros-Méndez S (2013) Battery separators based on vinylidene fluoride (VDF) polymers and copolymers for lithium ion battery applications. *RSC Adv* 3:11404–11417. https://doi.org/10.1039/c3ra40732b
21. Sousa RE, Ferreira JCC, Costa CM, et al. (2015) Tailoring poly(vinylidene fluoride-co-chlorotrifluoroethylene) microstructure and physicochemical properties by exploring its binary phase diagram with dimethylformamide. *J Polym Sci Part B Polym Phys* 53:761–773. https://doi.org/10.1002/polb.23692
22. Costa CM, Rodrigues LC, Sencadas V, et al. (2012) Effect of degree of porosity on the properties of poly(vinylidene fluoride–trifluorethylene) for Li-ion battery separators. *J Memb Sci* 407–408:193–201. https://doi.org/10.1016/j.memsci.2012.03.044
23. Kim JF, Jung JT, Wang HH, et al. (2016) Microporous PVDF membranes via thermally induced phase separation (TIPS) and stretching methods. *J Memb Sci* 509:94–104. https://doi.org/10.1016/j.memsci.2016.02.050
24. Ribeiro C, Costa CM, Correia DM, et al. (2018) Electroactive poly(vinylidene fluoride)-based structures for advanced applications. *Nat Protoc* 13:681–704. https://doi.org/10.1038/nprot.2017.157

25. Martins P, Lopes AC, Lanceros-Mendez S (2014) Electroactive phases of poly(vinylidene fluoride): Determination, processing and applications. *Prog Polym Sci* 39:683–706. https://doi.org/10.1016/j.progpolymsci.2013.07.006
26. Edmonds EA, Acharya UR, Bonjour E, et al. (2007) Founding editor. 193. https://doi.org/10.1016/S0950-7051(18)30220-X
27. Du Pasquier A, Warren PC, Culver D, et al. (2000) Plastic PVDF-HFP electrolyte laminates prepared by a phase-inversion process. *Solid State Ionics* 135:249–257. https://doi.org/10.1016/S0167-2738(00)00371-4
28. Li ZH, Xiao QZ, Zhang P, et al. (2008) Porous nanocomposite polymer electrolyte prepared by a non-solvent induced phase separation process. *Funct Mater Lett* 1:139–143. https://doi.org/10.1142/S1793604708000253
29. Iijima T, Toyoguchi Y, Eda N (1985) Quasi-solid organic electrolytes gelatinized with polymethyl-methacrylate and their applications for lithium batteries. *Denki Kagaku* 53:619–623.
30. Bohnke O, Frand G, Rezrazi M, et al. (1993) Fast ion transport in new lithium electrolytes gelled with PMMA. 1. Influence of polymer concentration. *Solid State Ionics* 66:97–104. https://doi.org/10.1016/0167-2738(93)90032-X
31. Prasanth R, Aravindan V, Srinivasan M (2012) Novel polymer electrolyte based on cob-web electrospun multi component polymer blend of polyacrylonitrile/poly(methyl methacrylate)/polystyrene for lithium ion batteries - preparation and electrochemical characterization. *J Power Sources* 202:299–307. https://doi.org/10.1016/j.jpowsour.2011.11.057
32. Rhoo H-J, Kim H-T, Park J-K, Hwang T-S (1997) Ionic conduction in plasticized PVCPMMA polymer blend electrolytes. *Electrochim Acta* 42:1571–1579. https://doi.org/10.1016/S0013-4686(96)00318-0
33. Tatsuma T, Taguchi M, Oyama N (2001) Inhibition effect of covalently cross-linked gel electrolytes on lithium dendrite formation. *Electrochim Acta* 46:1201–1205. https://doi.org/10.1016/S0013-4686(00)00706-4
34. Gray FM, MacCallum JR, Vincent CA (1986) Poly(ethylene oxide) - LiCF3SO3 - polystyrene electrolyte systems. *Solid State Ionics* 18–19:282–286. https://doi.org/10.1016/0167-2738(86)90127-X
35. Mahant YP, Kondawar SB, Bhute M, Nandanwar DV (2015) Electrospun Poly (Vinylidene Fluoride)/Poly (Methyl Methacrylate) Composite Nanofibers Polymer Electrolyte for Batteries. *Procedia Mater Sci* 10:595–602. https://doi.org/10.1016/j.mspro.2015.06.011
36. Mahant YP, Kondawar SB, Nandanwar DV, Koinkar P (2018) Poly(methyl methacrylate) reinforced poly(vinylidene fluoride) composites electrospun nanofibrous polymer electrolytes as potential separator for lithium ion batteries. *Mater Renew Sustain Energy* 7:1–9. https://doi.org/10.1007/s40243-018-0115-y
37. Ding Y, Zhang P, Long Z, et al. (2009) The ionic conductivity and mechanical property of electrospun P(VdF–HFP)/PMMA membranes for lithium ion batteries. *J Memb Sci* 329:56–59. https://doi.org/10.1016/j.memsci.2008.12.024
38. Rajendran S, Kannan R, Mahendran O (2001) An electrochemical investigation on PMMA/PVdF blend-based polymer electrolytes. *Mater Lett* 49:172–179. https://doi.org/10.1016/S0167-577X(00)00363-3
39. Jiang Z, Carroll B, Abraham KM (1997) Studies of some poly (vinylidene fluoride) electrolytes. *Electrochimica Acta* 42(17):2667–2677
40. Nicotera I, Coppola L, Oliviero C, et al. (2006) Investigation of ionic conduction and mechanical properties of PMMA – PVdF blend-based polymer electrolytes. 177:581–588. https://doi.org/10.1016/j.ssi.2005.12.028
41. Sundaram NTK, Musthafa OTM, Lokesh KS, Subramania A (2008) Effect of porosity on PVdF- co -HFP – PMMA-based electrolyte. 110:11–16. https://doi.org/10.1016/j.matchemphys.2007.12.024

42. Gohel K, Kanchan DK (2018) Ionic conductivity and relaxation studies in PVDF-HFP: PMMA-based gel polymer blend electrolyte with LiClO 4 salt. 8:1–13. https://doi.org/10.1142/S2010135X18500054
43. Ma T, Cui Z, Wu Y, et al. (2013) Preparation of PVDF based blend microporous membranes for lithium ion batteries by thermally induced phase separation: I. Effect of PMMA on the membrane formation process and the properties. *J Memb Sci* 444:213–222. https://doi.org/10.1016/j.memsci.2013.05.028
44. Polu AR, Rhee H-W (2015) Nanocomposite solid polymer electrolytes based on poly(ethylene oxide)/POSS-PEG (n=13.3) hybrid nanoparticles for lithium ion batteries. *J Ind Eng Chem* 31:323–329. https://doi.org/10.1016/j.jiec.2015.07.005
45. Shi J, Yang Y, Shao H (2018) Co-polymerization and blending based PEO/PMMA/P(VDF-HFP) gel polymer electrolyte for rechargeable lithium metal batteries. *J Memb Sci* 547:1–10. https://doi.org/10.1016/j.memsci.2017.10.033
46. Ramrakhiani M, Nogriya V (2012) Synthesis and characterization of zinc sulfide nanocrystals and zinc sulfide/polyvinyl alcohol nanocomposites for luminescence applications. *Polym Process Charact* 394:109–138. https://doi.org/10.1201/b13105
47. Xi J, Qiu X, Li J, et al. (2006) PVDF-PEO blends based microporous polymer electrolyte: Effect of PEO on pore configurations and ionic conductivity. *J Power Sources* 157:501–506. https://doi.org/10.1016/j.jpowsour.2005.08.009
48. Prasanth R, Shubha N, Hng HH, Srinivasan M (2014) Effect of poly(ethylene oxide) on ionic conductivity and electrochemical properties of poly(vinylidenefluoride) based gel polymer electrolytes prepared by electrospinning for lithium ion batteries. *J Power Sources* 245:283–291. https://doi.org/10.1016/j.jpowsour.2013.05.178
49. Shubha N, Prasanth R, Hng HH, Srinivasan M (2014) Study on effect of poly (ethylene oxide) addition and in-situ porosity generation on poly (vinylidene fluoride)-glass ceramic composite membranes for lithium polymer batteries. *J Power Sources* 267:48–57. https://doi.org/10.1016/j.jpowsour.2014.05.074
50. Ramesh S, Winie T, Arof AK (2007) Investigation of mechanical properties of polyvinyl chloride–polyethylene oxide (PVC–PEO) based polymer electrolytes for lithium polymer cells. *Eur Polym J* 43:1963–1968. https://doi.org/10.1016/j.eurpolymj.2007.02.006
51. Rajendran S, Babu RS, Sivakumar P (2008) Investigations on PVC/PAN composite polymer electrolytes. *J Memb Sci* 315:67–73. https://doi.org/10.1016/j.memsci.2008.02.007
52. Periasamy P, Tatsumi K, Shikano M, et al. (1999) An electrochemical investigation on polyvinylidene fluoride-based gel polymer electrolytes. *Solid State Ionics* 126:285–292. https://doi.org/10.1016/s0167-2738(99)00234-9
53. Rajendran S, Sivakumar P (2008) An investigation of PVdF/PVC-based blend electrolytes with EC/PC as plasticizers in lithium battery applications. *Phys B Condens Matter* 403:509–516. https://doi.org/10.1016/j.physb.2007.06.012
54. Vickraman P, Ramamurthy S (2006) A study on the blending effect of PVDF in the ionic transport mechanism of plasticized PVC–LiBF4 polymer electrolyte. *Mater Lett* 60:3431–3436. https://doi.org/10.1016/j.matlet.2006.03.028
55. Mohamed NHB and Mohamed NS (2009) Conductivity studies and dielectric behaviour of PVDF-HFP- PVC-LiClO 4 solid polymer electrolyte Nor Hazlizaaini Basri and N.S. Mohamed. *Solid State Sci Tech* 17:63–72
56. Basri NH, Ibrahim S, Mohamed NS (2011) PVDF-HFP/PVC blend based lithium ion conducting polymer electrolytes. *Adv Mater Res* 287–290:100–103. https://doi.org/10.4028/www.scientific.net/AMR.287-290.100
57. Zhong Z, Cao Q, Jing B, et al. (2012) Novel electrospun PAN-PVC composite fibrous membranes as polymer electrolytes for polymer lithium-ion batteries. *Ionics (Kiel)* 18:853–859. https://doi.org/10.1007/s11581-012-0682-3

58. Gopalan AI, Santhosh P, Manesh KM, et al. (2008) Development of electrospun PVdF-PAN membrane-based polymer electrolytes for lithium batteries. *J Memb Sci* 325:683–690. https://doi.org/10.1016/j.memsci.2008.08.047
59. Gopalan AI, Lee K-P, Manesh KM, Santhosh P (2008) Poly(vinylidene fluoride)–polydiphenylamine composite electrospun membrane as high-performance polymer electrolyte for lithium batteries. *J Memb Sci* 318:422–428. https://doi.org/10.1016/j.memsci.2008.03.007
60. Zhong Z, Cao Q, Jing B, et al. (2012) Electrospun PVdF – PVC nanofibrous polymer electrolytes for polymer lithium-ion batteries. *Mater Sci Eng B* 177:86–91. https://doi.org/10.1016/j.mseb.2011.09.008

9 Polymer Clay Nanocomposite Electrolytes for Lithium-Ion Batteries

*Jishnu N. S, Krishnan M. A., Akhila Das,
Neethu T. M. Balakrishnan, Jou-Hyeon Ahn,
Jabeen Fatima M. J., and Prasanth Raghavan*

CONTENTS

9.1 Introduction 187
9.2 Ion Transport in Polymer/Clay Nanocomposites 188
9.3 Polyvinylidene Difluoride (PVdF)-Clay Composite Polymer Electrolytes 192
9.4 PVdF-*co*-HFP/Clay Composite Polymer Electrolytes 199
9.5 Polyacrylonitrile/Clay Composite Polymer Electrolytes 204
9.6 Polymethyl Methacrylate/Clay Composite Polymer Electrolytes 207
9.7 Conclusion 209
Acknowledgment 210
References 210

9.1 INTRODUCTION

The demand for microelectronic devices has led to extensive research on energy storage devices, like lithium-ion batteries (LIBs), supercapacitors etc. Among the different energy storage solutions, LIBs are the most promising energy storage devices, exhibiting higher energy and power density than other electrochemical energy storage devices. In order to avoid the safety concerns caused by the use of liquid electrolytes, polymer-based electrolytes have been widely explored for LIBs. Different polymer matrices, such as polyethylene oxide (PEO) [1], polyacrylonitrile (PAN) [2], polymethyl methacrylate (PMMA) [3], polyvinylidene difluoride (PVdF) [4] and its copolymer poly(vinylidene difluoride-*co*-hexafluoro propylene) (PVdF-*co*-HFP) [5, 6], have been extensively studied as host matrices for polymer electrolytes, that can offer good electrochemical and physical properties to the polymer electrolytes. Even though the polymer-based electrolytes can enhance the safety of the battery, the electrochemical performance offered by them is inferior to that of their liquid counterparts. As a consequence, different modification methods have been proposed, in which blending with other polymers [7–10] and additions of fillers [11–13], to form composites, are considered to be more simple and effective methods than chemical modifications [14–17], cross-linking [18] or grafting [19], to improve electrochemical properties and battery performance. Blending with other polymers can confer the synergistic advantages of the individual polymers and thereby can improve the performance of the gel polymer electrolyte. While considering the composite formations, different fillers have been examined for the composite polymer electrolyte. The organic compounds, such as cellulose [20–23], cellulose derivatives [24], kraft lignin [25], chitin [26], chitosan [27], carbon nanotubes (CNTs) [28], graphene oxide (GO) [29] and inorganic/ceramic fillers such as aluminum oxide (Al_2O_3) [30–34], titania (TiO_2) [35–39], silica (SiO_2) [40–44], barium titanate ($BaTiO_3$), zirconia (ZrO_2) nickel oxide (NiO) [45], glass ceramics, such as lithium aluminum germanium phosphate (LAGP) [10] and lithium aluminum titanium phosphate (LATP) [46, 47], have also been tried out as fillers-cum-ion conduction promoters in electrolytes. Among various inorganic fillers, nanoclay, consisting of layers of silicates, are less-studied fillers in the polymer electrolytes. The unique structure of nanoclay readily allows intercalation of polymers into two-dimensional lattices, which can confer the advantageous properties to the resulting nanocomposite material. Montmorillonite (MMT) clays, with high aspect ratio, high cation-exchange capacity, large specific surface area and appropriate interlayer charges, have been explored as fillers for polymer composite electrolytes [11–13] The unique, layered structure of clay can provide the exfoliation and intercalation mechanism that is beneficial for imparting increased mechanical, thermal, and dimensional stability, along with improved electrochemical performance within the polymer electrolyte. This chapter basically discusses the electrochemical and physical properties of clay-based composite polymer gel electrolytes fabricated by various methods, different polymer matrices and transport mechanisms in polymer clay nanocomposite electrolytes.

9.2 ION TRANSPORT IN POLYMER/CLAY NANOCOMPOSITES

Clay is a natural compound formed as a result of the disintegration of igneous rocks. It exhibits a layered fine texture of particle size <2 mm. Two types of clay are known, based on the chemical structure, namely 2:1 clay (montmorillonite/bentonite) and 1:1 clay (kaolinite). Among the different nanoclays, montmorillonite/bentonite clay has attracted greater attention, due to its wide range of applications. A schematic representation of the special arrangement of layers in 2:1 clay and 1:1 clay is shown in Figure 9.1. The planes of cations in the clay minerals may be tetrahedrally or octahedrally coordinated with oxygen. In the montmorillonite/bentonite clay, the planes are composed of two tetrahedral and one octahedral sheet, whereas, in kaolinite, the planes are composed of alternating tetrahedral and octahedral sheets. The polymer composite electrolytes based on montmorillonite (MMT) clays have been investigated for the preparation of polymer electrolytes in LIBs, because of their ability to exhibit particular sizes in terms of length and thickness, along with a sandwich-type structure. MMT clays consist of silicate platelet stacks, of 10 Å thickness and

Polymer Clay Nanocomposite Electrolytes

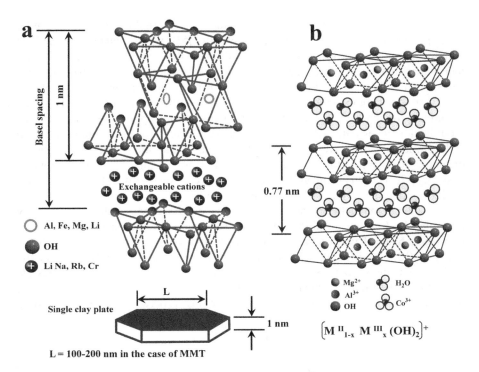

FIGURE 9.1 Images representing layered silicates/nanoclay and its constituent elements: (a) 2:1 clay (montmorillonite/bentonite) and (b) 1:1 clay (kaolinite). Adapted and reproduced with permission from Ref. [53]. Copyright © 2013 Elsevier.

2800 Å length. The mineral of this group, montmorillonite, has an ideal chemical formula of $R_{0.33}(Al_{1.67}Mg_{0.33})Si_4O_{10}(OH)_2$. They are a type of smectite, charged 2:1 phyllosilicate, consisting of interlayered cations that join the layers by van der Waals forces, electrostatic forces or hydrogen bonding [48–53]. Due to the presence of loosely bonded silicate layers of the MMT clay, they can exhibit intercalation and swelling properties, which are important when being used in polymer-based systems. The polymer clay nanocomposites are generally termed intercalated nanocomposites or delaminated nanocomposites. The intercalation or exfoliation of clay helps to form the composite structure with high thermal and mechanical stability along with good electrochemical properties, even at very low levels of clay loading [53, 54]. The polymer electrolytes based on the clay/polymer composites incorporating nanoclay into the polymeric matrix are popularly prepared by solution-mixing or melt-mixing processes. Unfortunately, the natural clays are hydrophilic in nature, owing to the presence of charges on the minerals, hence there are compatibility issues between the hydrophobic clay and the organophilic polymer matrix. Due to the hydrophobic nature of the clay, it is difficult to incorporate the clay uniformly within the organophilic matrix medium (polymer). Hence, the uniform dispersion of clay in the polymer matrix is achieved by surface modification of the clay system, where the hydrophilic nature of the clay surface is converted to being organophilic by using surfactants or organo-modified systems [53], which makes it compatible with the organophilic polymer matrix. In addition to the compatibility with the polymer matrix, the electrolytes used in energy storage applications prefer lyophilic solvents, such as EC, DMC, PC, etc., rather than hydrophilic (aqueous) solvents. In order to make it compatible with these organic solvents, the clay has to be modified organically by ion-exchange reactions for energy storage applications. The organic modification is usually achieved by using surfactant moieties, containing both hydrophilic as well as lyophilic parts, for example, cationic surfactants including primary, secondary, tertiary and quaternary alkyl ammonium or alkyl phosphonium salts [53–55]. The surfactant system modifies the clay surface with lyophilic chains, similar to those involved in micelle formation [50].

The interaction between polymeric chains on the clay structure was clearly explained by Prasanth et al. [53, 54]. The schematic illustration of the intercalation mechanism of various types of clay platelets with polymeric chains is shown in Figure 9.2. The differences between the synthesis methods strongly influence the performance and properties of the composite electrolyte. Figure 9.2a describes the structure of the microcomposite clay structure being stacked together, decreasing the possibility of intercalation of polymeric chains in between the layers, leading

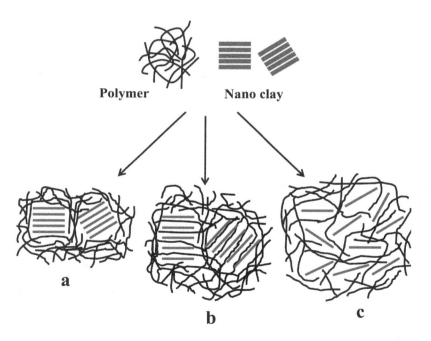

FIGURE 9.2 Different types of composite morphologies arising from the intercalation of layered silicates with the polymer: (a) phase-separated microcomposite, (b) intercalated nanocomposite and (c) exfoliated nanocomposite. Adapted and reproduced with permission from Ref. [53]. Copyright © 2013 Elsevier.

Polymer Clay Nanocomposite Electrolytes

to a phase separation in the composite. Figure 9.2b shows the intercalation of the polymeric chains into the stacked clay platelets, which increase the gap between the layers of the clay (clay gallery), where the layered structure of the clay is being maintained by some weak binding forces. Figure 9.2c shows a well-defined formation of an interlinked composite structure of exfoliated clay platelets with polymer chains, where the clay lost their inter-gallery relations, which, in turn, provide the maximum available surface area of the clay to interact with polymer chains, leading to the formation of polymer/clay nanocomposites with highest composite properties. The exfoliation of the clay results from the pulling apart of the clay platelets by the intercalated polymer chains ("pull apart") or the sliding of the clay platelets (shear apart) by the higher shear stress exerted by the mechanical force during mixing. The ultrasonication or mechanical mixing causes the de-stacking of the clay layers. The E-SEM images (Figure 9.3) displays the variation in the surface morphology on the structural variation of clay in DMAc solvent in response to differential sonication time (0 and 48 hours) [53]. Ultrasonication of the clay for a prolonged period of 48 hours leads to the swelling of clay bundles followed by primary to secondary aggregation of clay platelets, to form larger-sized particles with increased sonication time, leading to the formation of uniform dispersion of clay in DMAc.

The mechanism of composite formation of the polymer clay system was studied by Shukla et al. [56] From the X-ray diffraction analysis, they analyzed the crystallographic variation involved in the clay structure during the intercalation process. The modification of the clay results in an increase in the width of the silicate galleries, from 2.4 Å to 6.1 Å, which allows for the ease of intercalation of the polymer chains in between the clay layers [56]. During the intercalation of the polymer between the layers of MMT, it remains in its ordered multilayer structure, sustaining the self-assembling characteristics. Delaminated clay structures differ somewhat from

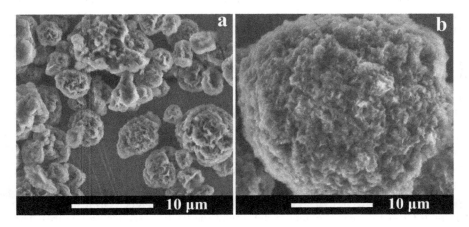

FIGURE 9.3 E-SEM (environmental-SEM) micrographs of the surface morphology of the nanoclay dispersion in DMAc in response to sonication time at room temperature: after (a) 0 h and (b) 48 h. Adapted and reproduced with permission from Ref. [53]. Copyright © 2013 Elsevier.

the exfoliated structure. In this structure, the silicate layers are separated and the interactions between the adjacent gallery cations are hindered. Hence, the silicate layers are viewed as being dispersed in the organic polymer. Completely exfoliated structures are seen to exhibit the best physico-mechanical and electrochemical performances. With increases in the clay concentration, transformation of the exfoliated phase into the intercalated phase was observed and, at a higher clay concentration, a greater number of clay channels allowed for the diffusion of polymer chains into it [56, 57]. Prasanth et al. [54] also investigated the effect of conversion of micron-sized clay tactoids to nanosized exfoliated clay, in response to an increase in sonication time, increasing the ionic conductivity of the composite polymer electrolyte. The micron-sized clay particles were initially broken down to nanosized clay and, on prolonged sonication for about 48 h in DMAc solvent, led to the exfoliation of the clay into platelets, which showed an increase in the interaction with the polymeric matrix (PVdF-co-HFP), leading to an increase in the ionic conductivity by the generation of an easier conduction path for the lithium ions. A schematic illustration of the effect of interaction on incorporation of various-sized clay particles into a polymeric matrix is shown in Figure 9.4.

Recently, though in only a few studies, the nanoclay has been utilized as the ceramic moiety to fabricate thermally and electrochemically a stable quasi-solid-state [58] (clay + Li-salt + room temperature ionic liquid [RTIL]) as well as a polymer gel electrolyte (polymer + clay + Li-salt + organic solvent/RTIL) [59] for lithium-ion batteries. Taking into account the thermal stability of the clay, a hybrid electrolyte separator, which can operate at high as well as at ambient temperatures, was fabricated by impregnating the lithiated room temperature ionic liquid (RTILs) into the bentonite clay structure. This electrolyte is stable up to 355°C, with a stable voltage window of 3V even at 120°C, which combines the excellent performance of the clay along with the RTIL [58]. The same research group later reported a quasi-solid electrolyte system by replacing the bentonite clay with hexagonal boron nitrile (h-BN), which exhibits a two-dimensional structure similar to that of bentonite clay. The cell fabricated by using this h-BN-based quasi solid-state electrolyte is shown to be efficient at a temperature range of 24 to 150°C, with this excellent performance being attributed to the presence of ceramic components in it [58].

9.3 POLYVINYLIDENE DIFLUORIDE (PVdF)-CLAY COMPOSITE POLYMER ELECTROLYTES

Polyvinylidene difluoride (PVdF) is a well-known matrix for polymer electrolyte in LIBs [53] [60] The high dielectric constant (ε = 8.4), and high mechanical and chemical stability ensures that PVdF membranes achieve the best performance of polymer electrolytes in LIBs. Furthermore, the presence of an electron-withdrawing group (–CF–) enhances the anodic stability of the PVdF-based electrolyte [61]. The ability to form homogeneous solutions with different organic solvents and the ability to exhibit high thermal stability (T_m=175°C) are the characteristic properties of PVdF which can be exploited for the preparation of polymer electrolytes [62]. The issues related to mechanical strength, as compared with that of ceramic or glass-ceramic

Polymer Clay Nanocomposite Electrolytes

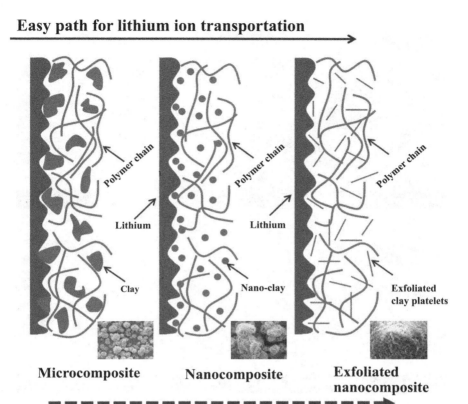

FIGURE 9.4 Schematic representation of the effect of ceramic filler particle size (clay is adapted as the representative ceramic filler) on composite morphology and lithium ion conduction channel formation in composite polymer electrolytes. Adapted and reproduced with permission from Ref. [54]. Copyright © 2013 Elsevier.

electrolytes, limits their applications in LIBs. In order to tackle this limitation, along with the electrochemical stability, composite electrolytes incorporating different fillers have been investigated. Among the different ceramic fillers, clay is a promising ceramic material to be utilized for the preparation of composite polymer electrolytes but one that has not yet been well studied [63, 64]. The layered structure of clay is unique in imparting a reinforcing effect in polymer composites. Analyzing the effect of clay on PVdF, it has been observed that an increase in mechanical properties with a similar porosity (80%), compared with that of pure PVdF membrane, has been reported [65]. The electrochemical performance of the PVdF/clay composite was examined by Deka et al. [66]. Organically modified MMT clays were incorporated into the PVdF matrix in different concentrations by weight (namely 1, 2.5 or 4 wt.%) by mechanical mixing followed by ultrasonication. The resultant slurry was then

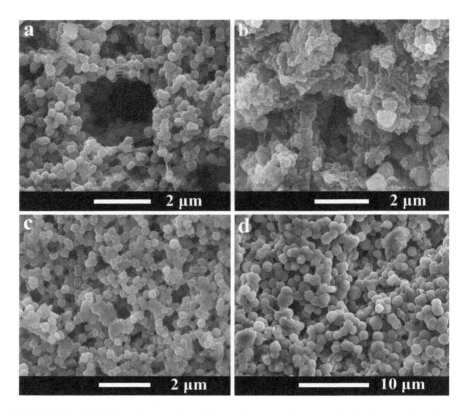

FIGURE 9.5 FE-SEM micrographs of the surface morphology of gel polymer electrolytes based on PVdF matrix with different loadings (wt.%) of organically modified MMT clay: (a) 0 wt.%, (b) 1 wt.%, (c) 2.5 wt.% or (d) 4 wt.%. Adapted and reproduced with permission from Ref. [66]. Copyright© 2011, Elsevier

cast (using the solvent-casting method) to form a polymeric composite membrane. A similar composite polymer electrolyte was reported by Prasanth et al. [53], fabricated using the phase-inversion technique. The concentrations of clay under investigation were chosen to be 0, 1, 2 or 4%. Both of these studies showed enhanced electrochemical performance of the composite membrane as compared with that of the pure one. At lower concentrations of modified MMT, the intercalation of the polymer may lead to the distortion of the layered clay structure. Also, compared with the pure MMT, an increase in gallery spacing (1-nm) was observed for MMT in the composite structure, a finding which clearly shows the interaction of polymer chains with the layered clay structure. Moreover, the addition of the filler can transform the crystalline phase of PVdF from the α phase to the β phase, the latter being electrochemically the more active phase of PVdF [67].

According to Deka et al. [66], the composite electrolyte has a spherical granular morphology; with an increase in the MMT clay content, the pore size becomes reduced and forms a smooth surface morphology (Figure 9.5). The morphology of

Polymer Clay Nanocomposite Electrolytes

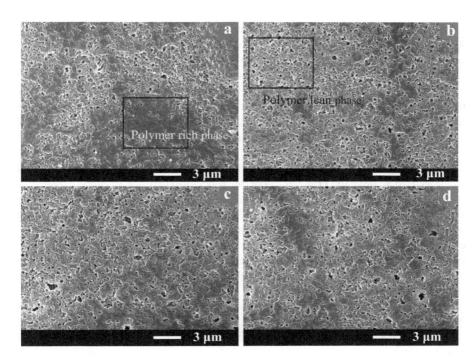

FIGURE 9.6 FE-SEM micrographs on the surface morphology (upper surface) of PVdF–clay nanocomposite membranes with different clay loadings (wt.%), prepared by the phase-inversion technique: (a) 0%, (b) 1 wt.%, (c) 2 wt.% or (d) 4 wt.%. Adapted and reproduced with permission from Ref. [53]. Copyright© 2013, Elsevier

the phase-inversion composite membrane, reported by Prasanth et al. [53], differs from that of the solvent-cast membrane [66]. The phase-inversion membrane shows a porous structure, with a blend of micro- as well as macrovoids (Figure 9.6). The membrane exhibits a closed cellular structure, with large numbers of micro- and macro-sized pores, which is beneficial for the delivery of a good electrochemical performance. The best honeycomb porous structure was observed with a system containing 2 wt.% MMT clay, which was suitable to achieve good absorption and retention of the electrolyte [53]. For the solvent-cast composite membrane, the highest concentration of clay tested (4 wt.%) is identified as being the best, as compared with that of lower MMT clay concentrations. However, the electrochemical performance of the phase-inversion membrane is shown to be better, as it exhibits better ionic conductivity and electrochemical performance, even at lower clay loading, compared with that of the solvent-cast polymer electrolyte [66].

The nanocomposite films are soaked with 1 M LiClO$_4$ in a 1:1 (v/v) solution of propylene carbonate (PC) and diethyl carbonate (DEC) to obtain the polymer gel electrolytes [66]. The solvent-cast composite electrolyte, containing 4 wt.% of nanoclay, exhibits an electrolyte uptake of 177%, which is the highest uptake reported. The same membrane also exhibits better electrochemical performance. Ionic conductivity exhibited by this electrolyte is 2.3×10^{-3} S cm^{-1} and is capable of exhibiting an electrochemical stability of 4.6 V. The higher ionic conductivity of membranes containing clay is attributed to their high dielectric constant, which facilitates the dissolution of more Li$^+$-ions. The phase-inversion composite membrane reported by Prasanth et al. [53] exhibits an ionic conductivity of 3.08×10^{-3} S cm^{-1} with a clay concentration of 2 wt.%, which is higher than that reported by Deka et al. [66] with a 4 wt.% clay content. The Li/LiMn$_2$O$_4$ cell fabricated by the polymer electrolyte is capable of exhibiting a stable cyclic performance with a discharge capacity of 114 mAh g^{-1}, even after 30 cycles [53]. The improved capacity of the electrolyte is attributed to the honeycomb porous structure, that can efficiently trap the liquid electrolyte to achieve the best electrochemical performance.

To achieve further enhancement of electrochemical performance, the porous structure of the polymer electrolyte can be modified, using suitable pore-forming agents. Polyvinylpyrrolidone (PVP) is a well-studied, pore-forming agent for PVdF-clay polymer composite electrolytes. By using PVP in the phase-inversion composite membrane, the pore structures are modified to obtain a uniform porous structure that is evident from the extremely high electrolyte uptake (802 wt.%) reported for the composite membrane consisting of 8 wt.% clay and 7 wt.% PVP (Figure 9.7), prepared by the phase-inversion technique. This is higher than the values reported by Prasanth et al. [53] and Deka et al. [66]. The cross-sectional morphology, depicted in Figure 9.7a–d, clearly pinpoints the increase in porosity of the electrolyte membranes containing the pore-forming agent (PVP) and the filler nanoclay. The surface morphology (Figure 9.7c and d) also confirms the role of poreforming agent and filler in the formation of uniform pores in the membrane. The PVdF/clay/PVP membrane exhibits an ionic conductivity of about 5.6×10^{-3} S cm^{-1} and a graphite/LiFePO$_4$ coin cell showed excellent discharge capacity of 127 mAh g^{-1} at a C rate of 0.2 C and a stable capacity for 50 cycles, with excellent Coulombic efficiency [68]. The cells also displayed good rate capability for the entire composite electrolyte and displayed a discharge capacity of about 75 mAh g^{-1} at a current density of 4 C. The cell performance is significantly better than in a cell with a conventional Celgard® separator, which shows a discharge capacity of 101 mAh g^{-1} at 0.2 C. Similarly, blends of PVdF with other polymers incorporated with clay have also been reported [69]. It has been reported that electrolytes with a polymer blend can achieve better electrochemical properties than those of pure polymer matrices. By incorporating clay into the blend, the advantage of both blends and composites can be achieved. Ma et al. [69] reported the fabrication of a PVdF/PVA blend polymer electrolyte incorporating MMT clay particles, using the solvent-casting method. Even though the blended polymer is chosen as the matrix into which to incorporate MMT, the ionic conductivity of the composite polymer electrolyte is lower than that reported by Prasanth et al. [53] or Deka et al. [66]. However, the composite electrolyte exhibited a discharge capacity of 123 mAh g^{-1} at 0.1 C rate in LiFePO$_4$/PVdF-PVA-MMT/Li cell, with a Coulombic efficiency of 97%, even after 100 cycles, which shows the exceptionally good stability of the electrolyte in LIBs.

Electrospinning is a well-known technique for the fabrication of polymer electrolytes for LIBs. It is a simple and versatile method to prepare fibrous membranes with

Polymer Clay Nanocomposite Electrolytes 197

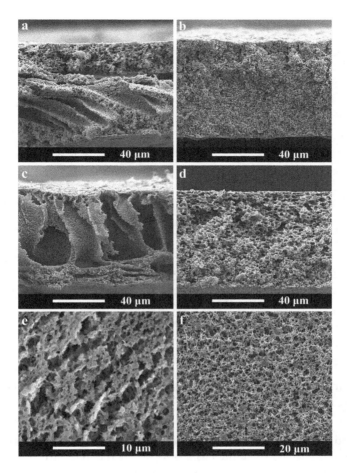

FIGURE 9.7 FE-SEM micrographs of the cross-sectional and surface morphology of polymer electrolyte membranes (PEMs) based on poly(vinylidene difluoride) (PVdF) with or without polyvinylpyrrolidone (PVP) pore-forming agent, Cross-sectional morphology: (a) pure PVdF (10 wt.%), (b) PVdF (10 wt.%) with nano-clay (8 wt.%), (c) PVdF (10 wt.%) with PVP (7 wt.%) and (d) PVdF (10 wt.%) with PVP (7 wt.%) and nano-clay (8 wt.%), Surface morphology: (e) pure PVdF (10 wt.%) and (f) PVdF (10 wt.%) with PVP (7 wt.%) and nano-clay (8 wt.%). Adapted and reproduced with permission from Ref. [68]. Copyright© 2018 Elsevier.

porosity values greater than 80% for the fabrication of polymer gel electrolytes. The schematic of the production of fibrous membranes by the electrospinning technique is shown in Figure 9.8. Essentially, the electrospinning set-up has a spinning head, consisting of an injection syringe fitted with a blunt steel needle, connected to a syringe pump or air pump to push the polymer solution/melt into the syringe, a high-voltage source connected to the tip of the steel needle and a collection target, typically a metal drum or plate. An electrospun polymer electrolyte, based on PVdF/organo-modified clay (OMC)/tripropylene glycol diacrylate (TPGDA) composite nanofibers,

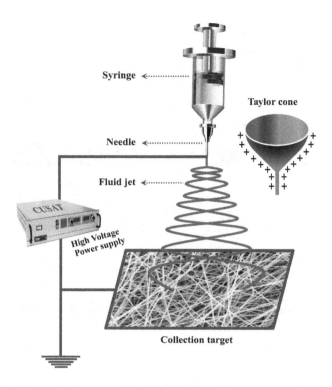

FIGURE 9.8 Schematic diagram of the visualization of the process and machine set-up of electrospinning.

was studied in LIBs [60]. The as-prepared electrospun polymer membrane consists of fully interconnected three-dimensional open microporous structures. The electrospun membrane was irradiated to graft onto and cross link between electrospun fibers. By incorporation of OMC and TPGDA, the mechanical and electrochemical properties are significantly enhanced by increasing the amounts of OMC and TPGDA. The tensile strength is increased by 156% by adding 5 wt.% of OC, whereas further addition of OC to 10 wt.% results in a rather lower increase (67%) in the tensile strength compared with neat PVdF. This might be due to a re-agglomeration of OMC particles [70]. A significantly higher ionic conductivity is observed for the electrolyte containing 5 wt.% OMC along with 10 wt.% TPGDA, which is about 1.7×10^{-3} S cm^{-1} compared to the pure PVdF electrolyte (PVdF nanofiber 1.1×10^{-4} S cm^{-1}, compared with 8.5×10^{-5} S cm^{-1} for the PVdF cast film) or 8.0×10^{-4} S cm^{-1} for the PVdF electrolyte containing 10 wt.% TPGDA nanofiber without OMC at room temperature. The electrochemical stability is slightly decreased by the incorporation of 5 wt.% clay and 10 wt.% TPGDA; however, the PVdF/xOMC/yTPGDA electrolytes displayed electrochemical stability above 5 V *versus* Li/Li$^+$, which is sufficient to be compatible with most common materials (limit *ca.* 4.5 V) used for lithium batteries. The charge/discharge studies in Li/sulfur batteries showed that the cycling

Polymer Clay Nanocomposite Electrolytes

performance of the PVdF electrolyte is significantly improved by incorporating OC and TPGDA, and the cell with composite electrolytes exhibits a discharge capacity of 1000 mAh g^{-1} of sulfur at a current density of 105 mA g^{-1} (equivalent to a C/15 cycle rate), indicating a higher proportion of sulfur utilization (ca. 60%). The discharge capacity of PVdF/TPGDA electrolytes with OC is higher than those without OC or for a purely PVdF electrolyte. A fast decay of discharge capacity is observed for all the electrolytes, with or without OC, and all the PVdF/xOC/yTPGDA electrolytes show a discharge capacity of >500 mAh g^{-1} of sulfur, whereas, for the pure PVdF electrolyte, the discharge capacity was <300 after 20 cycles.

9.4 PVdF-co-HFP/CLAY COMPOSITE POLYMER ELECTROLYTES

PVdF-co-HFP is a semi-crystalline copolymer [71], consisting of a crystalline PVdF part and an amorphous HFP part. By the copolymerization of vinylidene fluoride (VdF) with hexafluoropropylene (HFP), the crystallinity of the polymer decreases, providing an easier conduction path for the lithium ions, and thereby improving the ionic conductivity and transport properties. Compared with PVdF, the flexibility of PVdF-co-HFP is higher, which is a side-effect of the reduced crystallinity. The crystalline region achieves a greater mechanical strength and dimensional stability, whereas, at the same time, the amorphous region could maintain low temperature flexibility and soft conduction channels for Li$^+$-ion transportation. For PVdF, the property that inhibits its practical applications is its high crystallinity, which reduces the ionic conductivity and charge discharge capacity, as well as rate capability, which all basically arise from the more crystalline domains present in PVdF. Hence, PVdF-co-HFP is extensively studied as a host matrix for polymer gel electrolytes in LIBs [72–74].

PVdF-co-HFP-based gel and solid electrolytes are studied as separator-cum-electrolyte for lithium-ion batteries. Different processing methods, such as solution casting [75], phase inversion [76] and electrospinning [77], are the major methods employed for the preparation of the electrolytes. However, the cast (<50% porosity) and phase-inverted (<70% porosity) membranes have poor values for porosity, mechanical properties and ionic conductivity [75, 76, 78]. The ionic conductivity and mechanical stability of these electrolytes are improved by incorporating ceramic fillers such as SiO$_2$ [79], TiO$_2$ [80] or Al$_2$O$_3$ [81, 82]. By incorporating the ceramic fillers in the PVdF-co-HFP, the ionic conductivity and electrochemical performance are significantly improved by strong Lewis acid-base interactions between the filler particles and the electrolyte polar groups, and between the filler particles and the polymer chains [11, 83–85]. The ceramic particles in the polymer matrix suppress the formation of spherulites or crystalline domains, leading to significant levels of reduction in crystallinity. The increased surface area and surface energy facilitate dissociation of more lithium salts, thereby improving the ionic conductivity. For instance, the PVdF-co-HFP/TiO$_2$ composite electrolyte, activated with 1 M LiPF$_6$ and dissolved in EC/DMC (1:1 w/w) electrolyte, delivers ionic conductivity of ~10^{-3} S cm^{-1} (at room temperature), compared with that of pure PVdF-co-HFP (~10^{-4} S cm^{-1}) [80].

PVdF-*co*-HFP/clay composites are attracting much interest, owing to their advantages over the conventional reinforcing composite structures. It is reported that there are three types of possible complex formation possible in P(VdF-*co*-HFP)-based electrolytes, when clay is incorporated as a filler-cum-ion conduction promoter as shown in Figure 9.9. When the membrane is activated with a liquid electrolyte, for instance 1 M LiPF$_6$, in aprotic solvents to transform the P(VdF-*co*-HFP) to a polymer gel electrolyte, the only possibility is the formation of Complex I (only with PVdF-*co*-HFP), whereas, in the polymer/clay nanocomposite electrolyte, Complex III (only with clay platelets) may form within the clay phase and Complex II (PVdF-*co*-HFP + clay platelets) at the interface of the P(VdF-*co*-HFP) molecule and the clay platelets, which plays a key role in ion conduction. The extent of the formation of Complexes II and III depends greatly on the amount of clay in the composite and the extent of exfoliation of the clay [54].

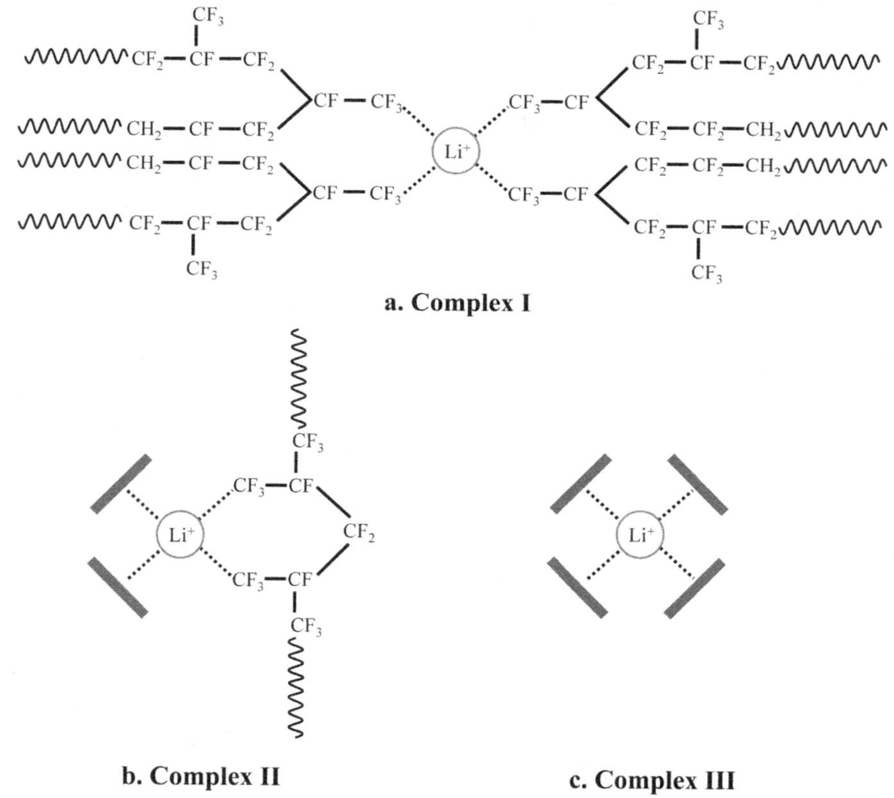

FIGURE 9.9 Graphical representation of complex formation of Li$^+$-cations with (a) P(VdF-*co*-HFP) chain (Complex I), (b) P(VdF-*co*-HFP) and clay platelets (Complex II), and (c) clay platelets (Complex III). Adapted and reproduced with permission from Ref. [54]. Copyright© 2013 Elsevier.

Polymer Clay Nanocomposite Electrolytes

Wang et al. [86, 87] fabricated the PVdF-*co*-HFP/clay composite polymer electrolyte by the conventional solvent-casting method, after modifying the clay layers with alkyl ammonium cations ($C_{16}H_{33}(CH_3)3N^+$). Ultrasonication is used to achieve the clay dispersion as well as the composite polymer solution in the study and is reported to be a good method to aid the delamination process and to prepare the polymer/clay nanocomposite [53–55, 88]. In order to avoid the pillar effect, caused by the MMT clay, the modified MMT clay is incorporated/dispersed as exfoliate sheets into the PVdF-*co*-HFP and the resulting nanocomposite exhibits a good film-forming property as well as high dimensional stability. Different concentrations, varying from 0 to 8 wt.% MMT clay, were incorporated into the PVdF-*co*-HFP to prepare solid polymer electrolytes, with lithium trifluoromethane sulfonate (LiTf) as the lithium salt [86] and the gel polymer electrolyte with LiTf and the plasticizer PC (40 wt.%) [86] or 1 M LiPF$_6$ in EC:DMC (1:1 v/v), and the ionic properties were compared. With the increasing concentration of the lithium salt, LiTf, the ionic conductivities of both the solid polymer electrolyte and the gel polymer electrolyte formed by incorporation of the plasticizer (PC) are seen to be increased. The system consists of 11.14 wt.% of LiTf, exhibiting an ionic conductivity of 5.5×10^{-6} S cm^{-1} in the solid electrolyte, which is better than the solid polymer electrolyte formed without MMT (1.07×10^{-8} S cm^{-1}) at 25°C. When the concentration of MMT reaches about 6 wt.%, the dissociation of LiTf increases and forms ion triplets and doublets, which can result in high ionic conductivity. It is worthwhile noting that the ionic conductivity of the gel polymer electrolyte covers a range about four orders of magnitude higher than that exhibited by the solid polymer electrolyte with MMT. The higher ionic conductivity displayed by the gel polymer electrolyte can be attributed to the following reasons. It is known that the dissociation field effect can be strengthened by the use of high dielectric constant plasticizers. Also, the plasticizer can be absorbed in clay galleries as well as soft the polymer chain due to the high solubility of PVdF-*co*-HFP in PC, which can increase the free volume of the polymer electrolyte. The room-temperature conductivities of the polymer gel electrolytes are even higher than 10^{-3} S cm^{-1} [86]. The nanocomposite polymer electrolytes, with exfoliated organophilic MMT (3 wt.%), show relatively low ionic conductivity, approximately 1.6×10^{-8} S cm^{-1} for the electrolyte without the plasticizer EC/DMC and 3.5×10^{-7} S cm^{-1} for the electrolyte with 40 wt.% EC/DMC (1:1 v/v), respectively, at 25°C, whereas the gel polymer electrolyte, which was made with 1 M LiPF$_6$ in EC:DMC, has a comparatively high ionic conductivity (about 2×10^{-3} S cm^{-1}). It is worthwhile noting that the liquid electrolyte 1 M LiPF$_6$ in EC: DMC (1:1 v/v) has an ionic conductivity at around 11.5×10^{-3} S cm^{-1} at room temperature, while the polymer clay nanocomposite film becomes brittle when the organophilic MMT content exceeds 5 wt.% [87]. The composite membranes show good electrolyte uptake, that reflects the affinity of the composite membrane toward the organic electrolyte, and display high cation-transference numbers, which are essential to achieving the best electrochemical performance. According to the method proposed by Vincent [89], the estimated Li$^+$-transference number is reported to be 0.36 to 0.64 for the gel polymer electrolyte incorporating 0 to 4 wt.% organophilic MMT [87]. The increasing transference number with clay loading illustrates that the electronegative platelets of silicate play an important role in cation transportation. In addition, all of the composite electrolytes show an anodic stability of about 4.2 V *versus* Li/Li⁺ [86].

The evaluation of electrochemical performance in meso-carbon micro-beads, MCMB/LiCoO$_2$ cells, displayed good cycling stability and rate capability with gel polymer electrolytes containing 3 wt.% MMT. At a higher current density of 1 C rate, the cell could deliver about 89% of its 0.5 C capacity. When the current density is pushed to 2 C rate, the cell could still deliver a discharge about 74% of that exhibited at a discharge rate of 0.5 C. On the other hand, the cell with a gel polymer electrolyte at 0 wt.% MMT displayed a reasonably good charge/discharge capacity at low C rates, but showed poorer capacity at higher C rates. The cycling stability studied on the cell charged at 0.1 C and discharged at 0.5 C showed a discharge capacity of about 120 mAh g^{-1} for the first few cycles, then the capacity started to fade gradually, retaining about 85% of the initial discharge capacity after 50 cycles, indicating that the interfaces between the composite electrolyte and the electrodes in the battery are stable under active conditions.

To enhance the electrochemical performance, especially the oxidative and thermal stability of the polymer electrolyte for high-voltage LIBs, incorporation of room-temperature ionic liquids, such as those based on organic cations like imidazolium [13, 90–95], pyrrolidinium [96–100] and piperidinium [101, 102], has proved to be among the most effective methods and is the alternative choice for replacing organic electrolytes, which have serious safety issues. Solvent-cast polymer electrolytes can be treated with the ionic liquids in order to improve their thermal, chemical and electrochemical properties. PVdF-*co*-HFP/MMT composite electrolytes, prepared with different concentrations of MMT loading (1.5 to 10 wt.% loading into PVdF-*co*-HFP on a weight basis) are treated with 1-butyl-3-methylimidazolium bromide (BMIMBr) as room-temperature ionic liquids, that can exhibit good electrochemical performance (anodic stability >5 V) even without the presence of organic solvents. A 5 wt.% MMT-containing system, treated with 1-butyl-3-methylimidazolium bromide, exhibits an ionic conductivity of 9.8×10^{-3} S cm^{-1} [103], which is higher than the previous reports [86, 87]. The PVdF-*co*-HFP containing 5 wt.% MMT exhibits a uniform porous structure that can easily entrap the ionic liquid (Figure 9.10). The same compositions achieve an electrochemical stability 5.5 V *versus* Li/Li⁺, which is high enough to exhibit improved electrochemical performance [103].

An electrospun PVdF-*co*-HFP/clay nanocomposite was fabricated by Shubha et al. [54] by dispersing different concentrations of organically modified MMT clay (0, 1 or 2 wt.%). The electrospinning was adopted for the preparation of fibrous host membranes by keeping a constant solution concentration of 16 wt.%, maintaining the gap between the collection drum and the tip of the needle to 20 cm, and keeping the applied voltage at 16–20 kV (depending on the viscosity and clay content of the composite solution) for a feed rate of 0.2 ml min^{-1} [54]. The surface morphology and average fiber diameter of the electrospun membrane is displayed in Figure 9.11a–c. The electrospun membrane consists of a multilayered fibrous morphology formed with ultrafine smooth fibers of bead-free morphology. The average fiber diameter (AFD) of the membrane, electrolyte uptake and electrolyte retention capacity are observed to increase with increasing clay content. Figure 9.11d shows the AFD of the PVdF-*co*-HFP membrane prepared with 1 wt.% clay content. The membrane

Polymer Clay Nanocomposite Electrolytes 203

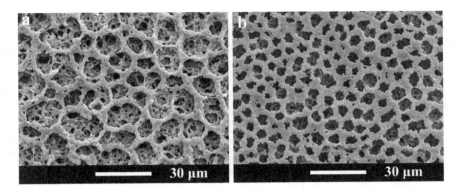

FIGURE 9.10 FE-SEM micrographs of the structural morphology of PVdF-*co*-HFP with 5 wt.% MMT composite electrolytes before and after treatment with 1-butyl-3-methylimidazolium bromide as room-temperature ionic liquid (RTIL): (a) before immersion in RTIL; (b) after immersion in RTIL. Adapted and reproduced with permission from Ref. [103]. Copyright © 2013 Springer.

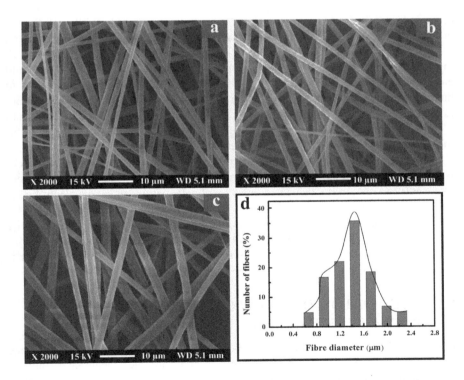

FIGURE 9.11 Surface morphology of electrospun PVdF-*co*-HFP/clay nanocomposite membranes with different clay concentrations (wt.%): (a) 0, (b) 1 and (c) 2 wt.%, and (d) fiber diameter distribution histogram of electrospun PVdF-*co*-HFP/clay nanocomposite membranes with 1 wt.% clay content. Adapted and reproduced with permission from Ref. [54]. Copyright © 2013, Elsevier.

exhibits an AFD of 1.6 μm and this clay concentration is optimal, exhibiting an electrochemical performance better than that of any other clay concentration. Along with the electrochemical performance, the thermal stability of the composite appears to be enhanced by increasing the clay content. This is attributed to the restricted chain movement in nanocomposites, that will hinder the chain scission. The presence of clay platelets with a high aspect ratio can further hamper the outward diffusion of the volatile products and thereby slow down the degradation of underlying material.

The electrolyte with 1 wt.% clay showed the highest ionic conductivity (5.5×10^{-2} mS cm^{-1}) at 30°C, with 2 wt.% clay showing a lower ionic conductivity (4.1 mS cm^{-1}), and, in turn, is higher than that of the pure PVdF-*co*-HFP (2.9 mS cm^{-1}). Compared to the electrolyte with 1 wt.% clay, the ionic conductivity of the electrolyte with 2 wt.% clay is lower even though it shows a higher electrolyte uptake. This unexpected phenomenon is attributed to the fact that the higher loading of clay results in the delamination of excess clay platelets, which may bind the Li$^+$-ions tightly, thus restricting the mobility of cations and reducing the conductivity. Also, the higher porosity and lower AFD may play a role in determining the higher ionic conductivity [104]. The earlier reports of the ionic conductivities (mS cm^{-1}) of electrospun PVdF-*co*-HFP-based composite electrolytes incorporating 6 wt.% different ceramic fillers (SiO_2, Al_2O_3 or $BaTiO_3$) activated with (i) 0.5 M LiTFSI in BMITFSI at 30°C, with ionic conductivities of 3.6 mS cm^{-1} (SiO_2), 4.4 mS cm^{-1} (Al_2O_3) and 5.2 mS cm^{-1} ($BaTiO_3$) [13] and (ii) 1 M LiPF$_6$ in EC/DMC (1:1, v/v) at 25°C as 5.92 mS cm^{-1} (Al_2O_3), 6.45 mS cm^{-1} (SiO_2) and 7.21 mS cm^{-1} ($BaTiO_3$) [12]. The liquid electrolytes, 1 M LiPF$_6$ in EC/DMC and 1 M LiPF$_6$ in EC/DEC are reported to have ionic conductivity of 10.7×10^{-3} S cm^{-1} [105, 106], at 25°C. The anodic stability increases from 4.6 V to 5.1 V *versus* Li/Li$^+$, by incorporation of 1 wt.% of clay into PVdF-*co*-HFP, which may due to the good affinity of the composite membrane for the carbonate-based liquid electrolyte solution, which partially swells the fibers.

Initial charge/discharge properties of composite electrolytes in a Li/LiFePO$_4$ cell, with or without clay, exhibit discharge capacities of 160 and 145 mAh g^{-1}, respectively at 0.1 C rate. Initial discharge capacity of cells with a composite electrolyte is 10% higher than in cells based on the pure PVdF-*co*-HFP, which corresponds to 94% utilization of the active material. In addition, the cells show better cycling stability and rate capability for all rate sweeps studied, up to 3 C. After 50 continuous charge/discharge cycles, the cell delivers a discharge capacity of 147 mAh g^{-1}, which is about 17% higher than the cell containing electrolyte without clay [54].

9.5 POLYACRYLONITRILE/CLAY COMPOSITE POLYMER ELECTROLYTES

Polyacrylonitrile (PAN) is a special kind of polymer with high processability, flame resistance, resistance to oxidative degradation and good electrochemical stability. The oxidative stability of PAN is very high, even at high temperatures, e.g., 300°C. Among the different polymeric materials used for the preparation of polymer electrolytes, PAN-based polymer electrolytes exhibit interesting characteristics, like high ionic conductivity, thermal stability, high affinity for liquid electrolytes, high electrolyte uptake and good compatibility with lithium metal electrodes, minimizing dendrite growth during the charging/discharging process of LIBs [29, 107]. The presence of polar –CN groups in PAN could interact with –CO groups of the liquid electrolytes, such as propylene carbonate (PC), ethylene carbonate (EC), etc., as well with lithium ions [8, 30, 108, 109], that will be beneficial for ionic conduction within the polymer electrolyte. Compared with other polymer electrolyte matrices, such as polyethylene oxide (PEO), PAN exhibits high mechanical properties and dimensional stability. The PAN/clay composite electrolyte system is not very well studied and there are few reports on the performance of this composite electrolyte. Wei et al. [110] reported the performance of a solid polymer electrolyte fabricated by using PAN/LiCF$_3$SO$_3$/MMT. The clay used in the study is cetylpyridinium chloride (CPC)-modified montmorillonite. The solid polymer electrolyte is fabricated by the solvent-casting technique, where dimethylformamide (DMF) is used as the solvent. The highly polar –CN group in PAN forms complexes with Li$^+$-ions, and the presence of MMT leads to the formation of ionic aggregates that can improve the performance of the composite electrolyte. As discussed in the aforementioned sections (Sections 9.3 and 9.4), the ionic conductivity of the composite electrolyte increases with increasing MMT content, with the highest ionic conductivity being observed for the composite electrolyte with 6 wt.% clay loading at 3.6×10^{-7} S cm^{-1} at 40°C, which is about 80-fold higher than that of the pure PAN/LiCF$_3$SO$_3$ system (PAN/LiCF$_3$SO$_3$ equivalent ratio = 8). At higher clay concentrations, the conductivity is seen to be reduced. At a temperature of 40°C, the ionic conductivity observed is about eight-fold higher than that of the electrolyte without clay. At clay concentrations greater than the optimum concentration, the clay particles can inhibit the performance, which may due to the ability of the excess clay particles to form strong bonds with lithium salt and thereby hinder the movement of the Li$^+$-ions. The study compared the performance of the electrolyte with untreated and treated clay against the pure PAN electrolyte and observed that the ionic conductivity is two- and eight-fold higher than for the polymer electrolyte with untreated and treated clay, respectively, than that without clay. As a consequence, the composite gel polymer electrolyte, formed with the addition of treated clay particles, exhibits better electrochemical performance [110].

A similar effect of modified clay on the electrochemical performance of a PAN-based polymer electrolyte is reported by Hwang et al. [111] (gel polymer electrolyte) and Yang et al. [112] (solid polymer electrolyte), in which the MMT clay used is modified with quaternary alkyl ammonium salts, with the electrolyte being prepared by solvent casting. The lithium salt used in both studies is LiClO$_4$. The polymer electrolyte is fabricated by using a simple solvent-casting technique, incorporating different concentrations of clay (1 to 9 wt.%). From the X-ray diffraction analysis (XRD), it has been reported that the interaction of modified clay with the PAN matrix leads to a disturbance of the order of the crystalline domain, thereby increasing the amorphous content within the PAN structure and resulting in an improved ionic conductivity. The highest ionic conductivity reported for the gel polymer electrolyte is about 1.0×10^{-2} S cm^{-1} at room temperature (the composite gel electrolyte having 3 wt.% clay, with PC/EC 1:1 (v/v) at 60% and F–0.35, where F is the concentration of lithium salt, which is expressed as the molar ratio of salt fed to a PAN repeat unit, i.e.; F = [LiClO$_4$]/[CH$_2$CH(CN) repeating unit in PAN]) which is the best result in solvent-cast gel polymer electrolyte. When the PC/EC content in the same sample is increased from 60 to 90%, the ionic conductivity at room temperature increases to 1.4×10^{-2} S cm^{-1}. Interestingly, it is reported that the ionic conductivity was increased up to 3- to 4-fold when the percentage of PC/EC cosolvent was increased from 40% to 90% [111]. Up to the concentration of 5 wt.% clay [111], the system exhibits both an exfoliation and an intercalation structure, which is similar to the observation reported by Wei et al. [110]. In addition to this, the composite membrane displayed an ability to hold higher concentrations of plasticizer, e.g., propylene carbonate (PC) or ethylene carbonate (EC), than did the electrolyte formed without the clay. At higher loadings of the plasticizer, the membrane shows greater smoothness; up to the loading of 70 wt.% plasticizer, the composite polymer gel electrolyte combines sufficient mechanical stability along with better electrochemical performance. Another important requirement for the proper electrochemical performance of the gel electrolyte is its dimensional stability. If the polymer electrolyte could not maintain its dimension under pressure or a higher temperature, this would result in a short-circuit of the battery. The composite gel polymer electrolyte, containing 3 wt.% organophilic clay, 15 wt.% PC/EC mixed solvents, and LiClO$_4$ (F- 0.25), exhibits a thermal expansion coefficient of 778×10^{-6} mm °C^{-1}, whereas for the gel electrolyte without the clay the thermal expansion coefficient at the same temperature is about 3400×10^{-6} mm °C^{-1}. This result indicates that the addition of clay can improve the dimensional stability of the polymer composite electrolyte. This is attributed to the presence of exfoliated clay layers, which become uniformly dispersed throughout the PAN matrix and maintain a stable inter-molecular interaction. In addition to this, the PAN/clay composite electrolyte can achieve a stable electrochemical stability of 4.2 V [111].

Using the similar fabrication (solvent-casting) method, the ionic conductivity reported by Yang et al. [112] for the PAN/clay composite solid electrolyte, using the same lithium salt (LlClO$_4$), is significantly lower than that reported by Hwang et al. [111]. The solid polymer electrolyte fabricated without clay can exhibit an ionic conductivity of 3.4×10^{-5} S cm^{-1} at 30°C. Addition of clay increases the conductivity, with the highest ionic conductivity being noted for the composite electrolyte containing 7 wt.% of clay (2.44×10^{-4} S cm^{-1} at 30°C), which is more than seven times higher than that obtained from CPE [112] but much lower than that reported by Hwang et al. [111] for the gel polymer electrolyte with a clay concentration of 3 wt.% (1.4×10^{-2} S cm^{-1} at room temperature). However, the ionic conductivity of the composite solid polymer electrolyte, with or without clay, even at 30°C [112] is significantly higher than that reported for the solid polymer electrolyte PAN/LiCF$_3$SO$_3$ system (PAN/LiCF$_3$SO] equivalent ratio = 8). with 6 wt.% clay loading at 40°C (3.6×10^{-7} S cm^{-1}) by Wei et al. [110]. This difference in ionic conductivity may be due to the difference in type and amount of lithium salt used in the different studies. The fractured surface (Figure 9.12) shows that the PAN membrane exhibits a brittle surface, whereas the addition of clay reduces the brittleness, indicating the ability of the filler to resist the fractures, which may be the direct effect of the increase in amorphous content. On evaluating the electrochemical performance, the cell fabricated by using this gel

Polymer Clay Nanocomposite Electrolytes

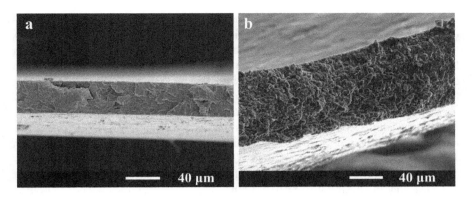

FIGURE 9.12 FE-SEM micrographs of fracture surface morphology of polyacrylonitrile (PAN)-based gel polymer electrolyte with hydrophobic clay prepared by ion-exchange reaction of MMT with a carboxy-terminal alkyl ammonium salt (ALA-MMT): (a) plain PAN, and (b) PAN/ALA–MMT (7 wt.%) composite. Adapted and reproduced with permission from Ref. [112]. Copyright© 2009, Elsevier

polymer electrolyte exhibits well-maintained anodic and cathodic peaks even after 50 cycles and a stable electrochemical stability and cyclability over a potential range of 0 to 4 V versus Li/Li[+] [112].

9.6 POLYMETHYL METHACRYLATE/CLAY COMPOSITE POLYMER ELECTROLYTES

Polymethyl methacrylate (PMMA), an amorphous thermoplastic polymer with water-like transparency, is a promising matrix for polymer electrolytes in LIBs. PMMA exhibits a good interfacial stability with the lithium electrode. In addition to this, the polymer exhibits high electrolyte uptake, ionic conductivity and good electrochemical stability. However, PMMA-based gel polymer electrolytes suffer from poor mechanical stability and toughness, that can be improved by either chemical cross-linking between the polymer chains or incorporating a reinforcing filler into the polymer matrix. Like other polymer matrices, such as PVdF, PVdF-co-HFP, PAN, PEO, etc., PMMA-based polymer composite electrolytes incorporated with clay are also less explored. There are only a few reports on the PMMA/clay composite electrolyte and, in all those, the electrolytes are fabricated by either pressing between glass sheets [113] or by solution casting [114, 115], and the lithium salt used is $LiClO_4$. For the preparation of gel electrolytes, the aprotic solvent used is either a mixture of propylene carbonate (PC) and ethylene carbonate (EC) [113] or a mixture of propylene carbonate (PC) and diethyl carbonate (DEC) [115]. Composite polymer electrolytes of high-molecular- weight PMMA/clay were synthesized by in-situ polymerization by Meneghetti et al. [113] or by solution mixing by Deka et al. [115]. The MMT clay was ion-exchanged with a zwitterionic surfactant (octadecyl dimethyl betaine), dispersed in methyl methacrylate (MMA) and polymerized to form the composite polymer [113] or mixed with PMMA solution (15% solid content in THF) by mechanical stirring [115]. The nanocomposite was dissolved in a mixture of EC/PC with LiClO$_4$, heated and either pressed between glass plates [113] or mixed with LiClO$_4$ and EC in dry acetonitrile [114], or the free-standing solvent-cast film is activated with 1 M LiClO$_4$ in a mixture of DEC/PC (1:1 v/v) [115] to obtain the polymer gel electrolyte. PMMA/clay with different clay concentrations, varying from 0 to 10 wt.%, is prepared by *in-situ* polymerization [113, 114], whereas PMMA/clay with 0 to 5 wt.% clay loading is prepared by the solvent-casting technique [115].

With the addition of the clay, the glass transition temperature (T_g) of the electrolyte increases, as measured by differential scanning calorimetry (DSC). The increase in T_g reflects the more amorphous nature of the polymer. The ionic conductivity exhibited by the polymer electrolyte without clay is 5×10^{-4} S cm^{-1}, with the addition of 1.5 wt.% clay increasing it to 8×10^{-4} S cm^{-1}, whereas, when the clay concentration reached 3 wt.%, the ionic conductivity decreased to 4.5×10^{-4} S cm^{-1}. From these results, it is evident that, at higher clay concentrations, the composite electrolyte is unable to exhibit the properties of even the plain electrolyte. This is due to the clay layers becoming trapped inside the polymer matrix, inhibiting the properties at higher concentrations. In addition to this, the fabricated PMMA/MMT composite gel polymer electrolyte can exhibit a stable interfacial stability over 3 weeks [115].

Dimensionally stable and ionically conducting, the PMMA/MMT composite electrolyte was synthesized by using MMT modified with dimethyldioctadecylammonium chloride (DDAC, D-clay) and the resulting structure of the complex, formed by Li$^+$ cations, PMMA chains and silicate layers, is depicted in Figure 9.13a [114]. The PMMA/clay composite polymer gel electrolyte is obtained by activating the composite membrane with EC containing the lithium salt LiClO$_4$. The property of the electrolyte was studied on the basis of the interacting mechanism (Figure 9.13b) taking place in it. The well-dispersed silicate layer in the polymer electrolyte carries negative charges that can act as dipoles. This well-distributed negative charge provides a better environment for the dissolution of the lithium salt. More lithium salt will get dissolved and the C=O group of PMMA will interact with the lithium ions to enhance the electrochemical performance. However, the ionic conductivity reported by this PMMA/Dclay composite electrolyte is 6×10^{-4} S cm^{-1} which is lower than that reported by Meneghetti et al. [113] for the PMMA/clay composite host membrane synthesized by *in-situ* polymerization and activated with 1 M LiClO$_4$ in the mixture of EC/PC [113].

Compared with Meneghetti et al. [113] and Chen et al. [114], much better results are reported by Deka et al. [115] for the partially exfoliated nanocomposite polymer electrolyte fabricated by using the solution-intercalation technique. The composite polymer gel electrolyte containing 5 wt.% clay exhibits high solvent absorption and thereby greater swelling than by any other clay concentrations. The same membrane exhibits the highest ionic conductivity of 1.3×10^{-3} S cm^{-1} at room temperature. When the temperature reaches 115°C, the conductivity of the clay-free electrolyte is reduced rapidly due to the softening of the polymer at high temperature. But, for the composite electrolyte, the conductivity is seen to increase up to 135°C, a phenomenon attributed to the presence of MMT. The MMT can limit the fluidity and thereby enhance the mechanical stability of the electrolyte. Hence, this polymer electrolyte

Polymer Clay Nanocomposite Electrolytes

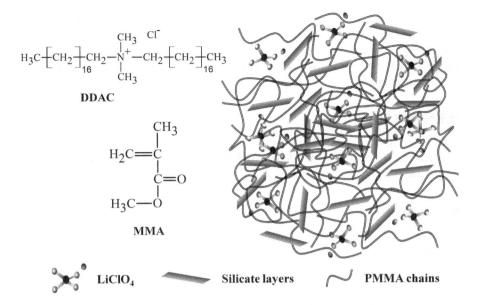

FIGURE 9.13 The chemical structure and schematic illustration of the complex lithium ion-transference channels formed in composite polymer electrolyte of MMT modified with dimethyldioctadecylammonium chloride (DDAC) (D-clay) with lithium salt: (a) Chemical structure of DDAC and MMA; (b) Schematic structure of the complex formed by Li⁺ cations, PMMA chains and silicate layers. Adapted and reproduced with permission from Ref. [114]. Copyright© 2009, Elsevier

is usable for the high-temperature applications as well. It is also reported that that the anodic limit of nanocomposite gel polymer electrolytes increases with increasing clay loading and reaches about 4.9 V for 5 wt.% clay loading, which is higher than that of the gel polymer electrolyte (about 4.5 V) [115]. Thus, there is a clear improvement in the voltage stability factor in the electrolyte films containing clay.

9.7 CONCLUSION

The composite gel polymer electrolytes are regarded as being promising for LIBs because of their high thermal and mechanical stabilities and their electrochemical properties. This chapter discusses the performance of composite gel polymer electrolytes formed with the incorporation of clay as the filler. Clay is an attractive host for intercalating the polymer chains and the resulting composite can exhibit the properties of both the intercalating guest and the host material. The MMT clays exhibiting excellent properties are chosen as the filler from among other clay minerals. Different polymer matrixes like PVdF, PVdF-*co*-HFP, PMMA and PAN are explored, as the matrix for the incorporation of the filler and each composite membrane exhibit better electrochemical properties than that of the gel polymer electrolyte without clay. The chapter discusses the mechanisms by which clay layers enhance the performance of the composite electrolyte, and the impact of the fabrication process on the improvement of thermal, mechanical and electrochemical properties is recognized. Even though clays are not well-studied fillers for the polymer electrolytes in LIBs, they can improve the mechanical, thermal and chemical stability of the gel polymer electrolyte to a great extent.

ACKNOWLEDGMENT

Authors Dr. Jabeen Fatima M. J. and Dr. Prasanth Raghavan would like to acknowledge Kerala State Council for Science, Technology and Environment (KSCSTE), Kerala, India, for financial assistance.

REFERENCES

1. Xue Z, He D, Xie X (2015) Poly(ethylene oxide)-based electrolytes for lithium-ion batteries. *J Mater Chem A* 3:19218–19253. https://doi.org/10.1039/c5ta03471j
2. Hu P, Chai J, Duan Y, et al. (2016) Progress in nitrile-based polymer electrolytes for high performance lithium batteries. *J Mater Chem A* 4:10070–10083. https://doi.org/10.1039/c6ta02907h
3. Tan CG, Siew WO, Pang WL, et al. (2007) The effects of ceramic fillers on the PMMA-based polymer electrolyte systems. *Ionics (Kiel)* 13:361–364. https://doi.org/10.1007/s11581-007-0126-7
4. Wang F, Li L, Yang X, et al. (2018) Influence of additives in a PVDF-based solid polymer electrolyte on conductivity and Li-ion battery performance. *Sustain Energy Fuels* 2:492–498. https://doi.org/10.1039/c7se00441a
5. Chen G, Zhang F, Zhou Z, et al. (2018) A flexible dual-ion battery based on PVDF-HFP-modified gel polymer electrolyte with excellent cycling performance and superior rate capability. *Adv Energy Mater* 8:1–7. https://doi.org/10.1002/aenm.201801219
6. Miao R, Liu B, Zhu Z, et al. (2008) PVDF-HFP-based porous polymer electrolyte membranes for lithium-ion batteries. *J Power Sources* 184:420–426. https://doi.org/10.1016/j.jpowsour.2008.03.045
7. Raghavan P, Zhao X, Shin C, et al. (2010) Preparation and electrochemical characterization of polymer electrolytes based on electrospun poly(vinylidene fluoride-co-hexafluoropropylene)/polyacrylonitrile blend/composite membranes for lithium batteries. *J Power Sources* 195:6088–6094. https://doi.org/10.1016/j.jpowsour.2009.11.098
8. Prasanth R, Aravindan V, Srinivasan M (2012) Novel polymer electrolyte based on cob-web electrospun multi component polymer blend of polyacrylonitrile/poly(methyl methacrylate)/polystyrene for lithium ion batteries - preparation and electrochemical characterization. *J Power Sources* 202:299–307. https://doi.org/10.1016/j.jpowsour.2011.11.057
9. Prasanth R, Shubha N, Hng HH, Srinivasan M (2014) Effect of poly(ethylene oxide) on ionic conductivity and electrochemical properties of poly(vinylidenefluoride) based polymer gel electrolytes prepared by electrospinning for lithium ion batteries. *J Power Sources* 245:283–291. https://doi.org/10.1016/j.jpowsour.2013.05.178
10. Shubha N, Prasanth R, Hng HH, Srinivasan M (2014) Study on effect of poly (ethylene oxide) addition and in-situ porosity generation on poly (vinylidene fluoride)-glass ceramic composite membranes for lithium polymer batteries. *J Power Sources* 267:48–57. https://doi.org/10.1016/j.jpowsour.2014.05.074

11. Raghavan P, Choi J-W, Ahn J-H, et al. (2008) Novel electrospun poly(vinylidene fluoride-co-hexafluoropropylene)–in situ SiO2 composite membrane-based polymer electrolyte for lithium batteries. *J Power Sources* 184:437–443. https://doi.org/10.1016/j.jpowsour.2008.03.027
12. Raghavan P, Zhao X, Kim JK, et al. (2008) Ionic conductivity and electrochemical properties of nanocomposite polymer electrolytes based on electrospun poly(vinylidene fluoride-co-hexafluoropropylene) with nano-sized ceramic fillers. *Electrochim Acta* 54:228–234. https://doi.org/10.1016/j.electacta.2008.08.007
13. Raghavan P, Zhao X, Manuel J, et al. (2010) Electrochemical performance of electrospun poly(vinylidene fluoride-co-hexafluoropropylene)-based nanocomposite polymer electrolytes incorporating ceramic fillers and room temperature ionic liquid. *Electrochim Acta* 55:1347–1354. https://doi.org/10.1016/j.electacta.2009.05.025
14. Pawlicka A, Machado GO, Guimaraes K V, Dragunski DC (2003) Solid polymeric electrolytes obtained from modified natural polymers. In: *Proc. SPIE*.
15. Xu Z, Li W, Chen Z, et al. (2019) Chemically modified polyvinyl butyral polymer membrane as a gel electrolyte for lithium ion battery applications. *Macromol Mater Eng* 304:1800477. https://doi.org/10.1002/mame.201800477
16. Takahashi T, Davis GT, Chiang CK, Harding CA (1986) Chemical modification of poly(ethylene imine) for polymeric electrolyte. *Solid State Ionics* 18–19:321–325. https://doi.org/10.1016/0167-2738(86)90134-7
17. Mobarak NN, Ahmad A, Abdullah MP, et al. (2013) Conductivity enhancement via chemical modification of chitosan based green polymer electrolyte. *Electrochim Acta* 92:161–167. https://doi.org/10.1016/j.electacta.2012.12.126
18. Karatas Y, Kaskhedikar N, Burjanadze M, Wiemhöfer H-D (2006) Synthesis of crosslinked comb polysiloxane for polymer electrolyte membranes. *Macromol Chem Phys* 207:419–425. https://doi.org/10.1002/macp.200500470
19. Guo M, Zhang M, He D, et al. (2017) Comb-like solid polymer electrolyte based on polyethylene glycol-grafted sulfonated polyether ether ketone. *Electrochim Acta* 255:396–404. https://doi.org/10.1016/j.electacta.2017.10.033
20. Leijonmarck S, Cornell A, Lindbergh G, Wågberg L (2013) Single-paper flexible Li-ion battery cells through a paper-making process based on nano-fibrillated cellulose. *J Mater Chem A* 1:4671–4677. https://doi.org/10.1039/c3ta01532g
21. Sheng J, Tong S, He Z, Yang R (2017) Recent developments of cellulose materials for lithium-ion battery separators. *Cellulose* 24:4103–4122. https://doi.org/10.1007/s10570-017-1421-8
22. Kim JH, Gu M, Lee DH, et al. (2016) Functionalized nanocellulose-integrated heterolayered nanomats toward smart battery separators. *Nano Lett* 16:5533–5541. https://doi.org/10.1021/acs.nanolett.6b02069
23. Lavoine N, Desloges I, Dufresne A, Bras J (2012) Microfibrillated cellulose - Its barrier properties and applications in cellulosic materials: A review. *Carbohydr Polym* 90:735–764. https://doi.org/10.1016/j.carbpol.2012.05.026
24. Ran Y, Yin Z, Ding Z, et al. (2013) A polymer electrolyte based on poly(vinylidene fluoride-hexafluoropropylene)/hydroxypropyl methyl cellulose blending for lithium-ion battery. *Ionics (Kiel)* 19:757–762. https://doi.org/10.1007/s11581-012-0808-7
25. Liu B, Huang Y, Cao H, et al. (2018) A high-performance and environment-friendly gel polymer electrolyte for lithium ion battery based on composited lignin membrane. *J Solid State Electrochem* 22:807–816. https://doi.org/10.1007/s10008-017-3814-x
26. Stephan AM, Kumar TP, Kulandainathan MA, Lakshmi NA (2009) Chitin-incorporated poly(ethylene oxide)-based nanocomposite electrolytes for lithium batteries. *J Phys Chem B* 113:1963–1971. https://doi.org/10.1021/jp808640j

27. Aziz NA, Majid SR, Yahya R, Arof AK (2011) Conductivity, structure, and thermal properties of chitosan-based polymer electrolytes with nanofillers. *Polym Adv Technol* 22:1345–1348. https://doi.org/10.1002/pat.1619
28. Tang C, Hackenberg K, Fu Q, et al. (2012) High ion conducting polymer nanocomposite electrolytes using hybrid nanofillers. *Nano Lett* 12:1152–1156. https://doi.org/10.1021/nl202692y
29. Kammoun M, Berg S, Ardebili H (2015) Flexible thin-film battery based on graphene-oxide embedded in solid polymer electrolyte. *Nanoscale* 7:17516–17522. https://doi.org/10.1039/c5nr04339e
30. Masoud EM, El-Bellihi A-A, Bayoumy WA, Mousa MA (2013) Organic–inorganic composite polymer electrolyte based on PEO–LiClO$_4$ and nano-Al$_2$O$_3$ filler for lithium polymer batteries: Dielectric and transport properties. *J Alloys Compd* 575:223–228. https://doi.org/10.1016/j.jallcom.2013.04.054
31. Liang B, Tang S, Jiang Q, et al. (2015) Preparation and characterization of PEO-PMMA polymer composite electrolytes doped with nano-Al$_2$O$_3$. *Electrochim Acta* 169:334–341. https://doi.org/10.1016/j.electacta.2015.04.039
32. Jeong H-S, Hong SC, Lee S-Y (2010) Effect of microporous structure on thermal shrinkage and electrochemical performance of Al$_2$O$_3$/poly(vinylidene fluoride-hexafluoropropylene) composite separators for lithium-ion batteries. *J Memb Sci* 364:177–182. https://doi.org/10.1016/j.memsci.2010.08.012
33. Liao YH, Li XP, Fu CH, et al. (2011) Polypropylene-supported and nano-Al$_2$O$_3$ doped poly(ethylene oxide)–poly(vinylidene fluoride-hexafluoropropylene)-based gel electrolyte for lithium ion batteries. *J Power Sources* 196:2115–2121. https://doi.org/10.1016/j.jpowsour.2010.10.062
34. Li Z, Su G, Wang X, Gao D (2005) Micro-porous P(VDF-HFP)-based polymer electrolyte filled with Al$_2$O$_3$ nanoparticles. *Solid State Ionics* 176:1903–1908. https://doi.org/10.1016/j.ssi.2005.05.006
35. Lin CW, Hung CL, Venkateswarlu M, Hwang BJ (2005) Influence of TiO$_2$ nano-particles on the transport properties of composite polymer electrolyte for lithium-ion batteries. *J Power Sources* 146:397–401. https://doi.org/10.1016/j.jpowsour.2005.03.028
36. Liu Y, Lee JY, Hong L (2003) Morphology, crystallinity, and electrochemical properties of in situ formed poly(ethylene oxide)/TiO$_2$ nanocomposite polymer electrolytes. *J Appl Polym Sci* 89:2815–2822. https://doi.org/10.1002/app.12487
37. Cui W-W, Tang D-Y, Gong Z-L (2013) Electrospun poly(vinylidene fluoride)/poly(methyl methacrylate) grafted TiO$_2$ composite nanofibrous membrane as polymer electrolyte for lithium-ion batteries. *J Power Sources* 223:206–213. https://doi.org/10.1016/j.jpowsour.2012.09.049
38. Zhou L, Wu N, Cao Q, et al. (2013) A novel electrospun PVDF/PMMA gel polymer electrolyte with in situ TiO$_2$ for Li-ion batteries. *Solid State Ionics* 249–250:93–97. https://doi.org/10.1016/j.ssi.2013.07.019
39. Cao J, Wang L, Shang Y, et al. (2013) Dispersibility of nano-TiO$_2$ on performance of composite polymer electrolytes for Li-ion batteries. *Electrochim Acta* 111:674–679. https://doi.org/10.1016/j.electacta.2013.08.048
40. Yang CL, Li ZH, Li WJ, et al. (2015) Batwing-like polymer membrane consisting of PMMA-grafted electrospun PVdF–SiO$_2$ nanocomposite fibers for lithium-ion batteries. *J Memb Sci* 495:341–350. https://doi.org/10.1016/j.memsci.2015.08.036
41. Lee Y-S, Ju SH, Kim J-H, et al. (2012) Composite gel polymer electrolytes containing core-shell structured SiO$_2$(Li$^+$) particles for lithium-ion polymer batteries. *Electrochem Commun* 17:18–21. https://doi.org/10.1016/j.elecom.2012.01.008

42. Park J-H, Cho J-H, Park W, et al. (2010) Close-packed SiO$_2$/poly(methyl methacrylate) binary nanoparticles-coated polyethylene separators for lithium-ion batteries. *J Power Sources* 195:8306–8310. https://doi.org/10.1016/j.jpowsour.2010.06.112
43. Zhang P, Yang LC, Li LL, et al. (2011) Enhanced electrochemical and mechanical properties of P(VDF-HFP)-based composite polymer electrolytes with SiO$_2$ nanowires. *J Memb Sci* 379:80–85. https://doi.org/10.1016/j.memsci.2011.05.043
44. Xie H, Liao Y, Sun P, et al. (2014) Investigation on polyethylene-supported and nano-SiO$_2$ doped poly(methyl methacrylate-co-butyl acrylate) based gel polymer electrolyte for high voltage lithium ion battery. *Electrochim Acta* 127:327–333. https://doi.org/10.1016/j.electacta.2014.02.038
45. Rajasudha G, Nancy AP, Paramasivam T, et al. (2011) Synthesis and characterization of polyindole–NiO-based composite polymer electrolyte with LiClO$_4$. *Int J Polym Mater Polym Biomater* 60:877–892. https://doi.org/10.1080/00914037.2010.551367
46. Liang Y, Lin Z, Qiu Y, Zhang X (2011) Fabrication and characterization of LATP/PAN composite fiber-based lithium-ion battery separators. *Electrochim Acta* 56:6474–6480. https://doi.org/10.1016/j.electacta.2011.05.007
47. Lin Y, Liu ke, Wu M, et al. (2020) Enabling solid-state li metal batteries by in-situ forming ionogel interlayers. *ACS Appl Energy Mater*. https://doi.org/10.1021/acsaem.0c00662
48. Giannees BEP (1996) Polymer layered silicate nanocomposites. *Adv Mater* 29–35.
49. Jeevanandam P, Vasudevan S (1998) Conductivity of a confined polymer electrolyte: Lithium-polypropylene glycol intercalated in layered CdPS3. *J Phys Chem B* 102:4753–4758. https://doi.org/10.1021/jp980357d
50. Yano K, Usuki A, Okada A (1997) Synthesis and properties of polyimide-clay hybrid films. *J Polym Sci Part A Polym Chem* 35:2289–2294. https://doi.org/10.1002/(SICI)1099-0518(199708)35:11<2289::AID-POLA20>3.0.CO;2-9
51. Uddin F (2018) Montmorillonite: An introduction to properties and utilization. *Curr Top Util Clay Ind Med Appl* 1.
52. Uddin F (2008) Clays, nanoclays, and montmorillonite minerals. *Metall Mater Trans A Phys Metall Mater Sci* 39:2804–2814. https://doi.org/10.1007/s11661-008-9603-5
53. Prasanth R, Shubha N, Hng HH, Srinivasan M (2013) Effect of nano-clay on ionic conductivity and electrochemical properties of poly(vinylidene fluoride) based nanocomposite porous polymer membranes and their application as polymer electrolyte in lithium ion batteries. *Eur Polym J* 49:307–318. https://doi.org/10.1016/j.eurpolymj.2012.10.033
54. Shubha N, Prasanth R, Hoon HH, Srinivasan M (2013) Dual phase polymer gel electrolyte based on non-woven poly(vinylidenefluoride-co-hexafluoropropylene)-layered clay nanocomposite fibrous membranes for lithium ion batteries. *Mater Res Bull* 48:526–537. https://doi.org/10.1016/j.materresbull.2012.11.002
55. Joshi M, Banerjee K, Prasanth R, Thakare V (2006) Polymer/clay nanocomposite based coatings for enhanced gas barrier property. *Indian J Fiber Text Res* 31:202–214
56. Shukla N, Thakur AK (2010) Ion transport model in exfoliated and intercalated polymer-clay nanocomposites. *Solid State Ionics* 181:921–932. https://doi.org/10.1016/j.ssi.2010.05.023
57. Jacob MME, Hackett E, Giannelis EP (2003) From nanocomposite to nanogel polymer electrolytes. *J Mater Chem* 13:1–5. https://doi.org/10.1039/b204458g
58. Kalaga K, Rodrigues MTF, Gullapalli H, et al. (2015) Quasi-solid electrolytes for high temperature lithium ion batteries. *ACS Appl Mater Interfaces* 7:25777–25783. https://doi.org/10.1021/acsami.5b07636

59. Rodrigues MTF, Kalaga K, Gullapalli H, et al. (2016) Hexagonal boron nitride-based electrolyte composite for Li-Ion battery operation from room temperature to 150°C. *Adv Energy Mater* 6:1–7. https://doi.org/10.1002/aenm.201600218
60. Jeong KU, Chae HD, Lim C Il, et al. (2010) Fabrication and characterization of electrolytemembranes based on organoclay/tripropyleneglycol diacrylate/poly(vinylidene fluoride) electrospun nanofiber composites. *Polym Int* 59:249–255. https://doi.org/10.1002/pi.2716
61. Muniyandi N, Kalaiselvi N, Periyasamy P, et al. (2001) Optimisation of PVdF-based polymer electrolytes. *J Power Sources* 96:14–19. https://doi.org/10.1016/S0378-7753(01)00562-6
62. Huang H, Wunder SL (2001) Preparation of microporous PVDF based polymer electrolytes. *J Power Sources* 97–98:649–653. https://doi.org/10.1016/S0378-7753(01)00579-1
63. Priya L, Jog JP (2002) Poly(vinylidene fluoride)/clay nanocomposites prepared by melt intercalation: Crystallization and dynamic mechanical behavior studies. *J Polym Sci Part B Polym Phys* 40:1682–1689. https://doi.org/10.1002/polb.10223
64. Priya L, Jog JP (2003) Polymorphism in intercalated poly(vinylidene fluoride)/clay nanocomposites. *J Appl Polym Sci* 89:2036–2040. https://doi.org/10.1002/app.12346
65. Hwang H-Y, Kim D-J, Kim H-J, et al. (2011) Effect of nanoclay on properties of porous PVdF membranes. *Trans Nonferrous Met Soc China* 21:s141–s147. https://doi.org/10.1016/S1003-6326(11)61078-9
66. Deka M, Kumar A (2011) Electrical and electrochemical studies of poly(vinylidene fluoride)-clay nanocomposite gel polymer electrolytes for Li-ion batteries. *J Power Sources* 196:1358–1364. https://doi.org/10.1016/j.jpowsour.2010.09.035
67. Koh MJ, Hwang HY, Kim DJ, et al. (2010) Preparation and characterization of porous PVdF-HFP / clay nanocomposite membranes. 26:633–638. https://doi.org/10.1016/S1005-0302(10)60098-9
68. Dyartanti ER, Purwanto A, Widiasa IN, Susanto H (2018) Ionic conductivity and cycling stability improvement of PVDF/Nano-clay using PVP as polymer electrolyte membranes for liFePo4 batteries. *Membranes (Basel)* 8:. https://doi.org/10.3390/membranes8030036
69. Ma Y, Li LB, Gao GX, et al. (2016) Effect of montmorillonite on the ionic conductivity and electrochemical properties of a composite solid polymer electrolyte based on polyvinylidenedifluoride/polyvinyl alcohol matrix for lithium ion batteries. *Electrochim Acta* 187:535–542. https://doi.org/10.1016/j.electacta.2015.11.099
70. Karasawa N, Goddard WA (1992) Force fields, structures, and properties of poly(vinylidene fluoride) crystals. *Macromolecules* 25:7268–7281. https://doi.org/10.1021/ma00052a031
71. Cardoso VF, Correia DM, Ribeiro C, et al. (2018) Fluorinated polymers as smart materials for advanced biomedical applications. *Polymers (Basel)* 10:1–26. https://doi.org/10.3390/polym10020161
72. Ruan L, Yao X, Chang Y, et al. (2018) Properties and applications of the β phase poly(vinylidene fluoride). *Polymer (Guildf)* 10:1–27. https://doi.org/10.3390/polym10030228
73. Dutta B, Kar E, Bose N, Mukherjee S (2015) Significant enhancement of the electroactive β-phase of PVDF by incorporating hydrothermally synthesized copper oxide nanoparticles. *RSC Adv* 5:105422–405434. https://doi.org/10.1039/c5ra21903e
74. Wang X, Sun F, Yin G, et al. (2018) Tactile-sensing based on flexible PVDF nanofibers via electrospinning : a review. *Sensors* 18:330. https://doi.org/10.3390/s18020330
75. Wu F, Feng T A study on PVDF-HFP gel polymer electrolyte for lithium-ion batteries. https://doi.org/10.1088/1757-899X/213/1/012036

76. Dawood Alamery HR, Hatim MI, Syarhabil M (2016) Investigating morphology of asymmetric PVDF-HFP membranes prepared by phase inversion. *Int J Eng Trends Technol* 37:149–155. https://doi.org/10.14445/22315381/ijett-v37p224
77. Kim JR, Choi SW, Jo SM, et al. (2005) Characterization and properties of P(VdF–HFP)-based fibrous polymer electrolyte membrane prepared by electrospinning. *J Electrochem Soc* 152:A295. https://doi.org/10.1149/1.1839531
78. Raghavan P, Zhao X, Manuel J, et al. (2010) Electrochemical studies on polymer electrolytes based on poly(vinylidene fluoride-co-hexafluoropropylene) membranes prepared by electrospinning and phase inversion-A comparative study. *Mater Res Bull* 45:362–366. https://doi.org/10.1016/j.materresbull.2009.12.001
79. Jeong HS, Noh JH, Hwang CG, et al. (2010) Effect of solvent-nonsolvent miscibility on morphology and electrochemical performance of SiO_2/PVdF-co-HFP-based composite separator membranes for safer lithium-ion batteries. *Macromol Chem Phys* 211:420–425. https://doi.org/10.1002/macp.200900490
80. Costa CM, Rodrigues HM, Gören A, et al. (2017) Preparation of poly(vinylidene fluoride) lithium-ion battery separators and their compatibilization with ionic liquid – A green solvent approach. *ChemistrySelect* 2:5394–5402. https://doi.org/10.1002/slct.201701028
81. Xie H, Tang Z, Li Z, et al. (2008) PVDF-HFP composite polymer electrolyte with excellent electrochemical properties for Li-ion batteries. *J Solid State Electrochem* 12:1497–1502. https://doi.org/10.1007/s10008-008-0511-9
82. Jeong HS, Kim DW, Jeong YU, Lee SY (2010) Effect of phase inversion on microporous structure development of Al2O3/poly(vinylidene fluoride-hexafluoropropylene)-based ceramic composite separators for lithium-ion batteries. *J Power Sources* 195:6116–6121. https://doi.org/10.1016/j.jpowsour.2009.10.085
83. Chung SH, Wang Y, Persi L, et al. (2001) Enhancement of ion transport in polymer electrolytes by addition of nanoscale inorganic oxides. *J Power Sources* 97–98:644–648. https://doi.org/10.1016/S0378-7753(01)00748-0
84. Croce F, Appetecchi GB, Persi L, Scrosati B (1998) Nanocomposite polymer electrolytes for lithium batteries. *Nature* 394:456–458. https://doi.org/10.1038/28818
85. Liu Y, Lee JY, Hong L (2004) In situ preparation of poly(ethylene oxide)–SiO2 composite polymer electrolytes. *J Power Sources* 129:303–311. https://doi.org/10.1016/j.jpowsour.2003.11.026
86. Wang M, Zhao F, Guo Z, Dong S (2004) clays nanocomposite lithium polymer electrolytes. 49:3595–3602. https://doi.org/10.1016/j.electacta.2004.03.028
87. Wang M, Dong S (2007) Enhanced electrochemical properties of nanocomposite polymer electrolyte based on copolymer with exfoliated clays. 170:425–432. https://doi.org/10.1016/j.jpowsour.2007.04.031
88. Yoonessi M, Toghiani H, Kingery WL, Pittman CU (2004) Preparation, characterization, and properties of exfoliated/delaminated organically modified clay/dicyclopentadiene resin nanocomposites. *Macromolecules* 37:2511–2518. https://doi.org/10.1021/ma0359483
89. Bruce PG, Vincent CA (1987) Steady state current flow in solid binary electrolyte cells. *J Electroanal Chem Interfacial Electrochem* 225:1–17. https://doi.org/10.1016/0022-0728(87)80001-3
90. Raghavan P, Zhao X, Choi H, et al. (2014) Electrochemical characterization of poly(vinylidene fluoride-co-hexafluoro propylene) based electrospun gel polymer electrolytes incorporating room temperature ionic liquids as green electrolytes for lithium batteries. *Solid State Ionics* 262:77–82. https://doi.org/10.1016/j.ssi.2013.10.044

91. Cheruvally G, Kim J-K, Choi J-W, et al. (2007) Electrospun polymer membrane activated with room temperature ionic liquid: Novel polymer electrolytes for lithium batteries. *J Power Sources* 172:863–869. https://doi.org/10.1016/j.jpowsour.2007.07.057
92. Choi JW, Cheruvally G, Kim YH, et al. (2007) Poly(ethylene oxide)-based polymer electrolyte incorporating room-temperature ionic liquid for lithium batteries. *Solid State Ionics* 178:1235–1241. https://doi.org/10.1016/j.ssi.2007.06.006
93. Garcia B, Lavallée S, Perron G, et al. (2004) Room temperature molten salts as lithium battery electrolyte. *Electrochim Acta* 49:4583–4588. https://doi.org/10.1016/j.electacta.2004.04.041
94. Nakagawa H, Izuchi S, Kuwana K, et al. (2003) Liquid and polymer gel electrolytes for lithium batteries composed of room-temperature molten salt doped by lithium salt. *J Electrochem Soc* 150:A695. https://doi.org/10.1149/1.1568939
95. Fung YS, Zhou RQ (1999) Room temperature molten salt as medium for lithium battery. *J Power Sources* 81–82:891–895. https://doi.org/10.1016/S0378-7753(99)00127-5
96. Shin J-H, Henderson WA, Passerini S (2005) PEO-based polymer electrolytes with ionic liquids and their use in lithium metal-polymer electrolyte batteries. *J Electrochem Soc* 152:A978. https://doi.org/10.1149/1.1890701
97. Shin J-H, Henderson WA, Scaccia S, et al. (2006) Solid-state Li/LiFePO₄ polymer electrolyte batteries incorporating an ionic liquid cycled at 40°C. *J Power Sources* 156:560–566. https://doi.org/10.1016/j.jpowsour.2005.06.026
98. Shin J-H, Henderson WA, Passerini S (2005) An elegant fix for polymer electrolytes. *Electrochem Solid-State Lett* 8:A125. https://doi.org/10.1149/1.1850387
99. Shin J-H, Henderson WA, Appetecchi GB, et al. (2005) Recent developments in the ENEA lithium metal battery project. *Electrochim Acta* 50:3859–3865. https://doi.org/10.1016/j.electacta.2005.02.049
100. Shin J-H, Henderson WA, Passerini S (2003) Ionic liquids to the rescue? Overcoming the ionic conductivity limitations of polymer electrolytes. *Electrochem Commun* 5:1016–1020. https://doi.org/10.1016/j.elecom.2003.09.017
101. Yuan LX, Feng JK, Ai XP, et al. (2006) Improved dischargeability and reversibility of sulfur cathode in a novel ionic liquid electrolyte. *Electrochem Commun* 8:610–614. https://doi.org/10.1016/j.elecom.2006.02.007
102. Sakaebe H, Matsumoto H (2003) N-Methyl-N-propylpiperidinium bis(trifluoromethanesulfonyl)imide (PP13–TFSI) – novel electrolyte base for Li battery. *Electrochem Commun* 5:594–598. https://doi.org/10.1016/S1388-2481(03)00137-1
103. Nath AK, Kumar A (2013) Ionic transport properties of PVdF-HFP-MMT intercalated nanocomposite electrolytes based on ionic liquid, 1-butyl-3- methylimidazolium bromide. 1393–1403. https://doi.org/10.1007/s11581-013-0878-1
104. Kim JR, Choi SW, Jo SM, et al. (2004) Electrospun PVdF-based fibrous polymer electrolytes for lithium ion polymer batteries. *Electrochim Acta* 50:69–75. https://doi.org/10.1016/j.electacta.2004.07.014
105. Schmidt M, Heider U, Kuehner A, et al. (2001) Lithium fluoroalkylphosphates: A new class of conducting salts for electrolytes for high energy lithium-ion batteries. *J Power Sources* 97–98:557–560. https://doi.org/10.1016/S0378-7753(01)00640-1
106. Hayashi K, Nemoto Y, Tobishima S, Yamaki J (1999) Mixed solvent electrolyte for high voltage lithium metal secondary cells. *Electrochim Acta* 44:2337–2344. https://doi.org/10.1016/S0013-4686(98)00374-0
107. Tsutsumi H, Matsuo A, Takase K, et al. (2000) Conductivity enhancement of polyacrylonitrile-based electrolytes by addition of cascade nitrile compounds. *J Power Sources* 90:33–38. https://doi.org/10.1016/S0378-7753(00)00444-4

108. Raghavan P, Zhao X, Shin C, et al. (2010) Preparation and electrochemical characterization of polymer electrolytes based on electrospun poly(vinylidene fluoride-co-hex afluoropropylene)/polyacrylonitrile blend/composite membranes for lithium batteries. *J Power Sources* 195:6088–6094. https://doi.org/10.1016/j.jpowsour.2009.11.098
109. Mohapatra A, Agarwal S, Genesereth M (1990) Li+-conductive solid polymer electrolytes with liquid-like conductivity. *J Electrochem Soc* 137:1657–1658. https://doi.org/10.1007/978-3-319-50127-7_24
110. Hsein-Wei Chen F-CC (2001) Interaction mechanism of a novel polymer electrolyte composed of poly(acrylonitrile), lithium triflate, and mineral clay. *J Polym Sci Part B Polym Phys* 39:2407–2419. https://doi.org/10.1002/polb.1212
111. Hwang JJ, Liu HJ (2002) Influence of organophilic clay on the morphology, plasticizer-maintaining ability, dimensional stability, and electrochemical properties of gel polyacrylonitrile (PAN) nanocomposite electrolytes. *Macromolecules* 35:7314–7319. https://doi.org/10.1021/ma020613r
112. Chen-Yang YW, Chen YT, Chen HC, et al. (2009) Effect of the addition of hydrophobic clay on the electrochemical property of polyacrylonitrile/LiClO$_4$ polymer electrolytes for lithium battery. *Polymer (Guildf)* 50:2856–2862. https://doi.org/10.1016/j.polymer.2009.04.023
113. Meneghetti P, Qutubuddin S, Webber A (2004) Synthesis of polymer gel electrolyte with high molecular weight poly(methyl methacrylate)-clay nanocomposite. *Electrochim Acta* 49:4923–4931. https://doi.org/10.1016/j.electacta.2004.06.023
114. Chen HW, Lin TP, Chang FC (2002) Ionic conductivity enhancement of the plasticized PMMA/LiClO4 polymer nanocomposite electrolyte containing clay. *Polymer (Guildf)* 43:5281–5288. https://doi.org/10.1016/S0032-3861(02)00339-7
115. Deka M, Kumar A (2010) Enhanced electrical and electrochemical properties of PMMA-clay nanocomposite gel polymer electrolytes. *Electrochim Acta* 55:1836–1842. https://doi.org/10.1016/j.electacta.2009.10.076

10 Polymer Silica Nanocomposite Gel Electrolytes for Lithium-Ion Batteries

Akhila Das, Anjumole P. Thomas, Neethu T. M. Balakrishnan, Nikhil Medhavi, Jou-Hyeon Ahn, Jabeen Fatima M. J., and Prasanth Raghavan

CONTENTS

10.1 Lithium-Ion Batteries (LIB): A Brief Introduction 219
10.2 Gel Polymer Electrolytes for Lithium-Ion Batteries 221
10.3 Silica-Based Gel Polymer Electrolytes for Lithium-Ion Batteries 222
    10.3.1 Fumed Silica-Based Gel Polymer Electrolytes 222
    10.3.2 Nanosilica-Based Gel Polymer Electrolytes 224
    10.3.3 *In-Situ*-Generated Silica-Based Gel Polymer Electrolytes 226
    10.3.4 Surface-Modified (Functionalized) Silica-Based Gel Polymer Electrolytes 227
10.4 Conclusion 230
Acknowledgment 231
References 231

10.1 LITHIUM-ION BATTERIES (LIB): A BRIEF INTRODUCTION

Lithium-ion batteries (LIB) were commercialized in 1990 by the Sony Corporation. The development of LIBs for the fabrication of portable energy storage devices resulted in the award of the 2019 Nobel Prize in Chemistry to three eminent scientists, namely Prof. John. B. Goodenough, Prof. Michael Stanley Whittingham and Prof Akira Yoshino (Figure 10.1). Prof. Goodenough laid the foundation for a cathode based on lithium cobalt oxide (LCO) by studying the intercalation property of these inorganic materials [1]. The chemistry of intercalation was explained by Prof. Stanley Wittingham during his investigation of the intercalation behavior of transition metal ions in certain chalcogenides, like TiS_2 [2]. Prof. Akira Yoshino was the person who created a prototype lithium-ion battery [3].

FIGURE 10.1 Images of Nobel laureates in Chemistry, 2019, from left to right: Prof. John B. Goodenough, Prof. M. Stanley Whittingham and Prof. Akira Yoshino. Adapted and reproduced with permission from The Royal Swedish Academy of Science [4]. Copyright © Nobel Media 2019. Illustration: Niklas Elmehed.

A typical lithium-ion battery consists of a cathode, an anode and an electrolyte (Figure 10.2). The material of the cathode employed in LIBs can include lithium cobalt oxide ($LiCoO_2$, LCO) or lithium iron phosphate ($LiFePO_4$, simply LFP). The anode mainly found in commercial lithium-ion batteries is graphite, which has the capability to intercalate lithium ions into the interstitial sites, which is first proposed by Prof. Rachid Yazami. The electrolyte is considered to be the heart of the battery. During the early stages of LIB development, liquid electrolytes containing lithium salts were used. These liquid electrolytes caused many serious safety issues, such as leakage and fires.

The most vital issues limiting the use of these energy storage devices are the safety issues. Even after decades of commercialization, safe portable energy storage devices are still a nightmare, owing to the incidence of safety issues generated and reported over the past few years. Several reports, regarding fires and explosions of lithium-ion batteries, have been reported all over the world, hence, the risk of explosion is the most worrying issue. The US Federal Aviation Administration (FAA) announced new aviation standards, published by the U.S. Department of Transportation (DOT) in July 2014 [5], in order to reinforce the safety conditions to facilitate the safe and reliable transport of the lithium-ion batteries, as these could cause fires, if not appropriately packaged, shipped and transported. Several news reports were published on the recall of the batteries owing to the risk of fires. One of the largest recalls occurred when the famous laptop provider, Dell, recalled about 4.1 million batteries of notebook computers in 2006 [6]. On 26 February 2013, the Ryobi lithium battery pack (18 V 4 Ah) was recalled due to some batteries overheating and catching fire during charging [7]. During 2016, Samsung was forced to

Polymer Silica Gel Electrolytes 221

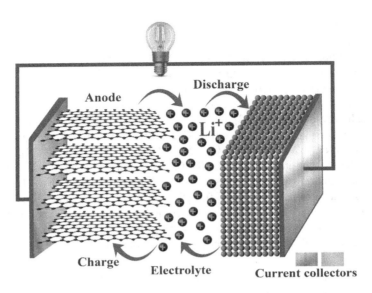

FIGURE 10.2 Schematic illustration of the structure and operating principles of lithium-ion batteries, including the movement of ions between electrodes during charge (forward arrow) and discharge (backward arrow) states.

recall the Dream mobile series Galaxy Note 7 as the batteries were reported to catch fire. These facts indicate that the batteries that we carried along with us were not at all safe. In addition to these portable devices, LIBs are being widely explored for use in electric vehicles (EVs) for a sustainable eco-friendly transportation system. Chevrolet's Volt EV was about to launch, but, during the crash test, the batteries was seriously affected, leading to them catching fire [8]. Many suggestions were put forward by scientists to overcome these issues, such as the use of safer electrolytes, like polymer electrolytes, and the inclusion of active flame-retardant agents in the electrolytes. Both these methods are being widely investigated by the researchers for commercialising safer and more efficient energy storage devices particularly for portable systems like mobile phones, laptops, tablets and EVs.

10.2 GEL POLYMER ELECTROLYTES FOR LITHIUM-ION BATTERIES

Gel polymer electrolytes are the most active topic of battery research, owing to the risk factors encountered in portable energy storage devices. These gel polymer electrolytes are synthesized by immobilizing liquid electrolytes, containing the active metal ion, in a polymer membrane. Such liquids act as a medium for active conduction of ions, whereas the polymers form a mechanical support for these active liquids and as a separator in LIBs. Among polymers, polyethylene oxide (PEO), polypropylene oxide (PPO), polymethyl methacrylate (PMMA), polyvinyl pyrrolidone (PVP), polyacrylonitrile (PAN), polyvinylidene difluoride (PVdF), polyvinylidene difluoride-*co*-hexafluoropropylene (PVdF-*co*-HFP) and polyethylene terephthalate (PET) are some of the polymer matrices used as electrolytes in LIBs. The major disadvantage of these polymer electrolytes is their lower ionic conductivity compared with conventional liquid electrolytes. Various methods have been adopted to increase the ionic conductivity of these electrolytes materials. The most effective strategy, and the one most widely adopted, is to increase the amorphous nature of these polymers by the incorporation of fillers. Several organic and inorganic fillers, such as cellulose [9–12], cellulose derivatives [13], chitin [14], chitosan [15], kraft lignin [16], carbon nanotubes (CNTs) [17], graphene oxide (GO) [18], aluminum oxide (Al_2O_3) [19–23], nickel oxide (NiO) [24], titania (TiO_2) [25–29] and silica (SiO_2) [30–34], have been studied. The present chapter investigates the role of silica as a ceramic filler in the performance of gel polymer electrolytes for lithium-ion batteries. The different forms of silica, such as nano-silica, fumed silica, *in-situ*-generated silica, functionalized silica, etc., have been evaluated. A detailed investigation will be carried out in this chapter on the synthesis and the properties (thermal, mechanical, ion transport and electrochemical) of silica incorporated into gel polymer electrolytes for LIBs.

10.3 SILICA-BASED GEL POLYMER ELECTROLYTES FOR LITHIUM-ION BATTERIES

Silica (SiO_2) is one the most abundant, low-cost and eco-friendly ceramic materials present in the Earth's crust. In silica, each silicon atom is surrounded by four oxygen atoms in a tetrahedral arrangement. The bond angle formed between O-Si-O is approximately 109°28′. Silica is a colorless solid with a molecular mass of 60.08 g mol^{-1}. Silica is thermally stable up to 1700°C. Considering the polymer electrolytes employed as electrolytes for LIBs, the crystallinity of silica is the highest. For Li$^+$-ion conduction in LIBs, the polymeric membrane should have both crystalline and amorphous phase. The crystalline domains in the polymer provides mechanical strength to the electrolyte, whereas the amorphous phase provides the medium for Li$^+$-ion conduction, which increases the ionic conductivity, as well as the Li$^+$-ion transference number. On incorporation of silica, these ceramic fillers penetrate the polymeric matrix, destroying the crystallinity of the polymer, inducing more amorphous regions, as shown in Figure 10.3. Silica is a unique material, with inertness, thermal stability, etc. It also facilitates the conductivity of the membrane by forming a networked structure.

10.3.1 Fumed Silica-Based Gel Polymer Electrolytes

Fumed silica, or pyrogenic silica, is a form of silica with a high surface area. This three-dimensional silica is formed by flame pyrolysis of silicon compounds. The silanol group (Si-O-H) in fumed silica imparts a hydrophilic nature to the material, facilitating the role of silica in many aqueous chemical reactions. In gel polymer electrolyte synthesis, this polar nature of the fumed silica is being utilized. In 1994, Saad et al. [35] introduced the concept of polymer composite electrolyte to enhance the ionic conductivity of polyethylene glycol (PEG) with fumed silica. In 1996, Tarascon et al. [36] reported increases in electrolyte uptake cyclability, and

Polymer Silica Gel Electrolytes

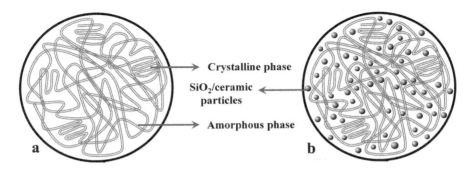

FIGURE 10.3 Schematic representation of crystalline and amorphous phases in gel polymer electrolytes and polymer silica composite gel electrolytes.

power density of LIBs constructed with graphite as the anode, lithium manganese oxide (LiMn$_2$O$_4$, LMO) as the cathode and the (PVdF-*co*-HFP/silica/1 M LiPF$_6$ (EC/DMC)) as the electrolyte. Polymethyl methacrylate was photopolymerized with ethyl glycol dimethacrylate as a crosslinker with lithium perchlorate (LiClO$_4$) and fumed silica in EC/PC solvent [37]. Detailed investigations of the electrochemical impedance of polymeric membranes with and without silica indicates that incorporation of silica increases the Li$^+$-ion conductivity and stability of the LIBs.

In 2003, the effect of fumed silica on the interfacial stability of the gel polymer electrolyte was proposed by Fu et al. [37]. The gel polymer electrolyte was fabricated with methyl methacrylate as a monomer, LiClO$_4$ as the doping salt and fumed silica (0–5 wt.%) as the filler. The minimum impedance was obtained for the gel electrolyte with 5 wt.% of silica with <500 Ω resistance, even after 1200-h storage time, indicating an improvement of the interfacial stability. In 2005, Ahmad et al. [38] reported the same gel polymer electrolyte to have the transparent property when prepared by the solvent-casting technique. The temperature-dependent ionic conductivity analysis of the samples indicates an increase in ionic conductivity in response to an increase in temperature from 20 to 70°C (Figure 10.4a). The highest ionic conductivity is observed in the electrolyte fabricated with 2 wt.% of fumed silica. The transference number of Li$^+$-ions increases from 0.19 to 0.25 on incorporation of silica, indicating a better pathway for the conduction of lithium ions, due to electro-osmotic behavior. The interaction of the solvent propylene carbonate with the silica molecules *via* hydrogen bonding, as shown in Figure 10.4b, facilitates the path for the conduction of Li$^+$-ions. In another report [39], addition of ammonium hexafluorophosphate (NH$_4$PF$_6$) into the composite electrolyte of PMMA and fumed silica significantly increases the ionic conductivity and the resulting polymer electrolyte exhibited an ionic conductivity almost equivalent to that of the liquid electrolyte (10^{-2} S cm^{-1}) under ambient conditions. Furthermore, the electrolyte was thermally stable up to 125°C. In another report [40], PVC was blended with PMMA, along with fumed silica as the filler and LiClO$_4$ as the doping salt. The polymeric blend electrolyte was compared with pure PVC electrolyte and the ionic conductivities were 1.1×10^{-3} S cm^{-1} at 25°C and 2.8×10^{-4} S cm^{-1} at −15°C, respectively.

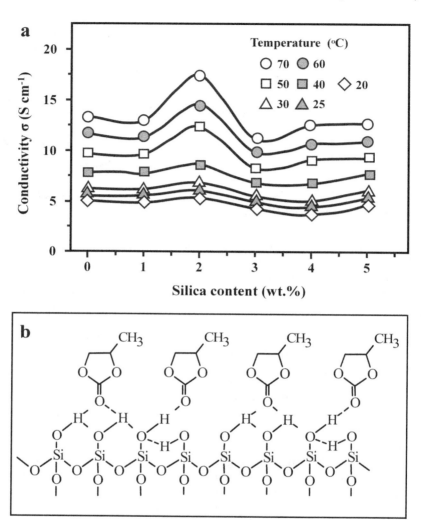

FIGURE 10.4 (a) Temperature-dependent ionic conductivity of gel polymer electrolytes: polymethyl methacrylate with LiClO$_4$ as the doping salt and fumed silica as the filler (0–5 wt.%). (b) Illustration of the interaction between the solvent propylene carbonate (PC) with the silica molecules *via* hydrogen bonding. Adapted and reproduced with permission from Ref. [38]. Copyright © 2005 Elsevier.

10.3.2 Nanosilica-Based Gel Polymer Electrolytes

The importance of nanoparticles was been widely investigated, following the famous talk given by Prof. Richard Feynman at the annual meeting of the American Physical Society held at Caltech. The famous quote of the speech was "There is plenty of room in the bottom" [41]. The surface-to-volume ratio increases for a nanoparticle, leading to an improvement in the surface properties of the material. In addition, the energy level of the material forms discrete levels, rather than bands, owing to the minimization of the number of atoms/molecules per particle. The change from the bulk to the nano regime thereby shows shifts in several electronic, physical and optical properties of the material, converting the system to form one which is more capable for various applications. In the present context of the investigation, the incorporation of nano-sized silica particles onto the polymer matrix is supposed to enhance the ionic conductivity of the polymer electrolyte. In 2012, Hu et al. [42] fabricated a composite gel polymer electrolyte, with polyvinyl alcohol (PVA) as the base polymer matrix, and incorporated hyperbranched polyamine ester grafted nano-silica (HBPAGS), followed by addition of lithium perchlorate through the mold-casting method. The HBPAGS: LiClO$_4$ ratio 15:54 wt.% resulted in an ionic conductivity of 1.51×10^{-4} S cm^{-1} at 25°C and 1.36×10^{-3} S cm^{-1} at 100°C. The effect of incorporation of nano-silica and titania in varying proportions into the PVA/PVdF polymer blend was investigated with the lithium triflate electrolyte [43]. The thermal, electrochemical mechanical and Li$^+$-ion transport properties of the composite polymer electrolyte were investigated. Changing the mole fraction of nanofiller resulted in a change in the bulk resistance. For 8 mole wt.% of silica-incorporated polymer composite electrolyte, exhibited an increased ionic conductivity, transport number and mechanical strength [43].

The polyethylene-supported poly(methyl methacrylate-co-butyl acrylate) (PMMA-co-BA/PE) copolymer was synthesized by the emulsion-polymerization method [34]. To enhance the performance of the pure polymer matrix, nanosilica (15 nm) was incorporated at varying concentrations from 0 wt.% to 15 wt.%. The morphological analysis shows that maximum porosity is obtained for polymer electrolyte having 5 wt.% silica, as shown in Figure 10.5. The studies revealed that the 5 wt.% nanosilica incorporation into (PMMA-co-BA/PE) copolymer based electrolyte, showing maximum improvement to their ionic conductivity and electrochemical properties. The electrolyte uptake was obtained at 200.8%, ionic conductivity at 2.26×10^{-3} S cm^{-1} at room temperature and thermal stability up to 320°C. The electrochemical stability of PMMA-co-BA/PE was 5.0 V, whereas (PMMA-co-BA/PE/SiO$_2$) was 5.6 V. The cell studies revealed that a capacity retention of only ~77% was obtained for the pure copolymer cell, whereas 99% retention was observed for the PMMA-co-BA/PE/SiO$_2$ cell. The study reveals that the various properties were improved markedly following incorporation of nano-silica, with increases in porosity, mechanical stability, thermal stability, Li$^+$-ion transport properties, electrochemical stability and specific capacitance [34].

In 2014, Li et al. [44] reported an ionic liquid-based gel polymer electrolyte (ILGE), into which nanosilica was incorporated as a filler and LiTFSI as the doping salt. The concentration of ionic liquid 1,2-dimethyl-3-butyl-imidazolium bis(trifluoromethanesulfonyl)imide (BMMIM-TFSI) was varied, keeping the remaining concentrations as a constant. The response of ionic conductivity of the samples to increasing concentrations of BMMIM-TFSI at different temperatures (25–70°C) indicated a marked swing between 10^{-5} and 10^{-3} S cm^{-1}. The performance of the electrolytes in batteries was analyzed using Li/ILGE/LiFePO$_4$ to obtain a maximum discharge capacity of 151 mAh g^{-1} at 0.1 C, with good capacity retention.

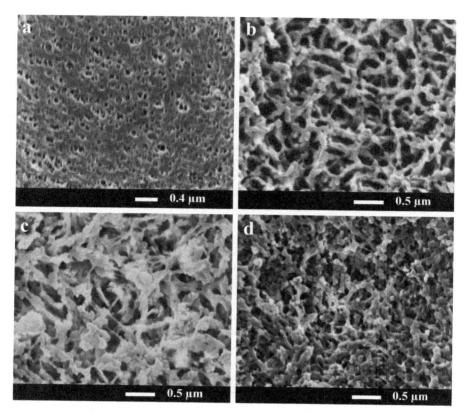

FIGURE 10.5 FE-SEM images on the surface morphology of PMMA-*co*-BA/PE membranes with different nanosized silica content: (a) 0 wt.%, (b) 5 wt.%, (c) 10 wt.%, and (d) 15 wt.%. Adapted and reproduced with permission from Ref. [34]. Copyright © 2014 Elsevier.

10.3.3 IN-SITU-GENERATED SILICA-BASED GEL POLYMER ELECTROLYTES

Uniformly dispersed silica in the polymeric matrix can be achieved from *in-situ*-generated silica. In the majority of reports, tetra ethoxy silane (TEOS) is used as the precursor. The polymeric electrolyte membrane is synthesized by various methods, such as incorporating TEOS followed by hydrolysis of the TEOS to form silica particles as per Equations 10.1 and 10.2. Because the silica particles are generated in the polymeric matrix, they are termed *in-situ*-generated silica. Recently, Xu et al. fabricated composite polymer electrolytes with *in-situ*-generated silica in the PEO matrix [45] *via* the solution-casting method. TEOS was used as the precursor for *in-situ* silica conversion (0–15 wt.%). Among the samples, the 10 wt.% silica-doped membrane showed the highest ionic conductivity of $\sim 1.1 \times 10^{-4}$ S cm^{-1} at 30°C, and Li/GPE/LiFePO$_4$ cell analysis revealing a specific capacity of 147 mAh g^{-1} under 0.3 C at 90°C. The samples were thus capable of withstanding elevated temperatures, with semi-transparent as well as flexible properties.

$$\text{Si}(\text{OC}_2\text{H}_5)_2 + 4\text{H}_2\text{O} \rightarrow \text{Si}(\text{OH})_4 + 4\text{C}_2\text{H}_5\text{OH} \quad (10.1)$$

$$\text{Si}(\text{OH})_4 \rightarrow \text{SiO}_2 + 2\text{H}_2\text{O} \quad (10.2)$$

Raghavan et al. [46] reported a comparative study on the effect of silica addition *via* direct (6 wt.%) and *in-situ*-generated (3, 6, 9 wt.%) silica in PVdF-*co*-HFP polymer electrolyte. The direct addition of fumed silica into the polymeric matrix resulted in lower measures of ionic conductivity and mechanical properties than did the *in-situ* synthesis of silica. The direct addition of silica causes aggregate formation, leading to islands of silica in the polymer matrix. Among the different concentrations, 6 wt.% of silica was identified as the optimum concentration, and this polymer electrolyte exhibited an ionic conductivity of 8.06 mS cm^{-1} [46]. The polymeric gel electrolyte was tested in a lithium-ion battery, with LiFePO$_4$ (LFP) as the cathode and lithium metal as the anode. The Li/GPE/LiFePO$_4$ cell delivered a discharge capacitance of 170 mAh g^{-1}, which is equal to the theoretical capacity of LFP, whereas direct addition of silica gave a capacity of 153 mAh g^{-1} [46].

In 2017, Wu et al. [47] fabricated high-temperature LIBs with ionogels prepared by the *in-situ* sol-gel method, forming organo-modified silica (epoxy-functionalized silane coupling agent (SCA)) from 3-glycidyloxy propyl trimethoxy silane with ionic liquid (lithium bis(trifluoromethanesulfonyl)imide (LiTFSI) (ILE) in *N*-propyl-*N*-methylpyrrolidinium bis(trifluoromethylsulfonyl)imide ([Py13][TFSI]) at different molar ratios. The morphology of ionogels fabricated at different concentrations of the ionic liquid is illustrated by FE-SEM in Figure 10.6, where Z=0.25, 0.5, 0.75, 1 or 1.25 wt.% of ionic liquid. The optimum ionogel concentration occurred with ionic liquid concentration of 0.75 wt.%, silica 1 wt.% and formic acid 5.9 wt.%, i.e., Z=0.75. Electrochemical studies reveal a stability window of 1–5 V and an ionic conductivity of 1.91×10^{-3} S cm^{-1} at 30°C and 4.70×10^{-3} S cm^{-1} at 60°C. The cell studies, with Li metal as anode and LiFePO$_4$ as cathode, gave a reversible capacity of 155 mAh g^{-1} at 0.1 C rate under 30°C, increasing to 160 and 168.6 mAh g^{-1} at 60 and 90°C, respectively [47]. In 2020, Chen et al. [48] reported ionogels based on LiTFSI, TEOS and 1-butyl-1-methylpyrrolidinium bis(trifluoromethylsulfonyl)imide (BMP-TFSI) *via* the sol-gel method, where TEOS was converted into silica *via* the *in-situ* hydrolysis reaction. According to the report, the ionic liquid was immobilized by hydrogen bonding onto the silanol groups of silica, providing a highly conductive medium for Li$^+$-ions similar to that of the liquid electrolyte. The electrochemical studies of the battery were tested using Li/ionogel/LiMn$_2$O$_4$ (LMO), delivering a stable cut-off voltage of 4.3 V for C rates varying between 1 and 20 C for over 20 cycles [48].

10.3.4 Surface-Modified (Functionalized) Silica-Based Gel Polymer Electrolytes

Silica particles were functionalized with various organic moieties to enhance the miscibility of the filler in the polymer matrix, in order to achieve greater

228 Polymer Electrolytes

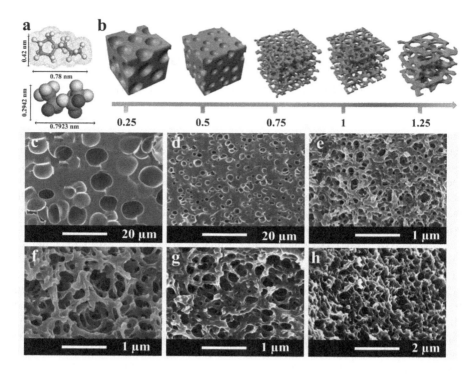

FIGURE 10.6 (a) Sizes of Py1+ and TFSI−; (b) Proposed structure of scaffolds with various epoxy-functionalized silane coupling agents (SCA) in an ionic liquid electrolyte (ILE) at different molar ratios; FE-SEM images on the morphologies of ionogels fabricated using organo-modified silica with different concentrations (x wt.%) of ionic liquid lithium bis(trifluoromethanesulfonyl)imide (LiTFSI) in *N*-propyl-*N*-methylpyrrolidinium bis(trifluoromethylsulfonyl)imide ([Py13][TFSI]): (c) x=0.25, (d) x =0.5, (e) x =0.75, (f) x =1, (g) x =1.25, and (h) x =0.75. Adapted and reproduced with permission from Ref. [46]. Copyright © 2017 Elsevier.

conductivity. In 2015, Li et al. [49] synthesized an ionic complex of silica nanoparticles by a chemical synthesis method in which polyacrylic acid (PAA) was complexed with silica, followed by neutralization using lithium hydroxide (LiOH) to form an ionic complex termed SiPAALi. The complex was used as filler in the PVdF matrix solution in a DMF/acetone mixture. The mixture was then electrospun and used as a gel polymer electrolyte for LIBs. The performance of these composite membranes was compared with the electrospun composite membrane prepared with an unmodified silica/PVdF composite membrane. The electrolyte uptake and electrochemical properties of the ionic complex were higher than those of the silica composite membrane. A dummy cell was fabricated with lithium metal as anode, lithium cobalt oxide (LiCoO$_2$) as cathode and nonwoven electrospun membrane as the electrolyte. A SiPAALi/PVdF membrane-containing

Polymer Silica Gel Electrolytes

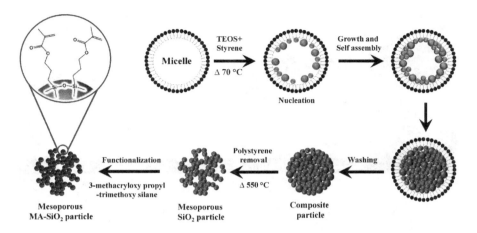

FIGURE 10.7 Schematic illustration of various steps and the nature of the silica (SiO$_2$) particles formed during the different stages of the synthesis procedure of mesoporous methacrylate functionalized silica nanoparticles (MA-SiO$_2$). Adapted and reproduced with permission from Ref. [50]. Copyright © 2016 Springer Nature.

cell delivered a specific capacitance of 156.5 mAh g^{-1} whereas SiO$_2$/PVdF delivered a specific capacitance of 152 mAh g^{-1} [49]. In another report [50], silica was modified with a commercially available pro-adhesive compound U-511 (3-methacryloxy propyl trimethoxy silane). PAN and PVdF-co-HFP were used as a polymeric matrix with LiPF$_6$ in tetra methoxy silane (TMS) as a lithiating compound. Modified and pure silica (3 wt.%) were compared to analyze the effect of modification on the electrochemical properties of the material. The ionic conductivity of the membranes was substantially increased from 10^{-4} to 10^{-2} S cm^{-1}, following the addition of modified silica.

Methacrylate was used as an organic modifier for mesoporous silica particles synthesized from TEOS, as shown in Figure 10.7 [51]. The report reveals that the non-porous silica particles block Li$^+$-ion transport. Thus, mesoporous silica was synthesized through surfactant-assisted synthesis. The polymeric electrolyte membrane was fabricated by electrospinning of PAN. followed by soaking in 5 wt.% of methacrylate-modified mesoporous silica (MA-SiO$_2$) and with nonporous silica particles, as morphological studies illustrated in Figure 10.8. The dried membrane was soaked in a 1.15 M LiPF$_6$ in ethylene carbonate/ethyl methyl carbonate (EC/DMC) solvent to form a gel polymer electrolyte and showed a room temperature ionic conductivity of 1.1 × 10^{-3} and 1.8 × 10^{-3} S cm^{-1} for non porous and mesoporous silica incorporated electrolyte membranes respectively. The cell was fabricated with graphite as the anode and LiNi$_{1/3}$Co$_{1/3}$Mn$_{1/3}$O$_2$ as the cathode, separated with PAN/silica polymer gel electrode delivered an initial charge capacity of ~180 mAh g^{-1}.

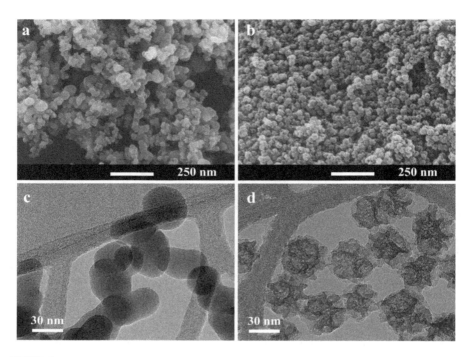

FIGURE 10.8 FE-SEM images of the structural morphology of (a) non-porous MA-SiO$_2$ nanoparticles and (b) mesoporous methacrylate functionalized silica nanoparticles (MA-SiO$_2$). Transmission electron microscopic (TEM) images of (c) non-porous MA-SiO$_2$ nanoparticles, and (d) mesoporous MA-SiO$_2$ nanoparticles. Adapted and reproduced with permission from Ref. [50]. Copyright © 2016 Springer Nature.

10.4 CONCLUSION

Gel polymer electrolytes are promising candidates for next-generation energy storage devices. The energy storage devices have emerged as an inevitable part of next-generation devices. In addition to the need for energy storage in portable devices, the global energy crisis also needs to store energy generated from intermittent or periodic renewable energy sources, such as solar energy, wind energy, hydro energy, etc. An efficient electrolyte is the heart of safe and efficient energy storage devices. Gel polymer electrolytes can overcome the problems associated with conventional liquid electrolytes. The ionic conductivity of the liquid electrolytes is good but continuous cycling can lead to batteries catching fire, with serious safety issues. The use of polymeric electrolytes overcomes these issues but the conductivity is decreased. The addition of ceramic fillers, like silica, leads to an increase in the amorphous areas in the polymeric matrix, causing improvements in ionic conductivity. The state of the polymer electrolyte in a gel electrolyte lies in between the solid and the liquid electrolytes, overcoming the issues generated by pure liquid as well as those associated with pure polymer electrolytes.

ACKNOWLEDGMENT

Authors Anjumole P. Thomas, Dr Jabeen Fatima M. J. and Dr Prasanth Raghavan would like to acknowledge the Department of Science and Technology (DST), India and Kerala State Council for Science, Technology and Environment (KSCSTE), Kerala, India for financial assistance.

REFERENCES

1. Mizushima K, Jones PC, Goodenough JB (1981) LixCoO2 (0<x<1): A new cathode material for batteries of high energy density. *Solid State Ionics* 4:171–174
2. Whittingham MS (1978) Chemistry of intercalation compounds: Metal guests in chalcogenide hosts. *Prog. Solid State Chem* 12:41–99.
3. Yoshino A (2012) The birth of the lithium-ion battery. *Angew Chemie - Int Ed* 51:5798–5800. https://doi.org/10.1002/anie.201105006
4. Prize N History. In: *Nobel Media*. www.nobelpeaceprize.org/History
5. News F (2014) New standards for the transport of lithium batteries. In: *News Rep.* www.faa.gov/news/updates/?newsId=78605
6. Darlin D (2006) Dell will recall batteries in PC's. In: *New York Times.* www.nytimes.com/2006/08/15/technology/15battery.html
7. CPSC (2013) One world technologies recalls ryobi cordless tool battery pack due to fire and burn hazards. In: *United States Consum. Prod. Saf. Comm.* www.cpsc.gov/Recalls/2013/one-world-technologies-recalls-ryobi-cordless-tool-battery-pack
8. Greenemeier L (2011) Could Chevy Volt lithium-ion battery fires burn out interest in EVs and hybrids? *In: Sci. Am.* https://blogs.scientificamerican.com/observations/could-chevy-volt-lithium-ion-battery-fires-burn-out-interest-in-evs-and-hybrids/
9. Leijonmarck S, Cornell A, Lindbergh G, Wågberg L (2013) Single-paper flexible Li-ion battery cells through a paper-making process based on nano-fibrillated cellulose. *J Mater Chem A* 1:4671–4677. https://doi.org/10.1039/c3ta01532g
10. Sheng J, Tong S, He Z, Yang R (2017) Recent developments of cellulose materials for lithium-ion battery separators. *Cellulose* 24:4103–4122. https://doi.org/10.1007/s10570-017-1421-8
11. Kim JH, Gu M, Lee DH, et al. (2016) Functionalized nanocellulose-integrated heterolayered nanomats toward smart battery separators. *Nano Lett* 16:5533–5541. https://doi.org/10.1021/acs.nanolett.6b02069
12. Lavoine N, Desloges I, Dufresne A, Bras J (2012) Microfibrillated cellulose - Its barrier properties and applications in cellulosic materials: A review. *Carbohydr Polym* 90:735–764. https://doi.org/10.1016/j.carbpol.2012.05.026
13. Ran Y, Yin Z, Ding Z, et al. (2013) A polymer electrolyte based on poly(vinylidene fluoride-hexafluoropropylene)/hydroxypropyl methyl cellulose blending for lithium-ion battery. *Ionics (Kiel)* 19:757–762. https://doi.org/10.1007/s11581-012-0808-7
14. Stephan AM, Kumar TP, Kulandainathan MA, Lakshmi NA (2009) Chitin-incorporated poly(ethylene oxide)-based nanocomposite electrolytes for lithium batteries. *J Phys Chem B* 113:1963–1971. https://doi.org/10.1021/jp808640j
15. Aziz NA, Majid SR, Yahya R, Arof AK (2011) Conductivity, structure, and thermal properties of chitosan-based polymer electrolytes with nanofillers. *Polym Adv Technol* 22:1345–1348. https://doi.org/10.1002/pat.1619
16. Liu B, Huang Y, Cao H, et al. (2018) A high-performance and environment-friendly gel polymer electrolyte for lithium ion battery based on composited lignin membrane. *J Solid State Electrochem* 22:807–816. https://doi.org/10.1007/s10008-017-3814-x

17. Tang C, Hackenberg K, Fu Q, et al. (2012) High ion conducting polymer nanocomposite electrolytes using hybrid nanofillers. *Nano Lett* 12:1152–1156. https://doi.org/10.1021/nl202692y
18. Kammoun M, Berg S, Ardebili H (2015) Flexible thin-film battery based on graphene-oxide embedded in solid polymer electrolyte. *Nanoscale* 7:17516–17522. https://doi.org/10.1039/c5nr04339e
19. Masoud EM, El-Bellihi A-A, Bayoumy WA, Mousa MA (2013) Organic–inorganic composite polymer electrolyte based on PEO–LiClO$_4$ and nano-Al$_2$O$_3$ filler for lithium polymer batteries: Dielectric and transport properties. *J Alloys Compd* 575:223–228. https://doi.org/10.1016/j.jallcom.2013.04.054
20. Liang B, Tang S, Jiang Q, et al. (2015) Preparation and characterization of PEO-PMMA polymer composite electrolytes doped with nano-Al$_2$O$_3$. *Electrochim Acta* 169:334–341. https://doi.org/10.1016/j.electacta.2015.04.039
21. Jeong H-S, Hong SC, Lee S-Y (2010) Effect of microporous structure on thermal shrinkage and electrochemical performance of Al$_2$O$_3$/poly(vinylidene fluoride-hexafluoropropylene) composite separators for lithium-ion batteries. *J Memb Sci* 364:177–182. https://doi.org/10.1016/j.memsci.2010.08.012
22. Liao YH, Li XP, Fu CH, et al. (2011) Polypropylene-supported and nano-Al$_2$O$_3$ doped poly(ethylene oxide)–poly(vinylidene fluoride-hexafluoropropylene)-based gel electrolyte for lithium ion batteries. *J Power Sources* 196:2115–2121. https://doi.org/10.1016/j.jpowsour.2010.10.062
23. Li Z, Su G, Wang X, Gao D (2005) Micro-porous P(VDF-HFP)-based polymer electrolyte filled with Al$_2$O$_3$ nanoparticles. *Solid State Ionics* 176:1903–1908. https://doi.org/10.1016/j.ssi.2005.05.006
24. Rajasudha G, Nancy AP, Paramasivam T, et al. (2011) Synthesis and characterization of polyindole–NiO-based composite polymer electrolyte with LiClO$_4$. *Int J Polym Mater Polym Biomater* 60:877–892. https://doi.org/10.1080/00914037.2010.551367
25. Lin CW, Hung CL, Venkateswarlu M, Hwang BJ (2005) Influence of TiO$_2$ nanoparticles on the transport properties of composite polymer electrolyte for lithium-ion batteries. *J Power Sources* 146:397–401. https://doi.org/10.1016/j.jpowsour.2005.03.028
26. Liu Y, Lee JY, Hong L (2003) Morphology, crystallinity, and electrochemical properties of in situ formed poly(ethylene oxide)/TiO$_2$ nanocomposite polymer electrolytes. *J Appl Polym Sci* 89:2815–2822. https://doi.org/10.1002/app.12487
27. Cui W-W, Tang D-Y, Gong Z-L (2013) Electrospun poly(vinylidene fluoride)/poly(methyl methacrylate) grafted TiO$_2$ composite nanofibrous membrane as polymer electrolyte for lithium-ion batteries. *J Power Sources* 223:206–213. https://doi.org/10.1016/j.jpowsour.2012.09.049
28. Zhou L, Wu N, Cao Q, et al. (2013) A novel electrospun PVDF/PMMA gel polymer electrolyte with in situ TiO$_2$ for Li-ion batteries. *Solid State Ionics* 249–250:93–97. https://doi.org/10.1016/j.ssi.2013.07.019
29. Cao J, Wang L, Shang Y, et al. (2013) Dispersibility of nano-TiO$_2$ on performance of composite polymer electrolytes for Li-ion batteries. *Electrochim Acta* 111:674–679. https://doi.org/10.1016/j.electacta.2013.08.048
30. Yang CL, Li ZH, Li WJ, et al. (2015) Batwing-like polymer membrane consisting of PMMA-grafted electrospun PVdF–SiO$_2$ nanocomposite fibers for lithium-ion batteries. *J Memb Sci* 495:341–350. https://doi.org/10.1016/j.memsci.2015.08.036
31. Lee Y-S, Ju SH, Kim J-H, et al. (2012) Composite gel polymer electrolytes containing core-shell structured SiO$_2$(Li$^+$) particles for lithium-ion polymer batteries. *Electrochem Commun* 17:18–21. https://doi.org/10.1016/j.elecom.2012.01.008

32. Park J-H, Cho J-H, Park W, et al. (2010) Close-packed SiO$_2$/poly(methyl methacrylate) binary nanoparticles-coated polyethylene separators for lithium-ion batteries. *J Power Sources* 195:8306–8310. https://doi.org/10.1016/j.jpowsour.2010.06.112
33. Zhang P, Yang LC, Li LL, et al. (2011) Enhanced electrochemical and mechanical properties of P(VDF-HFP)-based composite polymer electrolytes with SiO$_2$ nanowires. *J Memb Sci* 379:80–85. https://doi.org/10.1016/j.memsci.2011.05.043
34. Xie H, Liao Y, Sun P, et al. (2014) Investigation on polyethylene-supported and nano-SiO$_2$ doped poly(methyl methacrylate-co-butyl acrylate) based gel polymer electrolyte for high voltage lithium ion battery. *Electrochim Acta* 127:327–333. https://doi.org/10.1016/j.electacta.2014.02.038
35. Khan SA, Baker GL, Colson S (1994) Composite polymer electrolytes using fumed silica fillers: Rheology and ionic conductivity. *Chem Mater* 6:2359–2363. https://doi.org/10.1021/cm00048a023
36. Tarascon JM, Gozdz AS, Schmutz C, et al. (1996) Performance of Bellcore's plastic rechargeable Li-ion batteries. *Solid State Ionics* 86–88:49–54. https://doi.org/10.1016/0167-2738(96)00330-X
37. Fu YB, Ma XH, Yang QH, Zong XF (2003) The effect of fumed silica on the interfacial stability in the gel polymer electrolyte. *Mater Lett* 57:1759–1764. https://doi.org/10.1016/S0167-577X(02)01065-0
38. Ahmad S, Ahmad S, Agnihotry SA (2005) Nanocomposite electrolytes with fumed silica in poly(methyl methacrylate): Thermal, rheological and conductivity studies. *J Power Sources* 140:151–156. https://doi.org/10.1016/j.jpowsour.2004.08.002
39. Sharma JP, Sekhon SS (2007) Nanodispersed gel polymer electrolytes: Conductivity modification with the addition of PMMA and fumed silica. *Solid State Ionics* 178:439–445. https://doi.org/10.1016/j.ssi.2007.01.017
40. Choi N.-S., Park J.-K. (2001) New polymer electrolytes based on PVC/PMMA blend for plastic lithium-ion batteries. *Electrochim Acta* 46:1453–1459.
41. Feynman RP (1960) There's plenty of room at the bottom. *Eng Sci* 23:22–36
42. Hu X, Hou G, Zhang M, et al. (2012) A new nanocomposite polymer electrolyte based on poly(vinyl alcohol) incorporating hypergrafted nano-silica. *J Mater Chem* 22:18961–18967. https://doi.org/10.1039/c2jm33156j
43. Hema M, Tamilselvi P (2016) Lithium ion conducting PVA:PVdF polymer electrolytes doped with nano SiO$_2$ and TiO$_2$ filler. *J Phys Chem Solids* 96–97:42–48. https://doi.org/10.1016/j.jpcs.2016.04.008
44. Li M, Wang L, Du T (2014) Preparation of polymer electrolytes based on the polymerized imidazolium ionic liquid and their applications in lithium batteries. *J Appl Polym Sci* 131:1–7. https://doi.org/10.1002/app.40928
45. Xu Z, Yang T, Chu X, et al. (2020) Strong Lewis acid-base and weak hydrogen bond synergistically enhancing ionic conductivity of poly (ethylene oxide)@SiO2 electrolytes for high rate-capability Li-metal battery. *ACS Appl Mater & Interfaces* https://doi.org/10.1021/acsami.9b20128
46. Raghavan P, Choi JW, Ahn JH, et al. (2008) Novel electrospun poly(vinylidene fluoride-co-hexafluoropropylene)-in situ SiO$_2$ composite membrane-based polymer electrolyte for lithium batteries. *J Power Sources* 184:437–443. https://doi.org/10.1016/j.jpowsour.2008.03.027
47. Wu F, Chen N, Chen R, et al. (2017) Organically modified silica-supported ionogels electrolyte for high temperature lithium-ion batteries. *Nano Energy* 31:9–18. https://doi.org/10.1016/j.nanoen.2016.10.060

48. Chen X, Put B, Sagara A, et al. (2020) Silica gel solid nanocomposite electrolytes with interfacial conductivity promotion exceeding the bulk Li-ion conductivity of the ionic liquid electrolyte filler. *Sci Adv* 6:. https://doi.org/10.1126/sciadv.aav3400
49. Li W, Xing Y, Wu Y, et al. (2015) Study the effect of ion-complex on the properties of composite gel polymer electrolyte based on electrospun PVdF nanofibrous membrane. *Electrochim Acta* 151:289–296. https://doi.org/10.1016/j.electacta.2014.11.083
50. Kurc B (2016) Composite gel polymer electrolyte with modified silica for $LiMn_2O_4$ positive electrode in lithium-ion battery. *Electrochim Acta* 190:780–789. https://doi.org/10.1016/j.electacta.2015.12.175
51. Shin WK, Cho J, Kannan AG, et al. (2016) Cross-linked composite Gel polymer electrolyte using mesoporous methacrylate-functionalized SiO2 nanoparticles for lithium-ion polymer batteries. *Sci Rep* 6:1–10. https://doi.org/10.1038/srep26332

11 Polymer-Ionic Liquid Gel Electrolytes for Lithium-Ion Batteries

Jayesh Cherusseri

CONTENTS

11.1 Introduction 235
11.2 Properties of Polymer-Ionic Liquid Gel Electrolytes (PILGEs) 236
11.3 Types of Polymer-Ionic Liquid Gel Electrolytes (PILGEs) 239
11.4 Conclusion and Future Perspectives 251
References 251

11.1 INTRODUCTION

A prolonged search for fast-charging and high-energy-density energy storage devices ended with rechargeable metal-ion batteries, such as lithium ion batteries (LIBs) [1–5]. The LIBs are electrochemical systems which have been thoroughly investigated in the recent past and attracted great interest in the field of batteries, due to their high specific capacity, energy density, and compactness. Although the power density of LIBs is always inferior to that of supercapacitors, their popularity has exceeded that of supercapacitors because supercapacitors exhibit very low energy density, which limits their applications [6–9]. The LIBs have out-performed a variety of other battery systems, such as lead-acid, nickel-cadmium, nickel-metal hydride, etc. They have dominated the battery economy, accounting for more than 70% of all sales, when compared with other battery systems. Although LIBs are exhibiting impressive market sales, major bottlenecks remain, including reactivity, high cost, flammability, environmental hazards, etc. A solution to reduce the reactivity of Li by replacing them with sodium is on the way but this design needs considerable research and trialing to achieve market success.

The development of wearable electric and electronic devices as a part of progression into digitalization necessitates flexible and wearable power supplies. The safety issue is a major concern when using a LIB with wearable electronic and other portable devices and it is associated with the organic solvent-based liquid-state electrolyte used in it. In a LIB, electrodes (cathode and anode) are the major parts, but electrolytes also play a significant role in determining the electrochemical performances of LIBs. Electrolytes are the liquid-state organic solutions with good ionic transport properties in which the cell voltage, flammability, leakage, etc., are determined. The reactivity of the electrolytes with the electrode surfaces is another hurdle associated with liquid electrolytes. Though the liquid electrolyte-based LIBs are not safe to use, their commercial application is not hindered. This is mainly due to the unavailability of alternative electrochemical energy storage systems that could potentially replace the existing liquid electrolyte-based LIBs.

A search for safe batteries ended up in research and development into LIBs, based on polymer electrolytes. Polymer electrolytes were later found to be potential candidates that can effectively replace the existing liquid-state organic electrolytes. They are membranes that have the same function as an ordinary liquid-state organic electrolyte, such as good ionic transport. Polymer electrolyte-based LIBs have attracted great interest in the recent past, mainly due to their safety when in use [10]. They are of two kinds, namely solid-state polymer electrolytes (SPEs) and gel polymer electrolytes (GPEs). The former exhibits poor ionic conductivities, whereas the latter offers high ionic conductivities. GPEs utilize low-viscosity liquids with high-concentration Li salts to achieve better wettability with the LIB electrodes [11, 12]. In earlier days, various solvents, such as ethylene carbonate, dimethyl carbonate, dimethoxymethane, etc., were used in the preparation of GPEs. The major demerits of using these liquid solvents are leakage, flammability, and evaporation [13].

The other disadvantages of using organic liquid electrolytes in LIBs include corrosion of the battery electrodes, issues related to hermetic sealing, the formation of Li dendrites and hence short-circuiting of the electrodes, limited temperature range of operation, etc. These issues cannot be avoided completely but should be minimized. To eradicate these issues, ideal solvents have been developed using ionic liquids (ILs), which are non-flammable and show low volatility [14]. Although the ILs are costly, their development has opened the door to a safe battery technology. The polymer electrolytes prepared using ILs open up wider electrochemical potential windows for the LIBs. This chapter focuses on the recent advances that have happened in the field of polymer IL (PIL) gel electrolytes (PILGEs) and their application to LIBs. A schematic illustration showing the various components of a PILGE and the salient features is shown in Figure 11.1. Preparation and performance evaluation of PILGEs for LIBs will be explained in detail. The challenges and future perspectives of developing PILGEs for the next-generation advanced LIBs will also be explained briefly.

11.2 PROPERTIES OF POLYMER-IONIC LIQUID GEL ELECTROLYTES (PILGES)

A sudden increase in the research activity to develop solid-state electrolytes for LIBs is taking place nowadays. Electrolytes developed in the recent past consisted of organic solvents, which were quite incompatible with the next-generation flexible wearable technologies. Solid-state electrolytes are being developed to tackle this problem. Solid-state electrolytes, consisting of ILs and polymers to achieve both flexibility and good electrochemical performance, are required to achieve widespread acceptance. The LIB structure includes two electrodes separated by an electrically

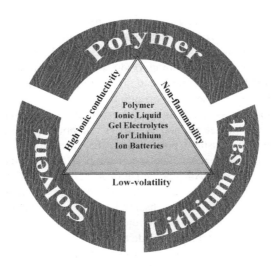

FIGURE 11.1 Schematic diagram of the various components of a polymer-ionic liquid gel electrolyte (PILGE).

insulating polymeric membrane. The electrodes interact with opposed electrolyte-ions and these electrolyte-ions travel through the polymer electrolyte membrane. Each component of a LIB is influential to its overall performance, but the most critical role is played by the electrolyte used. Electrolytes play a significant role in determining the electrochemical performance of a LIB. Therefore, most of LIB research is focused on electrolytes and is progressing along similar lines to that involved in developing electrodes. In a PILGE, PIL is the matrix that provides not only the ionic conduction pathways but also the mechanical robustness. A block copolymer approach is well established to achieve good mechanical properties as well as high ionic conductivity [15]. The best PILGE should have excellent ion transport characteristics coupled with mechanical robustness. PILGEs provide enough space in the chemistry research to design and tune the ionic conductivity while achieving appropriate mechanical properties. To develop a reliable rechargeable LIB, the PILGEs must possess the following properties.

- *Lithium-ion conductivity*: Ionic conductivity is the property of liquid electrolyte solutions which can be defined as a measure of its tendency towards ionic conduction. This involves the movement of electrolyte-ions from one site to another through the electrolyte. The organic solvent-based liquid electrolytes have advantages of high ionic conductivity. The ionic conductivity of a typical organic electrolyte is $\geq 10^{-2}$ S cm^{-1}. For a reliable PILGE to be used in a LIB, the ionic conductivity should be $\geq 10^{-4}$ S cm^{-1}. The mobility of electrolyte-ions in a PILGE depends on various factors, such as the nature of the cation and anion, the type of spacer used, the nature of the polymer chain, molecular weight, glass transition temperature, humidity, purity, etc.

- *Ion-transference number*: The ion-transference number, otherwise known as the ion-transport number, is defined as the total electric current carried in an electrolyte by a given electrolyte-ion. It can be either a cation or an anion. For example, in sodium chloride solution, less than half of the current is carried by the Na^+ ions (cations) and the remainder by the Cl^- ions (anions). The sum of the transport numbers for all the electrolyte-ions in a given solution always equals unity. The ion-transference numbers for Li^+-ion electrolytes can be determined by various methods, such as potentiostatic polarization, galvanostatic polarization, the electromotive force method, and nuclear magnetic resonance spectroscopy [16]. But this is different in the case of a PILGE. An ion-transference number of ~1 is desirable for the PILGE. In typical polymer electrolytes, a single ion-conducting system (such as cation conduction) is observed. Many of the PILGEs reported in the literature show a cation-transference number ≤ 0.5, which indicates that only half of the ions can move inside a PILGE. When this value is closer to unity, a greater number of cations will be able to move through the polymer electrolyte and hence the power density of the LIB becomes very high.
- *Compatibility with electrodes*: A good compatibility of the PILGE membrane is needed for the fabrication of a LIB, since the preparation involves sandwiching anode and cathode materials together with the electrolyte membrane in-between. The electrochemical performance of the LIB depends mainly on the synchronization of the electrode material with the electrolyte. A loss of compatibility will lead to low-capacity, low rate capability, feeble cycle life, etc.
- *Chemical stability*: PILGE membranes should have good chemical stability to achieve a long cycle life for the LIB. Any cross-reaction between the electrode and the electrolyte membrane may lead to either performance degradation or damage to the battery.
- *Thermal stability*: PILGE membranes should be able to withstand a wide temperature window. LIBs with a wide temperature window are much in demand for commercial applications. A shrinkage or breakage of the PILGE membrane as a result of thermal shock may lead to shorting of the anode and cathode that, in turn, leads to battery failure.
- *Electrochemical stability*: Fast electrochemical reactions are needed to achieve high power density. The electrochemical reaction during the charging phase leads to Li-intercalation, whereas, during the discharging phase, the de-intercalation of the Li^+-ions occurs. The continuous intercalation/de-intercalation results in electrode volume expansion due to swelling. This volume change should be minimized to achieve reversibility in the charging/discharging process of the LIB and should not affect the microstructure and morphology of the PILGE membrane. A disruption of the electrolyte membrane, as a result of volume expansion of the electrodes, results in shorting out the anode and cathode, which ultimately leads to battery failure.

- *Mechanical strength*: The durability of the PILGE membrane is an important factor in assembling as well as in the safe operation of LIBs. A brittle electrolyte membrane is not useful since the application of pressure during cell assembly or pouch cell fabrication can damage the electrolyte, leading to the failure or short-circuiting of the LIB..

11.3 TYPES OF POLYMER-IONIC LIQUID GEL ELECTROLYTES (PILGEs)

When compared with the conventional SPEs, IL-based SPEs exhibit several advantages, such as high ionic conductivities and wider operable voltage windows. But the liquid nature of the typical ILs leads to electrolyte leakage as well as making the LIB fabrication process difficult. This was the root cause of the development of a new type of electrolyte, called PILGE. Electrolyte leakage is a major bottleneck in the battery industry. Improper sealing and leakage due to the nature of the solvent used are equally detrimental for any best-performing LIB. In typical, PILGEs are single-ion conductors, although the preparation process of PILs with Li$^+$-ions is a great challenge. Zwitterions are used in the synthesis process, where they act as Li$^+$-ion dissociators. There are various types of PILGEs used in developing advanced LIBs. This section briefly discusses the various types of PILGEs used in LIBs, their synthesis, characterization, and applications in improving the electrochemical performance of the LIBs. After the successful implementation of PILGEs in LIBs, the performance, as well as the packaging capabilities, improve further. This has boosted its potential for developing novel and safe LIBs for commercial applications.

Lithium bis (trifluoromethane sulfonyl)imide (LiTFSI) is a hydrophilic salt with the chemical formula LiC$_2$F$_6$NO$_4$S$_2$. It is a widely accepted Li$^+$-ion source in commercially available LIBs. After the introduction of LiTFSI, lithium hexafluorophosphate became less attractive because of its poorer performance when compared with LiTFSI. Free-standing PILGEs, consisting of a PIL, poly[diallyldimethylammonium] bis-trifluoromethane sulfonimide] (PDAD-MATFSI), an IL, 1-ethyl-3-methylimidazolium bis-trifluoromethane sulfonimide and a Li salt for LIB application, was developed by Safa et al. [17]. In a typical procedure, the pyrrolidinium-based PIL, PDAD-MATFSI, was synthesized by a simple anion-exchange reaction between the chlorinated polymer and LiTFSI. The as-synthesized PIL was insoluble in water but readily soluble in acetone. Furthermore, the IL, 1-ethyl-3-methyl imidazolium bis (trifluoromethane sulfonyl)imide (EMIM-TFSI), was synthesized by reacting [EMIM][Cl] and LiTFSI in de-ionized water. The as-synthesized IL was colorless, odorless, and a fluid, which was dried at 60°C in a vacuum oven for two hours and immediately stored in an oxygen-free and humidity-free glove-box until later use. Furthermore, the GPE was synthesized in a procedure consisting of mixing an already prepared 1 M LiTFSI in [EMIM][TFSI] with 20 wt.% of PIL (final composition 80:20 electrolyte: PIL by weight). This mixture was dissolved in acetone and stirred (or sonicated) until completely dissolved. The final solution was drop-cast on a 0.0127 m circular polydimethylsiloxane (PDMS) template and enough time was given to evaporate the acetone from it, before it was vacuum dried at 90°C for 72 hours. The PILGE film developed in this way was transparent and free-standing. The PILGE film was thermally stable and contained IL and salt contents of up to 80 wt.%. This high concentration was found to enhance the ionic conductivity of the electrolyte films. A high ionic conductivity of 3.35×10^{-3} S cm^{-1} was obtained for the electrolyte films. Combining PIL with the salt component has allowed the LIB to exhibit a wide electrochemical stability window of −0.1 to 4.9 V. An increased Li$^+$-ion-transference number was also achieved. In addition, the PILGEs were found to suppress the formation of Li dendrites when compared with the pure IL component without the polymer.

Galvanostatic charge/discharge cycling of Li/GPE/lithium iron phosphate (LiFePO$_4$) cells have displayed discharge capacities of 169.3 mAh g^{-1} at a current rate of C/10 and 126.8 mAh g^{-1} at 1 C. These PILGEs also provide good capacity retention of 40 charge/discharge cycles at a current rate of 5 C at an operating temperature of 22°C. The IL, 1-methyl-3-propyl pyrrolidinium-TFSI(Py$_{13}$TFSI)/polyvinylidene difluoride-co-hexafluoropropylene (PVdF-co-HFP), blends with LiTFSI were used to prepare PILGE [18]. The Py$_{13}$TFSI/PVdF-co-HFP/1 M LiTFSI-based PILGE exhibited good thermal stability of 128°C. The LIB cell fabricated with this PILGE was able to operate at a voltage of 5.75 V vs. Li$^+$/Li. The authors have investigated the effect of ethylene carbonate (EC) on the PILGE and found that it can improve Li$^+$-ion-transport number, ionic conductivity, and the Li$^+$-ion-transport kinetics. But the addition of EC in the PILGE caused a small decrease in the anodic stability as well as in the thermal stability.

Fuller and co-workers [19] have developed PILGEs, using a room-temperature IL and PVdF-co-HFP. The ILs used in the preparation were 1-ethyl-3-methylimidazolium salts of triflate (CF$_3$SO$_3^-$) and BF$_4^-$. The PILGEs developed in this way were free-standing and flexible. The PILGE films exhibited room-temperature ionic conductivities ranging from 1.1×10^{-3} to 5.8×10^{-3} S cm^{-1}. The ILs and PVdF-co-HFP were non-volatile and thermally stable, so that the PILGE films could operate at elevated temperatures, with no loss of the electrochemical performance being noticed. A triflate IL-PVdF-co-HFP-based PILGE film was developed which exhibited an ionic conductivity of 41×10^{-3} S cm^{-1} at an operating temperature of 205°C.

The preparation of a PILGE with strong ion-dipole interactions, between an imidazolium-based IL and a fluorinated co-gel polymer, was reported by Chen et al. [20]. This PILGE was developed to obtain a stable and dendrite-free Li$^+$ plating/stripping. They have found that the selection of imidazolium-based IL leads to the formation of a tightly cross-linked gel framework with tethered anions and provides improved mechanical strength, high ionic conductivity, good heat resistance, and favorable self-healing capability. The PILGE is superior to the pure IL-based electrolyte in terms of Li dendrite formation as it enables a uniform Li$^+$ flux, which is schematically shown in Figure 11.2. A stable voltage of 4.5 V vs. Li$^+$/Li was obtained that could satisfy the demand of high-voltage cathodes. An optical micrograph and cross-sectional scanning electron microscope (SEM) image of the PILGE, prepared using IL-PVdF-co-HFP, are shown in Figure 11.3. The PILGE membrane has enabled dendrite-free Li deposition and hence achieved a stable cycling durability for 1000

Polymer-Ionic Liquid Gel Electrolyte 241

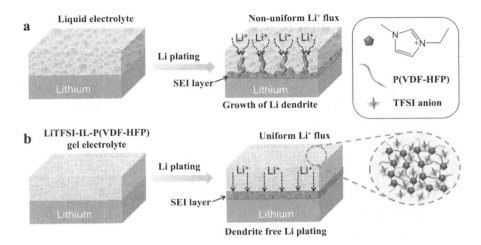

FIGURE 11.2 Schematic illustration of the electrochemical deposition behavior of Li anodes with (a) liquid-state organic electrolyte and (b) LiTFSI-IL-PVdF-*co*-HFP gel electrolyte. Adapted and reproduced with permission from Ref. [20]. Copyright © 2018 Elsevier.

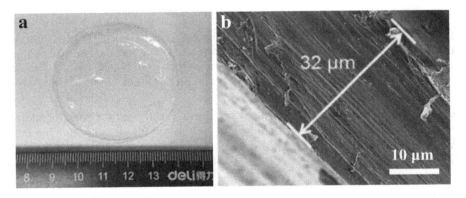

FIGURE 11.3 (a) Optical photograph, and (b) cross-sectional morphology (FE-SEM image) of the PVdF-*co*-HFP-based PILGE membrane. Adapted and reproduced with permission from Ref. [20]. Copyright © 2018 Elsevier.

hours when tested at a current density of 0.5 mA cm^{-2}. The LiFePO$_4$/Li LIB cell was found to exhibit good cycling stability and rate performance that was attributed to the performance of the PILGE membrane developed. The authors have assembled Li-sulfur batteries, using this PILGE membrane, and observed that this membrane efficiently suppresses the polysulfide shuttling and self-discharge, with the LIB cell exhibiting high specific capacity and long cycling life.

IL-organic solvent mixture-based PILGE, with a high Li concentration, was developed for LIBs by Lahiri et al. [21] recently. This PILGE was based on PVdF-*co*-HFP, 1-butyl-1-methylpyrrolidinium bis(fluorosulfonyl)imide [(Py1,4)FSI] and Li

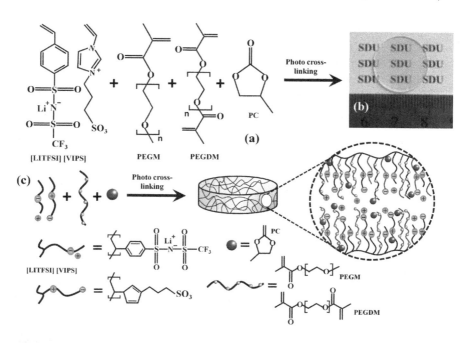

FIGURE 11.4 (a) Chemical structures of monomeric ionic liquid [LiSTFSI][VIPS], PEGM, PEGDM and PC; (b) A photograph of the polymer electrolyte film obtained after photo-cross-linking; (c) Schematic illustration of the strategy for preparation of PILGE film. Adapted and reproduced with permission from Ref. [22]. Copyright © 2018 Royal Society of Chemistry.

bis(fluorosulfonyl)imide (LiFSI). The authors have investigated the effect of adding acetonitrile and different concentrations of LiFSI to the PILGE and observed that, when compared with 1 M LiFSI, the addition of 4 M LiFSI resulted in higher Li storage capacity. The thickness of the solid-electrolyte interfacial layer was found to vary in response to the Li-salt concentration and the presence of acetonitrile in the PILGE. The authors hypothesized that this might be the reason for the greater Li storage capacity. This study shows that the addition of an organic solvent and high concentrations of a Li salt in a PILGE improves the performance of LIBs.

Recently, Yu et al. [22] have reported a promising method to synthesize PILGE for applications in LIBs. Initially, a room-temperature IL was synthesized using an equimolar monomer mixture of zwitterion, 3-(1-vinyl-3-imidazolic) propane sulfonate (VIPS) and 4-styrene sulfonyl-TFSI (LiSTFSI) *via* intermolecular electrostatic interactions. Later, *in-situ* photopolymerization was carried out with flexible chain poly(ethylene glycol) (PEG) methyl ether methacrylate (PEGM) and cross-linker PEG-di-methacrylate (PEGDM) in the presence of propylene carbonate (PC), generating free-standing PILGE films. The chemical structures of monomeric IL [LiSTFSI] [VIPS], PEGM, PEGDM, and PC are depicted in Figure 11.4(a). A photograph of the as-synthesized PILGE membrane is shown in Figure 11.4(b). A photo-cross-linking strategy was adopted for the synthesis of the PILGE, which

Polymer-Ionic Liquid Gel Electrolyte

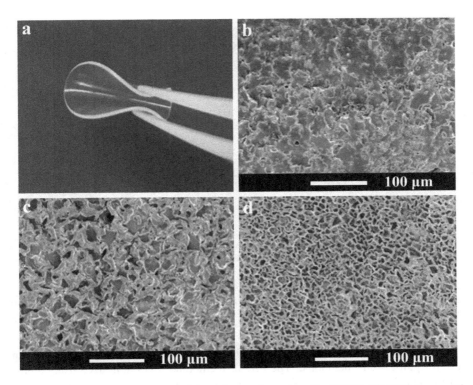

FIGURE 11.5 (a) A photograph of the polymer electrolyte film SIPE8. FE-SEM images of (b) SIPE2, (c) SIPE5, and (d) SIPE8. Adapted and reproduced with permission from Ref. [22]. Copyright © 2018 Royal Society of Chemistry.

is shown schematically in Figure 11.4(c). This PILGE film is enriched with Li$^+$-ion channels and exhibited a Li$^+$-ion-transference number of 0.93 at room temperature. The flexible electrolyte film exhibited an ionic conductivity of 1.31×10^{-4} S cm^{-1} at an operating temperature of 30°C. As depicted in Figure 11.5, the polymer matrix exhibited a sponge-like structure after completely evaporating away the PC. It was observed that the porosity of the polymer electrolyte increased with an increase in the PC concentration. Hence, a high concentration of PC can be accommodated within the polymer matrix. The sample named 'SIPE8' exhibited a more uniform pore distribution when compared with other samples, such as 'SIPE2' and 'SIPE5'. The electrochemical performance of this PILGE, when used in a LIB, was examined by fabricating a Li/LiFePO$_4$ half-cell. It was observed that the discharge capacity of this LIB half-cell during 100 charge/discharge cycles maintained a steady-state value of 120 mAh g^{-1}. The Coulombic efficiency of the LIB half-cell was found to be ~93%, which confirms the potential of the PILGE membrane for the manufacture of commercial LIBs.

The electrospinning strategy was adopted to synthesize a PILGE membrane by Cheng et al. [23]. A nanostructured IL, silica (SiO$_2$) nanoparticle-tethered

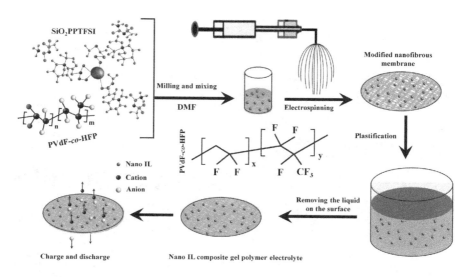

FIGURE 11.6 Schematic illustration of the strategic approach for the preparation of nanostructured ionic liquid-based PILGE membranes. Adapted and reproduced with permission from Ref. [23]. Copyright © 2018 Royal Society of Chemistry.

1-methyl-1-propyl piperidinium-TFSI (SiO$_2$PPTFSI) was used in the synthesis of an electrospun PILGE membrane. The PILGE membrane was synthesized using a protocol carried out as follows. Initially, the IL was introduced into a GPE matrix, PVdF-co-HFP, that was prepared at first by electrospinning. Furthermore, plasticization in a fluoroethylene carbonate (FEC) electrolyte was carried out. A schematic diagram, showing the preparation of this nanostructured IL-based PILGE membrane, is depicted in Figure 11.6. This novel PILGE has exhibited good mechanical properties and an increased electrolyte uptake of 552 wt.%. An ionic conductivity of 0.64×10^{-3} S cm^{-1} and a Li$^+$-ion-transference number of 0.6 were obtained for this novel PILGE membrane. In addition to these excellent features, the membrane has helped by suppressing the formation of Li dendrites and exhibited stable plating/striping cycles over a period of 1200 hours when tested in a symmetrical LIB cell. The PILGE membrane was able to operate in a stable voltage of 5.1 V (vs. Li/Li$^+$), showing its potential in commercial applications. The Li/LiNi$_{0.5}$Mn$_{1.5}$O$_4$ LIB cell fabricated with this PILGE membrane exhibited a discharge capacity of 119 mAh g^{-1} at a current rate of 1 C with a capacity retention of 92.1%, after undergoing 460 charge/discharge cycles. A reversible discharge capacity of 74 mAh g^{-1} was also obtained at a current rate of 6 C.

Libo Li et al. [24] have introduced a short-cut method to prepare high-performance PILGEs. The PILGE membranes were prepared *via* ultraviolet (UV) cross-linking of polyurethane acrylate (PUA), methyl methacrylate (MMA), Py$_{13}$TFSI, LiTFSI, ethylene glycol di-methacrylate (EGDMA), and benzoyl peroxide (BPO). The Py$_{13}$TFSI-based IL was prepared initially by mixing *N*-methyl-*N*-propyl pyrrolidinium bromide (Py$_{13}$Br) and LiTFSI. The addition of Py$_{13}$TFSI during the preparation

Polymer-Ionic Liquid Gel Electrolyte   245

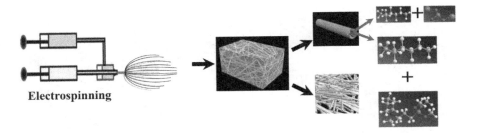

FIGURE 11.7 Schematic illustration of the synthesis of the PHP@PHL-based PILGE membrane. Adapted and reproduced from Ref. [25]. Copyright © 2019 Liu, Ren, Zhang and Zhang.

of the PILGE led to the formation of network structures *via* chain cross-linking. The as-prepared PILGE membrane exhibited an ionic conductivity of 1.37×10^{-3} S cm^{-1} at room temperature. A Li$^+$-ion-transference number of 0.22 was achieved for this electrolyte membrane, with an electrochemical stability window of 4.8 V vs. Li$^+$/Li. The charge/discharge characteristics of the Li/PILGE/LiFePO$_4$ LIB half-cell assembled delivered a discharge capacity of 151.9 mAh g^{-1} with a Coulombic efficiency of 87.9%. Although the discharge capacity fell to 131.9 mAh g^{-1}, the Coulombic efficiency was found to increase after completing 80 charge/discharge cycles. A functional IL-modified core-shell structured, fibrous GPE-based PILGE was reported in the recent past [25]. The authors have claimed that the PILGE developed in this way is suitable for fast-charging LIBs. The co-axial electrospinning technique was employed in the preparation of a core-shell structured, three-dimensional (3D) porous nanofiber membrane. In this work, a piperidine IL, 1-methyl-1-propyl piperidinium chloride (PPCl), and Li$_2$SiO$_3$ (LSO) nanoparticles were used with the PVdF-*co*-HFP matrix. The nanostructured PILGE membrane, consisting of PVdF-*co*-HFP, PVdF-*co*-HFP-PPCl (PHP), and PVdF-*co*-HFP-LSO (PHL), was prepared by mixing PVdF-*co*-HFP in dimethylformamide (DMF) (20:80), PPCl, and PVdF-*co*-HFP in DMF (1:19:80) or LSO and PVdF-*co*-HFP in DMF (2:18:80), respectively. A schematic diagram, showing the preparation of PHP@PHL membrane, is depicted in Figure 11.7. The SEM images of various separators, such as Celgard-2325, nanofibrous PVdF-HFP, PHL, and PHP@PHL are depicted in Figure 11.8(a)-(d) [25]. The PHP@PHL membrane exhibits high-temperature stability when compared with its pure counterparts and the commercially available Celgard®-2325. The electrolyte uptake and porosity of the PHP@PHL nanofiber membrane were 597% and 74%, respectively. An ionic conductivity of 4.05×10^{-3} S cm^{-1} was obtained for the PHP@PHL nanofiber membrane, with a Li$^+$-ion-transference number of 0.62. Symmetric-type LIB cells, fabricated with a PHP@PHL nanofiber membrane, exhibited good plating/stripping cycling stability for a period of 1000 hours without any possible short-circuiting. The LIB cell fabricated in this way exhibits good rate capability with a reversible capacity of 65 mAh g^{-1} at a current rate of 20 C. It should be noted here that the discharge capacity for the LIB cell fabricated with the commercially available Celgard®-2325 was only 5 mAh g^{-1}.

FIGURE 11.8 SEM images of the surface morphology of the various separators: (a) Celgard®-2325, (b) nanofibrous PVdF-*co*-HFP, (c) nanofibrous PHL, and (d) nanofibrous PHP@PHL electrospun membrane. Adapted and reproduced from Ref. [25]. Copyright © 2019 Liu, Ren, Zhang, and Zhang.

Zhang et al. [26] have reported the synthesis of IL-doped GPEs for use in flexible LIBs. The PILGE samples containing IL at different volume percentages were synthesized and their electrochemical performances in LIBs were investigated. This study has shown that the use of excess IL during the preparation of PILGE can damage the internal structure of the LIB and invite unwanted electrochemical reactions. They observed that the PILGE samples containing 40–50 vol.% IL exhibited lower internal resistance and superior electrochemical properties when compared with those prepared with lower or higher concentration of ILs. The PILGEs prepared using phosphonium IL, composed of the trihexyl(tetradecyl)phosphonium cation combined with TFSI counter-anions and Li salt, and an epoxy pre-polymer, were reported [27]. The PILGEs developed exhibit good thermal stability of more than 300°C, along with good thermo-mechanical properties. An ionic conductivity of 0.13×10^{-3} S cm^{-1} was achieved at an operating temperature of 100°C. An electrochemical potential window of 3.95 V vs. Li^0/Li$^+$ was achieved. A LIB cell fabricated using Li//LiFePO$_4$ electrodes exhibited a cycle life of 30 cycles. Taghavikish et al. [28] reported the preparation of a PILGE by chemically cross-linking 2-hydroxyethyl methacrylate (HEMA) monomer and a polymerizable IL, 1,4-di(vinylimidazolium)butane bis bromide (DVIMBr) in an IL, 1-butyl-3-methylimidazolium hexafluorophosphate, as the polymerization solvent. The authors claimed that *in-situ* entrapment of the IL took place in the gel during polymerization, which enabled the cross-linking of the polymer. This DVIMBr cross-linked HEMA polymer was thermally stable up to an operating temperature of 300°C, which is attributed to the high thermal stability of the DVIMBr.

Safa et al. [29] have reported the preparation of a composite PILGE, which consists of a PIL, an imidazolium cation-based IL as a solvent, LiTFSI as the Li salt, and glass fillers at various concentrations. This composite PILGE, prepared with 1 wt.% glass filler, showed a high ionic conductivity of 4.08×10^{-3} S cm^{-1} and a Li$^+$-ion-transference number of 0.44. These compositions improved the electrochemical performance of the composite PILGE, when compared with the one without glass fillers, which was attributed to an improved ion-pair dissociation of LiTFSI, which, in turn, caused an improvement in the Li$^+$-ion mobility. The galvanostatic charge/discharge measurement of the LIB cell, fabricated using a binder-free LiFePO$_4$/carbon cathode and a Li anode, exhibited a good cycling performance for 100 charge/discharge cycles at various current rates.

Poly(diallyldimethylammonium) (PDADMA) is a well-known host for developing PILGEs. A free-standing flexible PILGE membrane was synthesized by mixing high concentrations of LiFSI, trimethylisobutylphosphonium bis(fluorosulfonyl) imide (P$_{111i4}$FSI), and PDADMA-FSI [30]. The as-synthesized PILGE membrane was found to be transparent. The PILGE with 60:40 wt.% electrolyte, containing 3.8 M LiFSI in P$_{111i4}$FSI (LiFSI/IL): PDADMA FSI, exhibited an ionic conductivity of 0.49×10^{-3} S cm^{-1} at 40 °C. A comparatively low Li$^+$-ion-transference number of 0.21 was exhibited by this PILGE at an operating temperature of 50 °C. An imidazolium-based PIL, poly(1-ethyl-3-vinylimidazolium-TFSI), was synthesized *via* a three-step process, including direct radical polymerization of the 1-vinylimidazole monomer, a quaternization reaction, and an anion-exchange procedure [31]. A PILGE was synthesized by blending PIL as the polymer host, an IL, and LiTFSI. The PILGE synthesized in this way exhibited excellent electrochemical properties when compared with the one prepared *via* the conventional route. The reason was a high loading of IL in the PILGE. Singh et al. [32] have reported the synthesis of a PILGE composed of PVdF-*co*-HFP, an IL, 1-ethyl-3-methylimidazolium-FSI (EMIMFSI), and LiTFSI. This PILGE was prepared by the solution-casting method and the as-prepared PILGE film was free-standing. The PILGE prepared with 80 wt.% IL showed a thermal stability up to ~200°C, with an ionic conductivity of 6.42×10^{-4} S cm^{-1}. A wide electrochemical stability window of ~4.1 V vs. Li/Li$^+$ was achieved for this PILGE membrane at an operating temperature of 30°C.

A copolymeric IL, poly[*N*-(1-vinylimidazolium-3-butyl)-ammonium TFSI]-*co*-PEG methyl ether methacrylate) (PVIMTFSI-*co*-PPEGMA) was synthesized initially and then used to prepare a flexible PILGE, PVIMTFSI-*co*-PPEGMA/LiTFSI [33]. This novel PILGE membrane was prepared by the solution-casting method. The electrochemical performance of the PVIMTFSI-*co*-PPEGMA/LiTFSI-based PILGE was compared with poly(*N*-vinylimidazole)-*co*-PEG-methyl ether methacrylate (PVIM-*co*-PPEGMA)/LiTFSI-based electrolyte, that contained non-ionized imidazole groups. The electrochemical potential window of the PVIMTFSI-*co*-PPEGMA/LiTFSI-based PILGE was found to be 5.3 V, whereas it was slightly higher (5.5 V) for the PILGE consisting of PVIM-*co*-PPEGMA/LiTFSI. The fact that the PIL segments constrained the formation of Li dendrites was the reason for obtaining the wider window. The LiFePO$_4$/Li LIB cell was fabricated using the PILGE consisting of PVIMTFSI-*co*-PPEGMA/LiTFSI, which exhibited a discharge capacity of 136 mAh g^{-1} at a current rate of 0.1 C. The LIB was found to maintain a capacity of 70 mAh g^{-1} at a current rate of 1 C at an operating temperature of 60 °C.

Li et al. [34] have reported the synthesis of 80% [(1-x)PIL-(x)SN]–20% LiTFSI, using a pyrrolidinium-based PIL, (DADMA)TFSI, as a polymer host, succinonitrile (SN) as a plastic crystal, and LiTFSI. This novel PILGE, composed of 80% [50% PIL–50% SN]–20% LiTFSI (50% SN), displayed an ionic conductivity of 5.74×10^{-4} S cm^{-1} at room temperature. This PILGE also exhibited a wide electrochemical potential window of 5.5 V, with a Young's modulus of 4.9 MPa, showing its excellent mechanical stability. The Li/LiFePO$_4$-based LIB cells assembled with this PILGE, containing 50% SN, obtained a discharge capacity of 150 mAh g^{-1} at a rate of 0.1 C at an operating temperature of 25°C. The LIB cell fabricated in this way exhibited excellent capacity retention of 131.8 and 121.2 mAh g^{-1} at different current rates of 0.5 C and 1 C, respectively. An organic ionic plastic crystal, *N*-ethyl-*N*-methylpyrrolidinium-FSI (P$_{12}$FSI), a pyrrolidinium-based PIL, and LiTFSI were used to prepare a novel PILGE, PIL-P$_{12}$FSI-LiTFSI, in the recent past [35]. The solid-state electrolyte system, PIL-P$_{12}$FSI-LiTFSI exhibited good mechanical properties with an ionic conductivity of 10^{-4} S cm^{-1}. This PILGE was found to suppress the growth of Li dendrites. A Li/LiFePO$_4$ LIB cell fabricated with PIL-P$_{12}$FSI-LiTFSI electrolyte exhibited a wide operating temperature range of 25–80°C.

Recently, a pyrrolidinium-based PIL, PDADMA-TFSI, PEG800, and LiTFSI were used to prepare a PILGE [36]. The interactions between PDADMA-TFSI, PEG800, and Li$^+$-ions helped to achieve good ionic conductivity and thermal stability. The Li/LiFePO$_4$ LIB cells, fabricated with the as-prepared PILGE, has been shown to exhibit a discharge capacity of 150 mAh g^{-1} at a current rate of 0.2 C at an operating temperature of 80°C. A PILGE was developed using a PIL, 1,2-dimethyl-3-butylimidazolium-TFSI (BMMIM-TFSI), LiTFSI, and nano-SiO$_2$ [37]. An ionic conductivity of 1.07×10^{-3} S cm^{-1} was achieved at an operating temperature of 60°C with a BMMIM-TFSI concentration of 60% (the weight ratio of BMMIM-TFSI/PIL). The Li/LiFePO$_4$ LIB cell fabricated with this electrolyte delivered a discharge capacity of 146 mAh g^{-1} at an operating temperature of 60°C. A novel imidazolium-tetra-alkylammonium-based di-cationic PIL, poly(*N*,*N*,*N*-trimethyl-*N*-(1-vinyl-imidazolium-3-ethyl)-ammonium TFSI was prepared by Yin et al. [38]. The as-synthesized di-cationic PIL was further used as the polymer host to synthesize a ternary PILGE by blending it with an IL, 1,2-dimethyl-3-ethoxyethyl imidazolium TFSI [IM(2o2)11TFSI] and LiTFSI. This novel electrolyte exhibited a low glass transition temperature of −54°C and was found to be thermally stable up to a temperature of 330°C. This ternary PILGE displayed an ion conductivity of 1×10^{-4} S cm^{-1}. The Li/LiFePO$_4$ LIB cells assembled with the LiTFSI-IM(2o2)11TFSI electrolyte delivered a discharge capacity of 160, 140, and 120 mAh g^{-1} at different operating temperatures of 40, 30, and 25°C, respectively. This LIB cell retained a capacity of 161.1 mAh g^{-1} after 50 charge/discharge cycles when tested at a temperature of 40°C. At 30°C, the LIB cell maintained a capacity of 141.7 mAh g^{-1} after completing 50 charge/discharge cycles. These results have shown that the LiTFSI-IM(2o2)11TFSI-based PILGEs are potential candidates for commercial use in LIBs. Kuo et al. [39] have introduced a new strategy to prepare oligomeric PILGEs for high-performance and non-flammable LIBs. The authors have also introduced a new strategy by which to synthesize oligomeric IL economically from a conventional phenolic epoxy resin. This oligomeric IL was further used to prepare non-flammable PILGE membranes *via* blending with PVdF-*co*-HFP and a liquid-state organic electrolyte. During the preparation, the liquid electrolyte content was kept below 50% and achieved ionic conductivities of 2 and 6.6×10^{-3} S cm^{-1} at 30 and 80°C, respectively. The PVdF-*co*-HFP/70% oligomeric IL-based PILGE, with an uptake of 13% liquid electrolyte, exhibited a discharge capacity of 152 and 141 mAh g^{-1}, respectively. The Li/LiFePO$_4$ LIB cells fabricated with this PILGE exhibited a maximum operating voltage of 3.4 V. The interfacial compatibility of the PVdF-*co*-HFP/30% oligomeric IL-based PILGE with the electrodes led to the formation of a passivation layer. The use of high concentrations of oligomeric IL was found to reduce the electrode/electrolyte interfacial resistance, allowing the easy transportation of Li$^+$-ions. The charge/discharge test showed that the LIB cell exhibited a good cycle life since no degradation was observed after 100 charge/discharge cycles. A Coulombic efficiency of 99% was achieved for this LIB cell. Due to the oligomeric IL content, the PILGE exhibited excellent dimensional stability such that, at a high temperature of 150°C, the dimensional change was less than 1%. An important fact about this oligomeric IL-based PILGE was that it achieved the flame-retardant requirement for safety in LIB cells under normal conditions. This has enabled the PILGE to function as both a high-safety Li$^+$-ion conductor and as a separator for LIBs. Lu et al. [40] have reported the synthesis of a novel nanostructured proton-transporting film by the self-assembly of VIPS and 4-dodecylbenzenesulfonic acid *via* intermolecular electrostatic interactions. A schematic diagram representing a strategy to prepare nanostructured proton-transporting films is shown in Figure 11.9. The polymerized film-H1 and film-Lα exhibited good ionic conductivities. They exhibit lower activation energies than the commercially available Nafion® membrane and hence have the potential for practical use in advanced LIBs.

Patel et al. [41] have reported the synthesis of PILGE for high-rate capability LIBs. The PILGE was prepared *via* free-radical polymerization of a vinyl monomer, acrylonitrile (AN) in *N,N*-methyl butyl pyrrolidinium-TFSI-LITFSI (Py$_{1,4}$-TFSI-LITFSI). The as-obtained PILGE, with a content of AN: Py$_{1,4}$-TFSI = 0.16–0.18 (w/w), has shown high mechanical strength and thermal stability when compared with pure LiTFSI-Py$_{1,4}$-TFSI. The ionic conductivity of this PILGE ranged from 1.1×10^{-3}–1.7×10^{-3} Ω^{-1} cm^{-1}, which is almost identical to that of the IL (1.8×10^{-3} Ω^{-1} cm^{-1}). No ageing effect was observed for this PILGE, whereas it was observed in GPEs. The Li//LiFePO$_4$ LIB cell assembled using this PILGE was able to operate within a voltage window of 0.5–5 V vs. a Li reference electrode. A LIB cell, fabricated using a multi-walled carbon nanotubes-LiFePO$_4$ composite, and a Li electrode,

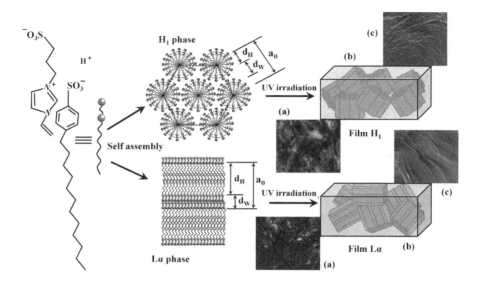

FIGURE 11.9 Schematic diagram depicting the strategy to prepare nanostructured proton-transporting films obtained by photopolymerization of H1 Phase and la(alpha) phase (a) POM images (b)schematic illustration and (c) scanning electron micrographs (SEM) of the cross sections. Adapted and reproduced with permission from Ref. [40]. Copyright © 2014 American Chemical Society.

delivered a discharge capacity of 141 mAh g^{-1}, which was found to be higher than that of the pure IL-based LIB cell (120 mAh g^{-1}). Dam et al. [42] proposed an idea of preparing PILGE membranes using semi-interpenetrating cross-linked polymer network (SICPN)-based polymer hosts. The SICPN polymer hosts were prepared *via* chemically cross-linking the polymer. Blending PEG-diacrylate-*co*-vinylene carbonate (PEGDA-*co*-VC) with PVdF-*co*-HFP and, later, carrying out UV-assisted cross-linking, was carried out between PEG-diacrylate (PEGDA) and vinylene carbonate (VC). The various steps involved in the preparation of PILGEs using this SICPN are as follows. Initially, a mixture of PVdF-*co*-HFP, PEGDA, VC, and 2-hydroxy-2-methylpropiophenone (HMPP) was prepared in a glass petri-dish, and it was irradiated with UV light for 5 minutes. The as-obtained stable membrane was washed and vacuum dried at a temperature of 49.85°C for 12 hours and stored inside a glovebox under an argon atmosphere. The weight ratio of VC to PEGDA was kept constant at 4:1, with PVdF-*co*-HFP: VC: PEGDA at a weight ratio of 1:x:x/4 with x= 2, 3, 4, 5, or 6. Later, a mixture of carbonate solvents [ethylene carbonate (EC): propylene carbonate (PC) = 1:1], Li salt (Li tetrafluoroborate) and 1-butyl-3-methylimidazolium tetrafluoroborate (BMIMBF$_4$)-based IL, was prepared. Finally, the SICPN membrane was immersed in this mixture for a period of 12 hours. A maximum ionic conductivity of 6.69×10^{-3} S cm^{-1} was achieved at a temperature of 29.85°C. The charge/discharge performance of the CR-2032 LIB coin-cell fabricated in a LiFePO$_4$/SICPN/Li configuration was carried out. The cell has exhibited discharge capacities of 112, 95, 78, 68, and 45 mAh g^{-1} when tested at current rates of C/10, C/8, C/5, C/3, and 1 C, respectively. A Coulombic efficiency of 99% was achieved at a rate of C/10 after completing 50 cycles. The LIB cell fabricated using SICPN-based PILGE was able to operate at a voltage of 4.3 V.

11.4 CONCLUSION AND FUTURE PERSPECTIVES

Organic electrolyte-based LIBs face critical issues associated with flammability, rigidity, and leakage. The recent developments of flexible and wearable devices necessitate flexible power supplies for their efficient working. To satisfy these requirements, solid-state electrolytes were developed, such as GPEs and PILGEs. In this chapter, various types of PILGEs to fabricate advanced LIBs have been discussed. Recent developments in the PILGEs were discussed and their advantages, when compared with their GPE counterparts, were prepared using organic solvents, were discussed. Strategies to develop flexible PILGE membranes were introduced from the relevant literature, and their electrochemical performances, when used in LIBs, were also explained.

REFERENCES

1. Fang X, Ge M, Rong J, Zhou C (2013) Graphene-oxide-coated LiNi 0.5 Mn 1.5 O 4 as high voltage cathode for lithium ion batteries with high energy density and long cycle life. *Journal of Materials Chemistry A* 1 (12):4083–4088
2. Landi BJ, Ganter MJ, Cress CD, DiLeo RA, Raffaelle RP (2009) Carbon nanotubes for lithium ion batteries. *Energy & Environmental Science* 2 (6):638–654
3. Chai J, Zhang J, Hu P, Ma J, Du H, Yue L, Zhao J, Wen H, Liu Z, Cui G (2016) A high-voltage poly (methylethyl α-cyanoacrylate) composite polymer electrolyte for 5 V lithium batteries. *Journal of Materials Chemistry A* 4 (14):5191–5197
4. Chen X, Lin H, Zheng X, Cai X, Xia P, Zhu Y, Li X, Li W (2015) Fabrication of core–shell porous nanocubic Mn 2 O 3 @ TiO 2 as a high-performance anode for lithium ion batteries. *Journal of Materials Chemistry A* 3 (35):18198–18206
5. Lv C, Peng Y, Yang J, Liu C, Duan X, Ma J, Wang T (2018) A free-standing Li 1.2 Mn 0.54 Ni 0.13 Co 0.13 O 2/MWCNT framework for high-energy lithium-ion batteries. *Inorganic Chemistry Frontiers* 5 (12):3053–3060
6. Cherusseri J, Kar KK (2015) Hierarchically mesoporous carbon nanopetal based electrodes for flexible supercapacitors with super-long cyclic stability. *Journal of Materials Chemistry A* 3 (43):21586–21598
7. Cherusseri J, Kar KK (2016) Ultra-flexible fibrous supercapacitors with carbon nanotube/polypyrrole brush-like electrodes. *Journal of Materials Chemistry A* 4 (25):9910–9922
8. Cherusseri J, Sharma R, Kar KK (2016) Helically coiled carbon nanotube electrodes for flexible supercapacitors. *Carbon* 105:113–125
9. Cherusseri J, Kar KK (2015) Self-standing carbon nanotube forest electrodes for flexible supercapacitors. *Rsc Advances* 5 (43):34335–34341
10. Fan L-Z, Wang X-L, Long F, Wang X (2008) Enhanced ionic conductivities in composite polymer electrolytes by using succinonitrile as a plasticizer. *Solid State Ionics* 179 (27–32):1772–1775

11. Hu J, Wang W, Zhou B, Feng Y, Xie X, Xue Z (2019) Poly (ethylene oxide)-based composite polymer electrolytes embedding with ionic bond modified nanoparticles for all-solid-state lithium-ion battery. *Journal of Membrane Science* 575:200–208
12. Rhoo H-J, Kim H-T, Park J-K, Hwang T-S (1997) Ionic conduction in plasticized PVCPMMA polymer blend electrolytes. *Electrochimica Acta* 42 (10):1571–1579
13. Tarascon J-M, Armand M (2011) Issues and challenges facing rechargeable lithium batteries. In: *Materials for Sustainable Energy: A Collection of Peer-Reviewed Research and Review Articles from Nature Publishing Group*. World Scientific, pp. 171–179
14. Watanabe M, Thomas ML, Zhang S, Ueno K, Yasuda T, Dokko K (2017) Application of ionic liquids to energy storage and conversion materials and devices. *Chemical Reviews* 117 (10):7190–7239
15. Forsyth M, Porcarelli L, Wang X, Goujon N, Mecerreyes D (2019) Innovative electrolytes based on ionic liquids and polymers for next-generation solid-state batteries. *Accounts of Chemical Research* 52 (3):686–694
16. Zugmann S, Fleischmann M, Amereller M, Gschwind RM, Wiemhöfer HD, Gores HJ (2011) Measurement of transference numbers for lithium ion electrolytes via four different methods, a comparative study. *Electrochimica Acta* 56 (11):3926–3933
17. Safa M, Chamaani A, Chawla N, El-Zahab B (2016) Polymeric ionic liquid gel electrolyte for room temperature lithium battery applications. *Electrochimica Acta* 213:587–593
18. Ye H, Huang J, Xu JJ, Khalfan A, Greenbaum SG (2007) Li ion conducting gel polymer electrolytes based on ionic liquid/PVDF-HFP blends. *Journal of the Electrochemical Society* 154 (11):A1048–A1057
19. Fuller J, Breda A, Carlin R (1997) Ionic liquid-gel polymer electrolytes. *Journal of the Electrochemical Society* 144 (4):L67–L70
20. Chen T, Kong W, Zhang Z, Wang L, Hu Y, Zhu G, Chen R, Ma L, Yan W, Wang Y (2018) Ionic liquid-immobilized gel polymer electrolyte with self-healing capability, high ionic conductivity and heat resistance for dendrite-free lithium metal batteries. *Nano Energy* 54:17–25
21. Lahiri A, Pulletikurthi G, Shapouri Ghazvini M, Höfft O, Li G, Endres F (2018) Ionic liquid–organic solvent mixture-based gel polymer electrolyte with high lithium concentration for Li-Ion batteries. *The Journal of Physical Chemistry C* 122 (43):24788–24800
22. Yu Y, Lu F, Sun N, Wu A, Pan W, Zheng L (2018) Single lithium-ion polymer electrolytes based on poly (ionic liquid) s for lithium-ion batteries. *Soft Matter* 14 (30):6313–6319
23. Cheng Y, Zhang L, Xu S, Zhang H, Ren B, Li T, Zhang S (2018) Ionic liquid functionalized electrospun gel polymer electrolyte for use in a high-performance lithium metal battery. *Journal of Materials Chemistry A* 6 (38):18479–18487
24. Li L, Yang X, Li J, Xu Y (2018) A novel and shortcut method to prepare ionic liquid gel polymer electrolyte membranes for lithium-ion battery. *Ionics* 24 (3):735–741
25. Zhang S, Zhang L, Liu X, Ren Y (2019) Functional ionic liquid modified core-shell structured fibrous gel polymer electrolyte for safe and efficient fast charging lithium-ion batteries. *Frontiers in Chemistry* 7:421
26. Zhang R, Chen Y, Montazami R (2015) Ionic liquid-doped gel polymer electrolyte for flexible lithium-ion polymer batteries. *Materials* 8 (5):2735–2748
27. Leclère M, Bernard L, Livi S, Bardet M, Guillermo A, Picard L, Duchet-Rumeau J (2018) Gelled electrolyte containing phosphonium ionic liquids for lithium-ion batteries. *Nanomaterials* 8 (6):435
28. Taghavikish M, Subianto S, Gu Y, Sun X, Zhao X, Choudhury NR (2018) A poly (ionic liquid) gel electrolyte for efficient all solid electrochemical double-layer capacitor. *Scientific Reports* 8 (1):1–10

29. Safa M, Adelowo E, Chamaani A, Chawla N, Baboukani AR, Herndon M, Wang C, El-Zahab B (2019) Poly (ionic liquid)-based composite gel electrolyte for lithium batteries. *ChemElectroChem* 6 (13):3319–3326
30. Yunis R, Girard GM, Wang X, Zhu H, Bhattacharyya AJ, Howlett P, MacFarlane DR, Forsyth M (2018) The anion effect in ternary electrolyte systems using poly (diallyldimethylammonium) and phosphonium-based ionic liquid with high lithium salt concentration. *Solid State Ionics* 327:83–92
31. Yin K, Zhang Z, Yang L, Hirano S-I (2014) An imidazolium-based polymerized ionic liquid via novel synthetic strategy as polymer electrolytes for lithium ion batteries. *Journal of Power Sources* 258:150–154
32. Singh SK, Gupta H, Balo L, Singh VK, Tripathi AK, Verma YL, Singh RK (2018) Electrochemical characterization of ionic liquid based gel polymer electrolyte for lithium battery application. *Ionics* 24 (7):1895–1906
33. Wang A, Liu X, Wang S, Chen J, Xu H, Xing Q, Zhang L (2018) Polymeric ionic liquid enhanced all-solid-state electrolyte membrane for high-performance lithium-ion batteries. *Electrochimica Acta* 276:184–193
34. Li X, Zhang Z, Li S, Yang L, Hirano S-i (2016) Polymeric ionic liquid-plastic crystal composite electrolytes for lithium ion batteries. *Journal of Power Sources* 307:678–683
35. Li X, Zhang Z, Li S, Yang K, Yang L (2017) Polymeric ionic liquid–ionic plastic crystal all-solid-state electrolytes for wide operating temperature range lithium metal batteries. *Journal of Materials Chemistry A* 5 (40):21362–21369
36. Li S, Zhang Z, Yang K, Yang L (2018) Polymeric ionic liquid-poly (ethylene glycol) composite polymer electrolytes for high-temperature lithium-ion batteries. *ChemElectroChem* 5 (2):328–334
37. Li M, Wang L, Du T (2014) Preparation of polymer electrolytes based on the polymerized imidazolium ionic liquid and their applications in lithium batteries. *Journal of Applied Polymer Science* 131 (20):40928.
38. Yin K, Zhang Z, Li X, Yang L, Tachibana K, Hirano S-i (2015) Polymer electrolytes based on dicationic polymeric ionic liquids: application in lithium metal batteries. *Journal of Materials Chemistry A* 3 (1):170–178
39. Kuo P-L, Tsao C-H, Hsu C-H, Chen S-T, Hsu H-M (2016) A new strategy for preparing oligomeric ionic liquid gel polymer electrolytes for high-performance and nonflammable lithium ion batteries. *Journal of Membrane Science* 499:462–469
40. Lu F, Gao X, Dong B, Sun P, Sun N, Xie S, Zheng L (2014) Nanostructured proton conductors formed via in situ polymerization of ionic liquid crystals. *ACS Applied Materials & Interfaces* 6 (24):21970–21977
41. Patel M, Gnanavel M, Bhattacharyya AJ (2011) Utilizing an ionic liquid for synthesizing a soft matter polymer "gel" electrolyte for high rate capability lithium-ion batteries. *Journal of Materials Chemistry* 21 (43):17419–17424
42. Dam T, Jena SS, Ghosh A (2019) Ion dynamics, rheology and electrochemical performance of UV cross-linked gel polymer electrolyte for Li-ion battery. *Journal of Applied Physics* 126 (10):105104

12 Biopolymer Electrolytes for Energy Storage Applications

S. Jayanthi and M. Ulaganathan

CONTENTS

12.1 Introduction 255
12.2 Polymer Electrolytes and Their Classifications...256
    12.2.1 Solvent-Free Polymer Salt Complexes...257
    12.2.2 Polyelectrolytes..257
    12.2.3 Gel Polymer Electrolytes 257
    12.2.4 Composite Polymer Electrolytes 257
12.3 Characteristics of the Polymer Electrolyte 258
12.4 Biopolymer-Based Polymer Electrolytes and Their Properties....................258
    12.4.1 Chitosan-Based Polymer Electrolytes 258
    12.4.2 Starch-Based Polymer Electrolytes 261
    12.4.3 Carrageenan-Based Polymer Electrolytes 261
12.5 Biopolymer-Based Electrolytes in Lithium Batteries 264
12.6 Biopolymer-Based Electrolytes for Supercapacitors 264
12.7 Biopolymer-Based Electrolytes for Fuel Cells..268
12.8 Conclusion 269
References..270

12.1 INTRODUCTION

In recent years, biopolymers have played a vital role in the field of polymer electrolytes, which are used in various electrochemical energy storage devices. Solid polymer electrolytes (SPEs) have many advantages over the liquid electrolytes which contain highly flammable organic solvents and inorganic molten salts [1]. The problems associated with liquid electrolytes, such as internal short-circuiting, leaks, and production of combustible reaction products at the electrode surfaces, have been eliminated by the utilization of a solid polymer electrolyte. The polymer electrolytes exhibit high ionic conductivity of the order of at least 10^{-3} S cm^{-1} to 10^{-2} S cm^{-1} at room temperature, and play the role of a separator, achieved by the liquids in liquid electrolytes. The polymer electrolytes should also exhibit good cycle lives, good low-temperature performances, and good thermal and mechanical strengths in order to withstand elevated internal temperatures and pressures, which build up during the battery operation. The polymer electrolytes, in general, being lightweight and consisting of non-combustible materials, can be fabricated to specific requirements of size and shape, thus offering a wide range of designs. Since stable, thin films of the polymer electrolytes can readily be made, high specific energy and high specific power batteries with polymer electrolytes can be expected for use in electrochemical devices and electric vehicles [2].

For the past few decades, interest in the naturally available polymers, such as polysaccharides (starch, carrageenan, and chitosan), has been increasing rapidly because they represent a renewable resource with a wide range of uses in nature, involved in energy storage, transport, signaling, and as structural components. Chitosan, a fundamental polysaccharide, can readily be prepared from the shells of crabs, shrimps, and prawns. Another polysaccharide biopolymer is starch, derived from various sources like corn (maize grain), rice, wheat, potato, etc. Carraggeenan is another example, which is extracted from edible red seaweeds. The above biopolymers are used in a wide variety of applications, including electrochemical energy storage technology. The biopolymers are mainly used as ion-conducting media, as well as separators, in energy storage and conversion devices, such as batteries, supercapacitors and in fuel cell applications. In this chapter, the authors aim to focus on the function of the various biopolymers used in the above-mentioned electrochemical energy-related devices. Importantly, this chapter also describes the ionic conductivities of the biopolymer-based electrolyte systems.

12.2 POLYMER ELECTROLYTES AND THEIR CLASSIFICATIONS

Polymer electrolytes are a highly specialized field, encompassing all the specialties of organic, inorganic, and polymer chemistry, along with electrochemistry. Polymer electrolytes are synthesized by dissolving a polymer in alkali metals, using some organic solvents, such as EC, PC, acetone, DMC etc. Polymer electrolytes offer several advantages over liquid inorganic and organic electrolytes, such as being inexpensive, non-volatile, with superior safety features, excellent processability, mechanical properties, ease of preparation in any size or shape, better electrochemical stability, ease of tuning of ionic conductivity and thermal stability, etc. Thus, polymer electrolytes are considered to be suitable candidates for the necessary innovations in batteries. However, the ionic conductivity of polymer electrolytes depend on several factors, including the dielectric constant of the polymer, the solubility of alkali salts, the total salt concentration in the polymer matrix, and so on. Polymer electrolytes are generally classified as solid polymer electrolytes (SPEs), gel polymer electrolytes (GPEs), or composite polymer electrolytes (CPEs). SPEs are those polymers consisting of a polymer matrix dissolved in a lithium salt, whereas GPEs are those polymer electrolytes synthesized by the addition of plasticizers or solvents into the polymer matrix. CPEs are those polymers synthesized by the incorporation of inorganic or ceramic fillers along with lithium salt in the polymer matrix. Based on the physical state, the polymer electrolytes have been classified as solvent-free polymer salt complexes, polyelectrolytes, gel polymer electrolytes, and composite polymer electrolytes.

12.2.1 SOLVENT-FREE POLYMER SALT COMPLEXES

Solvent-free polymer electrolytes are obtained by dissolving a salt in a high-molecular-weight polar polymer matrix. The choice of the polymer and its dopant salt is critical. The polymer should have a flexible chain, with a large number of polar groups (O, N, S, etc.) and a low glass transition temperature. The polymer must be highly solvating and be able to overcome the crystal lattice forces of the salt. The lattice energy of the salt should be low enough to facilitate a high level of dissociation of the salt. The main disadvantages with most of these conventional polymer electrolyte systems are: (i) high crystallinity (~70%) with very low chain flexibility at room temperature, (ii) conduction through both cations and anions, which is undesirable for device applications, and (iii) T_m<100°C, limiting the temperature range of operation of these materials.

12.2.2 POLYELECTROLYTES

Researchers have synthesized polymers in which anions are covalently bonded to the polymer backbone in order to increase the cation-transference number. These polymers are known as 'polyelectrolytes'. By virtue of the anions being effectively immobilized, the total ionic conductivity is due to cationic transport. Nevertheless, such materials are usually not sufficiently flexible, exhibiting ambient-temperature conductivities of only 10^{-6} S cm^{-1} or less. Unlike polymer electrolytes, polyelectrolytes are not susceptible to the build-up of potentially resistive layers of high or low salt concentrations at the electrode/electrolyte interfaces during charging and discharging. Polystyrene sulphonate and perfluorosulfonated polymers, like Nafion, belong to this group. The conductivity of these polymers is very low (~10^{-1} –10^{-15} S cm^{-1}) under dry conditions, but they can achieve high conductivities in the presence of a high dielectric constant solvent, such as water.

12.2.3 GEL POLYMER ELECTROLYTES

Generally, a polymeric gel is defined as a system with a polymer network swollen with solvent. Owing to their unique hybrid network structures, gels always possess both the cohesive properties of solids and the diffusive transport properties of liquids. Some gel polymer electrolytes have high ionic conductivities of 10^{-3} S cm^{-1} at room temperature. However, their soft morphology, poor mechanical properties, and considerable viscosity may lead to internal short-circuits and make gel polymer electrolytes unsuitable for high-speed manufacturing processes.

12.2.4 COMPOSITE POLYMER ELECTROLYTES

Composite polymer electrolytes (CPEs) are defined as a type of polymer electrolyte with inorganic or organic fillers in the polymer matrix. These are prepared by the addition of high-surface-area inorganic fillers, such as Al_2O_3, SiO_2, MgO, $LiAlO_2$, TiO_2, $BaTiO_3$, and zeolite powders. The mechanical strength and stiffness of the complex systems are improved appreciably when the fillers are incorporated into the polymer matrix. However, the main advantages of the composite polymer electrolytes are the enhancement of room-temperature ionic conductivity and an improved stability at the electrode/electrolyte interface.

12.3 CHARACTERISTICS OF THE POLYMER ELECTROLYTE

The following are the characteristics necessary to act as a successful polymer host for polymer electrolytes.

- A polymer repeat unit should have a donor group (an atom with at least one lone pair of electrons) to form coordinate bonds with cations.
- Low barriers to bond relations should be exhibited, so that segmental motion of the polymer chain can take place readily
- A suitable distance should exist between coordinating centers, because the formation of multiple intra-polymer ion bonds appears to be important.

Almost all of the polymer electrolytes have been optimized with respect to the ionic conductivity of the electrolytes. The ionic conductivity of the polymer electrolyte depends mainly on the structure and nature of the base polymer, the nature of the ionic salt, the concentration of the salt, the dissolution, ionic radius, etc. In the subsequent sections of this chapter (Sections 12.4–12.7), the role of biopolymers on the ionic conductivity of the polymer electrolytes will be discussed in detail.

12.4 BIOPOLYMER-BASED POLYMER ELECTROLYTES AND THEIR PROPERTIES

Biopolymers are biodegradable polymers which are found in nature or can be synthesized from non-biodegradable monomers. Carbohydrates and proteins are examples of natural biopolymers, whereas polylactic acid (PLA) and polycaprolactone (PCL) belong to synthetic biopolymers. Derivatives of biopolymers are being produced commercially nowadays and find applications in biomedical uses, tissue engineering, and the packaging industry. Some of the biopolymers are extensively used as electrolytes in energy storage devices. Biopolymer-based electrolytes have overcome the main drawbacks of synthetic electrolytes, such as high cost and environmentally unfriendly nature. These biopolymer-based electrolytes exhibit high ionic conductivity, low cost, and good dimensional and mechanical stability. Chitosan-based polymers, starch-based derivatives, carrageenan-based biopolymers, cellulose-based compounds, polyethylene glycol (PEG), polycaprolactone (PCL), etc. are biopolymer electrolytes widely used in energy storage devices.

12.4.1 CHITOSAN-BASED POLYMER ELECTROLYTES

Chitosan is a linear polysaccharide composed of randomly distributed β-(1→4)-linked D-glucosamine (deacetylated unit) and N-acetyl-D-glucosamine (acetylated unit). It is made by treating the chitin in shells of shrimps and other crustaceans with an alkaline substance, like sodium hydroxide. Table 12.1 lists some of the lithium ion and proton-conducting polymer electrolytes based on chitosan.

TABLE 12.1
Room-Temperature (RT) Ionic Conductivity of Various Chitosan-Based Polymer Electrolytes

Materials	Optimized Composition	Conductivity (S cm^{-1}) (RT)	References
Chitosan (CH), methylcellulose (MC), lithium tetrafluoroborate (LiBF$_4$)	60 wt.% CH–MC (75:25) and LiBF$_4$ (40 wt.%)	3.747×10^{-6}	[3]
Chitosan (CH), polyethylene oxide (PEO), lithium bis(trifluoromethane sulfonyl) imide (LiTFSI)	1 wt.% CH–1 wt.% PEO–30 wt.% LiTFSI	1.40×10^{-6}	[4]
Chitosan (CH), lithium triflate (LiCF$_3$SO$_3$)	40 wt.% LiCF$_3$SO$_3$	~10^{-5}	[5]
Chitosan (CH), lithium triflate (LiCF$_3$SO$_3$), ethylene carbonate (EC)	70 wt.% CH–30 wt.% EC + LiCF$_3$SO$_3$	2.75×10^{-5}	[6]
Chitosan (CH), dextran, ammonium thiocyanate (NH$_4$SCN)	40 wt.% dextran–60 wt.% chitosan–40 wt.% NH$_4$SCN	1.28×10^{-4}	[7]
Chitosan (CH), oxalic acid	60 wt.% chitosan–40 wt.% oxalic acid	4.95×10^{-7}	[8]
Chitosan, ammonium nitrate, acetic acid	1.9 wt.% CH–0.17 wt.% ammonium nitrate– 96.3 wt.% acetic acid	1.46×10^{-1}	[9]
Chitosan (CH), acetic acid, glycerol	-	2.9×10^{-5}	[10]
Chitosan–LiOAc–oleic acid	-	10^{-5}	[11]
Chitosan (CH), ammonium acetate	1g (CH) + 40 wt.% ammonium acetate	2.87×10^{-4}	[12]
Chitosan (CH), ammonium chloride	1g (CH) + 20 wt.% ammonium chloride	5.35×10^{-3}	[13]
Chitosan (CH), ammonium iodide (NH$_4$I), Ethylene carbonate (EC)	55 wt.% chitosan–45 wt.% NH$_4$I–40 wt.% EC	7.60×10^{-6}	[14]
Chitosan (CH), ammonium thiocyanate (NH$_4$SCN), aluminum titanate (Al$_2$TiO$_5$)	57 wt.% chitosan –38 wt.% NH$_4$SCN –5 wt.% Al$_2$TiO$_5$	2.10×10^{-3}	[15]
Chitosan (CH), ammonium thiocyanate (NH$_4$SCN), alumina (Al$_2$O$_3$)	60 wt.% chitosan–40 wt.% NH$_4$SCN 60 wt.% chitosan–40 wt.% NH$_4$SCN–6 wt.% Al$_2$O$_3$	1.29×10^{-4} 5.86×10^{-4}	[16]
Poly(vinyl alcohol) (PVA), chitosan (CH), phosphoric acid (H$_3$PO$_4$), niobium oxide (Nb$_2$O$_5$)	PVA:CS (80:20)–40% phosphoric acid–6% of Nb$_2$O$_5$	3.00×10^{-2}	[17]
Chitosan (CH), poly(vinyl alcohol), ammonium nitrate (NH$_4$NO$_3$), ethylene carbonate (EC)	7 wt.% chitosan–11 wt.% PVA–12 wt.% NH$_4$NO$_3$–70 wt.% EC	1.60×10^{-3}	[18]

(Continued)

TABLE 12.1 (CONTINUED)
Room-Temperature (RT) Ionic Conductivity of Various Chitosan-Based Polymer Electrolytes

Materials	Optimized Composition	Conductivity (S cm^{-1}) (RT)	References
Chitosan (CH), iota-carrageenan, phosphoric acid (H$_3$PO$_4$), poly(ethylene glycol)	37.50 wt.% chitosan–37.50 wt.% iota- carrageenan–18.75 wt.%, H$_3$PO$_4$–6.25 wt.% PEG.	6.29×10^{-4}	[19]
Chitosan (CH), kappa- carrageenan, ammonium nitrate (NH$_4$NO$_3$)	42 wt.% CH–42 wt.% k- carrageenan–16 wt.% NH$_4$NO$_3$	2.39×10^{-4}	[20]
Chitosan (CH), ammonium iodide (NH$_4$I), 1-butyl-3-methylimidazolium-iodide (BMII)	27.5 wt.% (CH)–22.5 wt.% NH$_4$I–50 wt.% BMII	3.43×10^{-5}	[21]
Chitosan (CH), poly(ethylene oxide) (PEO), ammonium iodide (NH$_4$I)	27 wt.% CH–27 wt.% PEO–44 wt.% NH$_4$I	4.32×10^{-6}	[22]
Chitosan (CH), poly(ethylene oxide) (PEO), ammonium iodide (NH$_4$I), 1-butyl-3-methylimidazolium-iodide (BMII)	30 wt.% C –70 wt.% PEO–9 wt.% NH$_4$I–80 wt.% BMII	5.52×10^{-4}	
Chitosan (CH), praseodymium (III) trifluoromethanesulfonate (PrTrif)	20 wt.% CH - 30 wt.% PrTrif	2.38×10^{-6} (at 90 °C)	[23]

Chitosan shows biocompatibility and biodegradability characteristics, which are suitable to create an environmentally friendly, inert, and flexible polymer for sensing and manipulating macromolecules and microorganisms in devices. In particular, the chitosan matrices provide a better environment for doping, blending, and grafting of acids, oxides, and salts to improve the conductance level to one comparable with synthetic ion-conducting polymers. In recent years, much research has been carried out on chitosan and its derivatives as components in batteries, supercapacitors, etc.

Chitosan is an odorless powder and its color varies from yellow to white. On the other hand, spray-dried chitosan salts exhibit a smooth texture and a pale color. The physico-chemical properties of chitosan are influenced by the degree of deacetylation and molecular weight. The process of deacetylation involves the removal of acetyl groups from the molecular chain of chitin, leaving behind a compound (chitosan) with a high degree of chemically reactive amino groups (-NH$_2$). This makes the degree of deacetylation (DD) an important property in chitosan production, as it affects the physicochemical properties, and, hence, determines its appropriate practical applications [24]. The degree of deacetylation of chitosan ranges from 50% to 99%, with an average of 80%, depending on the crustacean species and the preparation methods. Chitin with a degree of deacetylation of 75% or above is generally known as chitosan.

Biopolymer Electrolytes

Chitosan is hydrophilic in nature, finding utility in high-temperature and low-relative-humidity environments. Chitosan is soluble in most organic and alkaline solvents, but is not soluble in water, which limits its applications. In order to improve its properties, various chemical modifications, like sulfonation [25], phosphorylation [26], quaternization [27], chemical crosslinking [28], phthaloylation [29] and acylation [30], have been employed. Linear sweep voltammetry and variation in the discharge capacity of chitosan-based iota-carrageenan is shown in Figure 12.1.

12.4.2 Starch-Based Polymer Electrolytes

Starch is a polymeric carbohydrate consisting of numerous glucose units joined by glycosidic bonds. It is a polysaccharide and produced by most green plants for energy storage. It is the most common carbohydrate in human diets and is contained in large amounts in staple foods, such as potatoes, wheat, maize (corn), rice, and cassava.

Starch is white in color and forms a tasteless and odorless powder. It is insoluble in cold water or ethanol. It consists of two types of molecules: the linear and helical amylose and the branched amylopectin. Depending on the plant, starch generally contains 20 to 25% amylose and 75 to 80% amylopectin by weight.

Starch has attractive processing characteristics, which include (i) considerable abundance in nature, (ii) biocompatibility, (iii) cost effectiveness, and (iv) high mechanical integrity. Both amylopectin and amylose have hydroxyl (-OH) groups. Hence, the lone pair of electrons in the oxygen atom favors the solvation of the charge carriers within the polymer network and forms transient coordination bonds. A vacant site is created when the coordination bond is broken. As a consequence, the neighboring ions from the adjacent site occupy this vacant site, resulting in the transportation of ions *via* the hopping process [31]. Table 12.2 lists some of the lithium ion and ammonium salts added to polymer electrolytes based on starch.

12.4.3 Carrageenan-Based Polymer Electrolytes

Carrageenans can be obtained from the red seaweeds of the class Rhodophyceae. Carrageenans are a group of linear galactans with an ester sulfate content of 15–40% (w/w) and containing alternating α-(1→3)-D and β-(1→4)-D-galactopyranosyl (or 3,6-anhydro-α-D-galactopyranosyl) linkages. It dissolves in hot water at room temperature, but in solvents, such as dimethylsulfoxide (DMSO), the solubility temperature is between 40 and 70°C. It is insoluble in ethanol, aceton, and some other organic solvents [52]. Kappa, iota and lambda are the three main commercial classes of carrageenan. The primary differences that influence the properties of kappa-, iota-, and lambda-carrageenan are the number and position of the ester sulfate groups on the repeating galactose units. Higher levels of ester sulfate lower the solubility temperature of the carrageenan and produce lower strength gels or contribute to gel inhibition (in lambda-carrageenan). Figure 12.2 shows the electrochemical characteristics of iota-carrageenan with ammonium nitrate [53]. Table 12.3 describes the polymer electrolytes composed of lithium and proton salts based on carrageenan.

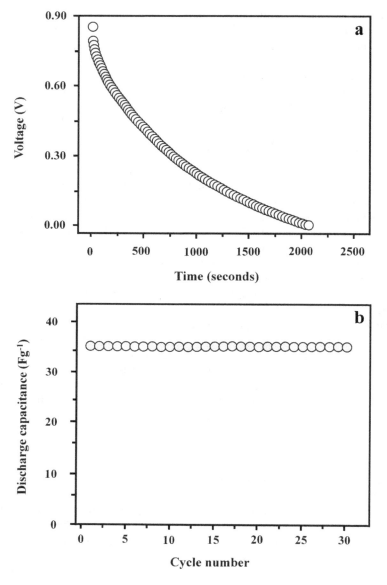

FIGURE 12.1 Electrochemical performance of EDLC with chitosan-based electrolytes sandwiched between activated carbon cloth electrodes: (a) Typical discharge characteristics of EDLC studied; (b) Variation in the discharge capacitance with cycle numbers for EDLC (iota- carrageenan/chitosan). Adapted and reproduced with permission from Ref. [18]. Copyright © 2011 Taylor and Francis.

TABLE 12.2
Room-Temperature (RT) Ionic Conductivity of Various Starch-Based Polymer Electrolytes

Materials	Optimized Composition	Conductivity (S cm^{-1}) (RT)	References
Corn starch (CS), lithium hexafluorophosphate (LiPF$_6$), 1-butyl-3-methylimidazolium hexafluorophosphate (BmImPF$_6$)	50 wt.% BmImPF$_6$	1.47×10^{-2}	[32]
Corn starch (CS), lithium hexafluorophosphate (LiPF$_6$), 1-butyl-3-methyl imidazolium trifluoromethanesulfonate (BmImTf)	80 wt.% of CS–20 wt.% of LiPF$_6$–80 wt.% BmImTf	3.21×10^{-4}	[33]
Corn starch (CS), lithium perchlorate (LiClO$_4$)	60 wt.% CS–40 wt.% LiClO$_4$	1.28×10^{-4}	[34]
Corn starch (CS), lithium bis(trifluoro methane sulfonyl)imide (LITFSI), deep eutectic solvent (DES)	14 wt.% CS–6 wt.% LiTFSI–80 wt.% DES	4.56×10^{-3}	[35]
Corn starch (CS), lithium iodide, glycerol	49 wt.% starch–21 wt.% LiI–30 wt.% glycerol	9.56×10^{-4}	[36]
Corn starch (CS), chitosan, ammonium iodide	40 wt.% NH$_4$I	3.04×10^{-4}	[37]
Corn starch, lithium perchlorate, nano silica	96 wt.%(CS-LiClO$_4$)–4 wt.% SiO$_2$	1.23×10^{-4}	[38]
Corn starch, ammonium bromide, glycerol	49 wt.% starch–21wt.% NH$_4$Br–30 wt.% glycerol	1.80×10^{-3}	[39]
Corn starch (CS), lithium bis(trifluoro methane sulfonyl)imide (LITFSI), 1-allyl-3-methylimidazolium chloride,[Amim] Cl	14 wt.% CS– 6 wt.% LiTFSI– 80 wt.% [Amim] Cl	4.18×10^{-2}	[40]
Corn starch (CS), 1-allyl-3-methylimidazolium chloride [Amim] Cl	5 g (CS)–30 wt.% [Amim] Cl–	10$^{-1.6}$	[41]
Corn starch (CS), lithium chloride, N,N-dimethyl acetamide (DMAc)	18 wt.% LiCl	10$^{-0.5}$	[42]
Rice starch (RS), lithium iodide (LI)	65 wt.% RS–35 wt.%LiI	4.68×10^{-5}	[43]
Rice starch (RS), lithium iodide (LI), 1-methyl-3-propylimidazolium iodide (MPII), titania (TiO$_2$)	44.2 wt.% RS–23.8 wt.%LiI–30.2 wt.%	3.63×10^{-4}	[44]
Potato starch (PS), ammonium iodide	-	2.4×10^{-4}	[45]
Potato starch (PS), chitosan, lithium trifluoromethanesulfonate (LiCF$_3$SO$_3$), glycerol	50 wt.% PS–50 wt.% chitosan–45 wt.% LiCF$_3$SO$_3$–30 wt.% glycerol	1.32×10^{-3}	[46]

(*Continued*)

TABLE 12.2 (CONTINUED)
Room-Temperature (RT) Ionic Conductivity of Various Starch-Based Polymer Electrolytes

Materials	Optimized Composition	Conductivity (S cm^{-1}) (RT)	References
Potato starch (PS), methylcellulose (MC), ammonium nitrate, glycerol	25 wt.% MC–17 wt.% PS–18 wt.% NH$_4$NO$_3$–40 wt.% glycerol	~10^{-3}	[47]
Potato starch (PS), graphene oxide (GO), lithium trifluoromethanesulfonate (LiCF$_3$SO$_3$), 1-butyl-3-methylimidazolium chloride ([Bmim][Cl])	PS-GO (80–20 wt.%)/0.333g LiCF$_3$SO$_3$/30wt.% [Bmim][Cl]	4.8×10^{-4}	[48]
Poly(vinyl alcohol) (PVA), starch, ammonium thiocyanate, glutaraldehyde	PVA:starch (50:50) + 0.4 ml of glutaraldehyde + 30% NH$_4$SCN	1.311×10^{-4}	[49]
Potato starch (PS), glycerol,	PS/Gly-0.5	5.2×10^{-5}	[50]
sago starch (SS), lithium chloride, glycerol	80 wt.% SS– 20 wt.% glycerol–8 wt.% LiCl.	~10^{-3}	[51]

12.5 BIOPOLYMER-BASED ELECTROLYTES IN LITHIUM BATTERIES

The lithium-ion polymer battery is a rechargeable battery of lithium-ion technology, using a polymer electrolyte, instead of a liquid electrolyte. These batteries provide higher specific energy than other commercially available lithium battery types. They are used in applications like mobile devices and radio-controlled aircraft [63].

Lithium polymer cells have evolved from lithium-ion and lithium-metal batteries. The primary difference is that, instead of using a liquid lithium-salt electrolyte (such as LiPF$_6$) mixed with an organic solvent (such as EC/DMC/DEC), the battery uses a solid polymer electrolyte (SPE), such as poly(ethyleneoxide) (PEO), poly(acrylonitrile) (PAN), poly(methyl methacrylate) (PMMA) or poly(vinylidene difluoride) (PVdF), etc.

During the electrochemical charge-discharge performance, at the first charge during the activation/formation stage, lithium from the cathode passes through the electrolyte and is stored (intercalated) in the graphite-based anode. It expands the gaps between layers and, hence, the cell thickness increases. During discharge of the cell, the cell thickness decreases again. The conductive salt most frequently used for the organic electrolyte medium is lithium hexafluorophosphate (LiPF$_6$) [64]. Table 12.4 shows the use of biopolymer electrolytes in lithium batteries.

12.6 BIOPOLYMER-BASED ELECTROLYTES FOR SUPERCAPACITORS

A supercapacitor is a type of capacitor that can store a large amount of energy, typically 10 to 100 times more energy per unit mass or volume than an electrolytic

Biopolymer Electrolytes

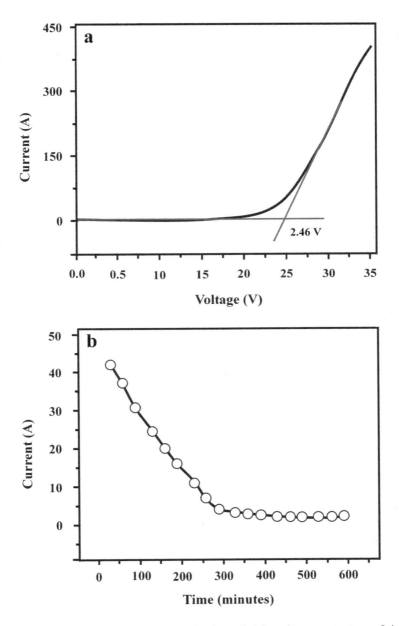

FIGURE 12.2 Electrochemical characterization of 1.0 g iota-carrageenan: 0.4 wt.% NH_4NO_3 polymer electrolyte: (a) Oxidative stability (linear sweep voltammetry) recorded at a scan rate of 1 mV s^{-1} at room temperature, and (b) polarization current vs. time plot. Adapted and reproduced with permission from Ref. [53]. Copyright © 2018 Elsevier.

TABLE 12.3
Room-Temperature (RT) Ionic Conductivity of Various Carrageenan-Based Polymer Electrolytes

Materials	Optimized Composition	Conductivity (S cm^{-1}) (RT)	References
Iota-carrageenan, ammonium nitrate	1.0 g iota-carrageenan– 0.4 wt.% NH$_4$NO$_3$–	1.46 × 10^{-3}	[53]
Carboxymethyl kappa-carrageenan, 1-butyl-3-methylimidazolium chloride ([Bmim]Cl)	30 wt.% ([Bmim]Cl)	5.76 × 10^{-3}	[54]
Iota-carrageenan, ammonium bromide	20 wt.% NH$_4$Br	5.76 × 10^{-3}	[55]
Iota-carrageenan, ammonium formate (NH$_4$HCO$_2$)	1.0 g iota-carrageenan and 0.4 wt.% NH$_4$HCO$_2$	1.13 × 10^{-3}	[56]
Iota-carrageenan, lithium chloride	1.0 g iota-carrageenan and 0.3 g LiCl	5.33 × 10^{-3}	[57]
Iota-carrageenan, lithium perchlorate, succinonitrile (SN)	1.0 g iota-carrageenan/0.5 wt.% LiClO$_4$/0.3 wt.% SN	3.33 × 10^{-3}	[58]
Kappa-carrageenan, ammonium bromide	20 wt.% NH$_4$Br	3.89 × 10^{-4}	[59]
Chitosan (CH), kappa-carrageenan, ammonium nitrate (NH$_4$NO$_3$)	42 wt.% CH–42 wt.% kappa-carrageenan–16 wt.% NH$_4$NO$_3$	2.39 × 10^{-4}	[60]
Kappa-carrageenan, 1-butyl-3-methylimidazolium chloride, [Bmim]Cl	120 wt.% of [Bmim]Cl	2.44× 10^{-3}	[61]
Kappa-carrageenan, lithium bromide	1g kappa-carrageenan – 0.5 wt.% LiBr	3.43× 10^{-3}	[62]

TABLE 12.4
Electrochemical Performance of Various Biopolymer-Based Electrolytes in Lithium Batteries

Biopolymer Electrolyte	Discharge Capacity	Open Circuit Voltage (OCV) V	References
Iota-carrageenan, lithium chloride (LiCl)	-	1.77	[58]
Kappa-carrageenan, lithium bromide (LiBr)	-	1.64	[63]
Chitosan-lithium bis (trifluoromethanesulfonyl) imide-succinonitrile	160 mAh g^{-1} at 17 mA g^{-1}	-	[65]

Biopolymer Electrolytes

capacitor. Unlike a battery, a supercapacitor has an unlimited life cycle, with little wear and tear following long-term use. Thus, it can be charged and discharged an unlimited number of times. A supercapacitor has many advantages. It can deliver high power and allow high load currents owing to its low resistance. Its charging mechanism is simple and fast and is not subject to overcharging. Compared to a battery, a supercapacitor has excellent high- and low-temperature charge and discharge performances. It is also highly reliable and has low impedance [66].

There are three types of capacitors, the most basic of which is the *electrostatic capacitor* with a dry separator. This classic capacitor has very low capacitance and is mainly used to tune radio frequencies and filtering. The size ranges from a few picofarads (pf) to low microfarad (µF). The *electrolytic capacitor* provides higher capacitance than the electrostatic capacitor and is rated in microfarads (µF), which is a million times higher than a picofarad. These capacitors deploy a moist separator and are used for filtering, buffering, and signal coupling. Similar to a battery, the electrostatic capacity has positive and negative charges that must be observed. The third type is the *supercapacitor*, which stores thousands of times more than the electrolytic capacitor. The supercapacitor is used for energy storage undergoing frequent charge and discharge cycles at high current and of short duration.

EDLC differs from conventional capacitors by the mechanism of energy storage. A conventional capacitor stores energy between the two electrode plates. EDLC stores electric energy in the electrochemical double layer formed at the electrode/electrolyte interface. Figure 12.3 represents the charge storage mechanism in EDLC and Table 12.5 shows the biopolymer electrolytes in EDLC applications.

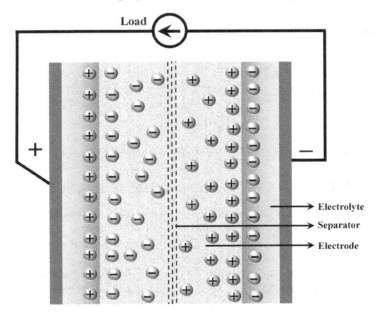

FIGURE 12.3 Schematic illustration of the structure and working principle of electrochemical double-layer capacitors (EDLC).

TABLE 12.5
Electrochemical Performance of Various Biopolymer-Based Electrolytes in Electrochemical Double-Layer Capacitors (EDLC)

Materials	Capacitance (Fg^{-1})	Energy Density (Wh kg^{-1})	Power Density (kW kg^{-1})	References
Chitosan (CH), poly(vinyl alcohol), ammonium nitrate (NH$_4$NO$_3$), ethylene carbonate (EC)	27.1	-	-	[18]
Chitosan (CH), iota- carrageenan, phosphoric acid (H$_3$PO$_4$), poly(ethylene glycol)	35	1.3–1.8	-	[19]
Kappa-carrageenan, chitosan, ammonium nitrate	13–18.5	-	-	[20]
Potato starch (PS), methylcellulose (MC), ammonium nitrate, glycerol	31	3.1	-	[48]
Chitosan, dextran, lithium perchlorate	8.7	1.21	685	[66]
Chitosan, dextran, ammonium fluoride (NH$_4$F)	12.4	1.4		[67]
Chitosan (CH), poly(ethylene oxide) (PEO), lithium perchlorate	6.88	0.94	305	[68]
Chitosan (CH), ammonium bromide, glycerol	7.5	-	-	[69]
Corn starch, citric acid	54	-	-	[70]
Corn starch (CS), lithium hexafluorophosphate (LiPF$_6$), 1-butyl-3-methylimidazolium hexafluorophosphate (BmImPF$_6$)	13.02	0.41	5.32	[71]
Corn starch, lithium perchlorate, nano silica	8.71	0.90	1.35	[72]
Corn starch, lithium acetate, glycerol	33.31	-	-	[73]

12.7 BIOPOLYMER-BASED ELECTROLYTES FOR FUEL CELLS

A fuel cell is a electrochemical conversion device which works due to the external supply of chemical energy, and can run indefinitely, as long as it is supplied with a source of hydrogen and oxygen (usually air) as negative and positive fuels. The source of hydrogen is generally referred to as the fuel and this gives the fuel cell its name, although there is no combustion involved. Oxidation of the hydrogen instead takes place electrochemically in a very efficient way. During oxidation, hydrogen atoms react with oxygen atoms to form water; in the process, electrons are released and flow through an external circuit as an electric current.

In general, a fuel cell consists of two electrodes, the anode and the cathode, separated by an electrolyte. A thin layer of platinum or other metal, depending on the

Biopolymer Electrolytes

TABLE 12.6
Electrochemical Performance of Various Biopolymer-Based Electrolytes in Fuel Cells

Materials	OCV (mV)	Current Density (mA cm^{-2})	Power Density (mW cm^{-2})	References
Iota-carrageeenan, ammonium nitrate,	442	-	-	[54]
Chitosan, phosphoric acid	-	3.9	1.61	[74]
Chitosan, phosphoric acid, ammonium nitrate	-	21.2	4.63	
Chitosan, phosphoric acid, ammonium nitrate, aluminum silicate	-	21.9	4.00	
Chitosan, phosphoric acid, aluminum silicate	-	21.0	5.40	
Chitosan, sulfuric acid	-	-	450	[75]
Chitosan, adenosine triphosphate, Nafion	-	616	243	[76]
Chitosan, hydroxyethyl cellulose, phosphotungstic acid	-	210	58	[77]
Chitosan, polyvinyl alcohol, sulfosuccinic acid, stabilized silicotungstic acid	-	400	156	[78]
Chitosan, polyvinyl alcohol, sulfosuccinic acid, nylon	-	177.2	54.08	[79]
Chitosan, 1-naphthalene acetic acid	-	115	16	[80]
Chitosan, 4-chlorophenoxy acetic acid	-	125	18	
Chitosan, 3-indole acetic acid	-	150	25	
Arrowroot starch, glutaraldehyde	-	85	16	[81]
Iota-carrageeenan, ammonium thiocyanate	503	-	-	[82]
Kappa-carrageeenan, ammonium thiocyanate	502	-	-	[60]

type of fuel cell, is coated on each electrode to activate the reaction between oxygen and hydrogen when they pass through the electrodes (Table 12.6). The overall reaction is shown by the equation below:

$$H_2 + \frac{1}{2}O_2 \rightarrow H_2O, \quad \Delta H = -287 \text{ KJmol}^{-1}$$

12.8 CONCLUSION

Polymer electrolytes based on biopolymers have attracted much attention. This is due to their cost-effectiveness, functionality, and their environmentally benign nature. Further studies are needed to improve the electrolytic performance of the electrochemical energy storage devices using biopolymer electrolytes. Biopolymer blends may allow more ion transport through the hopping conduction and also provide greater mechanical strength for easy of access during device fabrication. In addition, biopolymers of different molecular weights can also help to improve the characteristics of the polymer electrolytes. From the above assessment, it is clear that biopolymers will be potential candidates in future electrochemical energy-related applications.

REFERENCES

1. Alamgir M. (1994) KMA, in: G. Pistoia (Ed.), *Lithium Batteries-New Materials, Developments and Perspectives*. Elsevier.
2. Armand M (1983) Polymer solid electrolytes - an overview. *Solid State Ionics* 9–10:745–754. https://doi.org/10.1016/0167-2738(83)90083-8
3. Abdullah OG, Hanna RR, Salman YAK, Aziz SB (2018) Characterization of lithium Ion-conducting blend biopolymer electrolyte based on CH–MC doped with LiBF4. *Journal of Inorganic and Organometallic Polymers and Materials* 28 (4):1432–1438. doi:10.1007/s10904-018-0802-2
4. Idris NH, Senin HB, Arof AK (2007) Dielectric spectra of LiTFSI-doped chitosan/PEO blends. *Ionics* 13 (4):213–217. doi:10.1007/s11581-007-0093-z
5. Sudaryanto, Yulianti E, Jodi H (2015) Studies of dielectric properties and conductivity of chitosan-lithium triflate electrolyte. *Polymer-Plastics Technology and Engineering* 54 (3):290–295. doi:10.1080/03602559.2014.977424
6. Winie T, Majid SR, Khiar ASA, Arof AK (2006) Ionic conductivity of chitosan membranes and application for electrochemical devices. *Polymers for Advanced Technologies* 17 (7–8):523–527. doi:10.1002/pat.744
7. Kadir MFZ, Hamsan MH (2018) Green electrolytes based on dextran-chitosan blend and the effect of NH4SCN as proton provider on the electrical response studies. *Ionics* 24 (8):2379–2398. doi:10.1007/s11581-017-2380-7
8. Fadzallah IA, Majid SR, Careem MA, Arof AK (2014) Relaxation process in chitosan–oxalic acid solid polymer electrolytes. *Ionics* 20 (7):969–975. doi:10.1007/s11581-013-1058-z
9. Jamaludin A, Mohamad AA (2010) Application of liquid gel polymer electrolyte based on chitosan–NH4NO3 for proton batteries. *Journal of Applied Polymer Science* 118 (2):1240–1243. doi:10.1002/app.32445
10. Pawlicka A, Mattos RI, Tambelli CE, Silva IDA, Magon CJ, Donoso JP (2013) Magnetic resonance study of chitosan bio-membranes with proton conductivity properties. *Journal of Membrane Science* 429:190–196. doi:https://doi.org/10.1016/j.memsci.2012.11.048
11. Yulianti E, Karo AK, Susita L, Sudaryanto (2012) Synthesis of electrolyte polymer based on natural polymer chitosan by ion implantation technique. *Procedia Chemistry* 4:202–207. doi:https://doi.org/10.1016/j.proche.2012.06.028
12. Du JF, Bai Y, Pan DA, Chu WY, Qiao LJ (2009) Characteristics of proton conducting polymer electrolyte based on chitosan acetate complexed with CH3COONH4. *Journal of Polymer Science Part B: Polymer Physics* 47 (6):549–554. doi:10.1002/polb.21656
13. Du JF, Bai Y, Chu WY, Qiao LJ (2010) The structure and electric characters of proton-conducting chitosan membranes with various ammonium salts as complexant. *Journal of Polymer Science Part B: Polymer Physics* 48 (8):880–885. doi:10.1002/polb.21973
14. Buraidah MH, Teo LP, Majid SR, Arof AK (2009) Ionic conductivity by correlated barrier hopping in NH4I doped chitosan solid electrolyte. *Physica B: Condensed Matter* 404 (8):1373–1379. doi:https://doi.org/10.1016/j.physb.2008.12.027

15. Hassan F, Woo HJ, Aziz NA, Kufian MZ, Majid SR (2013) Synthesis of Al2TiO5 and its effect on the properties of chitosan–NH4SCN polymer electrolytes. *Ionics* 19 (3):483–489. doi:10.1007/s11581-012-0763-3
16. Aziz NA, Majid SR, Yahya R, Arof AK (2011) Conductivity, structure, and thermal properties of chitosan-based polymer electrolytes with nanofillers. *Polymers for Advanced Technologies* 22 (9):1345–1348. doi:10.1002/pat.1619
17. Quintana D, Baca E, Mosquera Vargas E, Vargas RA, Diosa J (2019) Improving the ionic conductivity in nanostructured membranes based on poly(vinyl alcohol) (PVA), chitosan (CS), phosphoric acid (H$_3$PO$_4$), and niobium oxide (Nb$_2$O$_5$). *Ionics* 25:1131–1136. doi:10.1007/s11581-018-2764-3
18. Kadir MFZ, Arof AK (2011) Application of PVA–chitosan polymer blend electrolyte membrane in electrical double layer capacitor. *Materials Research Innovations* 15 (2):s217–s220. doi:10.1179/143307511X13031890749299
19. Arof AK, Shuhaimi N, Alias N, Kufian MZ, Majid SR (2010) Application of chitosan/iota-carrageenan polymer electrolytes in electrical double layer capacitor (EDLC). *Journal of Solid State Electrochemistry* 14:2145–2152. doi:10.1007/s10008-010-1050-8
20. Shuhaimi NEA, Alias NA, Majid SR, Arof AK (2008) Electrical double layer capacitor with proton conducting κ-carrageenan–chitosan electrolytes. *Functional Materials Letters* 1 (3):195–201. doi:10.1142/S1793604708000423
21. Buraidah MH, Teo LP, Majid SR, Yahya R, Taha RM, Arof AK (2010). Characterizations of Chitosan-Based Polymer Electrolyte Photovoltaic Cells, 2010: 805836 https://doi.org/10.1155/2010/805836
22. Mohamad SA, Yahya R, Ibrahim ZA, Arof AK (2007) Photovoltaic activity in a ZnTe/PEO–chitosan blend electrolyte junction. *Solar Energy Materials and Solar Cells* 91 (13):1194–1198. https://doi.org/10.1016/j.solmat.2007.04.002
23. Alves R, Sabadini RC, Gonçalves TS, de Camargo ASS, Pawlicka A, Silva MM (2019) Structural, morphological, thermal and electrochemical characteristics of chitosan: praseodymium triflate based solid polymer electrolytes. *International Journal of Green Energy* 16 (15):1602–1610. doi:10.1080/15435075.2019.1677239
24. Rout Sandeep Kumar (2001) "Physicochemical, Functional and Spectroscopic Analysis of Crawfish Chitin and Chitosan as Affected by Process Modification." *LSU Historical Dissertations and Theses*. 432. https://digitalcommons.lsu.edu/gradschool_disstheses/432
25. Thanou M, Florea BI, Geldof M, Junginger HE, Borchard G (2002) Quaternized chitosan oligomers as novel gene delivery vectors in epithelial cell lines. *Biomaterials* 23 (1):153–159. https://doi.org/10.1016/S0142-9612(01)00090-4
26. Chiou MS, Li HY (2003) Adsorption behavior of reactive dye in aqueous solution on chemical cross-linked chitosan beads. *Chemosphere* 50 (8):1095–1105. https://doi.org/10.1016/S0045-6535(02)00636-7
27. Feng H, Dong C-M (2007) Synthesis and characterization of phthaloyl-chitosan-g-poly(l-lactide) using an organic catalyst. *Carbohydrate Polymers* 70 (3):258–264. https://doi.org/10.1016/j.carbpol.2007.04.004
28. Ming-chun LI, Chao LIU, Mei-hua XIN, Huang ZHAO, Min WANG, Zhen FENG, Xiao-li SUN (2005) Preparation and Characterization of Acylated Chitosan. Chemical research in chinese universities. 21(1):114–116. ID: 1005-9040(2005)-01-114-03; http://crcu.jlu.edu.cn/EN/Y2005/V21/I1/114
29. Fadzallah IA, Majid SR, Careem MA, Arof AK (2014) A study on ionic interactions in chitosan–oxalic acid polymer electrolyte membranes. *Journal of Membrane Science* 463:65–72. https://doi.org/10.1016/j.memsci.2014.03.044

30. Aziz NA, Majid SR, Arof AK (2012) Synthesis and characterizations of phthaloyl chitosan-based polymer electrolytes. *Journal of Non-Crystalline Solids* 358 (12):1581–1590. https://doi.org/10.1016/j.jnoncrysol.2012.04.019
31. Agnieszka Pawlicka, Aline C. Sabadini, Ellen Raphael, & Douglas C. Dragunski. (2008) Ionic Conductivity Thermogravimetry Measurements of Starch-Based Polymeric Electrolytes. *Molecular Crystals and Liquid Crystals*. 2008/04/17;485(1):804–816. https://www.tandfonline.com/doi/full/10.1080/15421400801918138.
32. Ramesh S, Liew C-W, Arof AK (2011) Ion conducting corn starch biopolymer electrolytes doped with ionic liquid 1-butyl-3-methylimidazolium hexafluorophosphate. *Journal of Non-Crystalline Solids* 357 (21):3654–3660. https://doi.org/10.1016/j.jnoncrysol.2011.06.030
33. Liew CW, Singh R (2013) Studies on ionic liquid-based corn starch biopolymer electrolytes coupling with high ionic transport number. *Cellulose* 20. doi:10.1007/s10570-013-0079-0
34. Teoh KH, Lim C-S, Ramesh S (2014) Lithium ion conduction in corn starch based solid polymer electrolytes. *Measurement* 48:87–95. https://doi.org/10.1016/j.measurement.2013.10.040
35. Ramesh S, Shanti R, Morris E (2012) Exerted influence of deep eutectic solvent concentration in the room temperature ionic conductivity and thermal behavior of corn starch based polymer electrolytes. *Journal of Molecular Liquids* 166:40–43. https://doi.org/10.1016/j.molliq.2011.11.010
36. Shukur MF, Ibrahim FM, Majid NA, Ithnin R, Kadir MFZ (2013) Electrical analysis of amorphous corn starch-based polymer electrolyte membranes doped with LiI. *Physica Scripta* 88 (2):025601. doi:10.1088/0031-8949/88/02/025601
37. Yusof YM, Shukur MF, Illias HA, Kadir MFZ (2014) Conductivity and electrical properties of corn starch–chitosan blend biopolymer electrolyte incorporated with ammonium iodide. *Physica Scripta* 89 (3):035701. doi:10.1088/0031-8949/89/03/035701
38. Teoh KH, Ramesh S, Arof AK (2012) Investigation on the effect of nanosilica towards corn starch–lithium perchlorate-based polymer electrolytes. *Journal of Solid State Electrochemistry* 16 (10):3165–3170. doi:10.1007/s10008-012-1741-4
39. Shukur MF, Kadir MFZ (2015) Electrical and transport properties of NH4Br-doped cornstarch-based solid biopolymer electrolyte. *Ionics* 21 (1):111–124. doi:10.1007/s11581-014-1157-5
40. Ramesh S, Shanti R, Morris E, Durairaj R (2011) Utilisation of corn starch in production of 'green' polymer electrolytes. *Materials Research Innovations* 15 (2):s13–s18. doi:10.1179/143307511X13031890747291
41. Ning W, Xingxiang Z, Haihui L, Benqiao H (2009) 1-Allyl-3-methylimidazolium chloride plasticized-corn starch as solid biopolymer electrolytes. *Carbohydrate Polymers* 76 (3):482–484. https://doi.org/10.1016/j.carbpol.2008.11.005
42. Ning W, Xingxiang Z, Haihui L, Jianping W (2009) N, N-dimethylacetamide/lithium chloride plasticized starch as solid biopolymer electrolytes. *Carbohydrate Polymers* 77 (3):607–611. https://doi.org/10.1016/j.carbpol.2009.02.002
43. Khanmirzaei MH (2013) Ionic transport and FTIR properties of lithium iodide doped biodegradable rice starch based polymer electrolytes. *International Journal of Electrochemistry Science* 8:9977–9991
44. Khanmirzaei MH, Ramesh S (2014) Nanocomposite polymer electrolyte based on rice starch/ionic liquid/TiO2 nanoparticles for solar cell application. *Measurement* 58:68–72. https://doi.org/10.1016/j.measurement.2014.08.009
45. Kumar M, Tiwari T, Srivastava N (2012) Electrical transport behaviour of bio-polymer electrolyte system: Potato starch+ammonium iodide. *Carbohydrate Polymers* 88 (1):54–60. https://doi.org/10.1016/j.carbpol.2011.11.059

46. Amran NNA, Manan NSA, Kadir MFZ (2016) The effect of LiCF3SO3 on the complexation with potato starch-chitosan polymer blend electrolytes. *Ionics* 22 (9):1647–1658. doi:10.1007/s11581-016-1684-3
47. Hamsan MH, Shukur MF, Kadir MFZ (2017) NH4NO3 as charge carrier contributor in glycerolized potato starch-methyl cellulose blend-based polymer electrolyte and the application in electrochemical double-layer capacitor. *Ionics* 23 (12):3429–3453. doi:10.1007/s11581-017-2155-1
48. Azli AA, Manan NSA, Kadir MFZ (2017) The development of Li+ conducting polymer electrolyte based on potato starch/graphene oxide blend. *Ionics* 23 (2):411–425. doi:10.1007/s11581-016-1874-z
49. Kulshrestha N, Gupta PN (2016) Structural and electrical characterizations of 50:50 PVA:starch blend complexed with ammonium thiocyanate. *Ionics* 22 (5):671–681. doi:10.1007/s11581-015-1588-7
50. Ayala Valencia G, Henao ACA, Zapata RAV (2014) Influence of glycerol content on the electrical properties of potato starch films. *Starch - Stärke* 66 (3–4):260–266. doi:10.1002/star.201300038
51. Pang SC, Tay CL, Chin SF (2014) Starch-based gel electrolyte thin films derived from native sago (Metroxylon sagu) starch. *Ionics* 20 (10):1455–1462. doi:10.1007/s11581-014-1092-5
52. Singh R, Polu AR, Bhattacharya B, Rhee H-W, Varlikli C, Singh P (2016) Perspectives for solid biopolymer electrolytes in dye sensitized solar cell and battery application. *Renewable and Sustainable Energy Reviews* 65:1098–1117. doi:10.1016/j.rser.2016.06.026
53. Moniha V, Alagar M, Selvasekarapandian S, Sundaresan B, Boopathi G (2018) Conductive bio-polymer electrolyte iota-carrageenan with ammonium nitrate for application in electrochemical devices. *Journal of Non-Crystalline Solids* 481:424–434. https://doi.org/10.1016/j.jnoncrysol.2017.11.027
54. Shamsudin I, Ahmad A, Hassan N, Kaddami H (2015) Biopolymer electrolytes based on carboxymethyl O-carrageenan and imidazolium ionic liquid. *Ionics* 22. doi:10.1007/s11581-015-1598-5
55. Karthikeyan S, Selvasekarapandian S, Premalatha M, Monisha S, Boopathi G, Aristatil G, Arun A, Madeswaran S (2017) Proton-conducting I-Carrageenan-based biopolymer electrolyte for fuel cell application. *Ionics* 23 (10):2775–2780. doi:10.1007/s11581-016-1901-0
56. Moniha V, Alagar M, Selvasekarapandian S, Sundaresan B, Hemalatha R (2019) Development and characterization of bio-polymer electrolyte iota-carrageenan with ammonium salt for: electrochemical application. *Materials Today: Proceedings* 8:449–455. https://doi.org/10.1016/j.matpr.2019.02.135
57. Chitra R, Sathya P, Selvasekarapandian S, Monisha S, Moniha V, Meyvel S (2019) Synthesis and characterization of iota-carrageenan solid biopolymer electrolytes for electrochemical applications. *Ionics* 25 (5):2147–2157. doi:10.1007/s11581-018-2687-z
58. Chitra R, Sathya P, Selvasekarapandian S, Meyvel S (2020) Synthesis and characterization of iota-carrageenan biopolymer electrolyte with lithium perchlorate and succinonitrile (plasticizer). *Polymer Bulletin* 77 (3):1555–1579. doi:10.1007/s00289-019-02822-y
59. Mobarak NN, Jumaah FN, Ghani MA, Abdullah MP, Ahmad A (2015) Carboxymethyl carrageenan based biopolymer electrolytes. *Electrochimica Acta* 175:224–231. https://doi.org/10.1016/j.electacta.2015.02.200
60. Christopher Selvin P, Perumal P, Selvasekarapandian S, Monisha S, Boopathi G, Leena Chandra MV (2018) Study of proton-conducting polymer electrolyte based on K-carrageenan and NH4SCN for electrochemical devices. *Ionics* 24 (11):3535–3542. doi:10.1007/s11581-018-2521-7

61. Shamsudin IJ, Hassan NH, Kaddami H (2014) Solid biopolymer electrolyte based on k-carrageenan for electrochemical devices application. *Asian Journal of Chemistry* 26:S77–S80. https://doi.org/10.14233/ajchem.2014.19019
62. Arockia Mary I, Selvanayagam S, Selvasekarapandian S, Srikumar SR, Ponraj T, Moniha V (2019) Lithium ion conducting membrane based on K-carrageenan complexed with lithium bromide and its electrochemical applications. *Ionics* 25 (12):5839–5855. doi:10.1007/s11581-019-03150-x
63. Bruno Scrosati (Editor), K. M. Abraham (Editor), Walter A. van Schalkwijk (Editor), Jusef Hassoun (Editor) (2013) Lithium Batteries: Advanced Technologies and Applications, ISBN: 978-1-118-18365-6, The ECS Series of Texts and Monographs, Wiley.
64. Jürgen Heydecke (2018) Introduction to lithium polymer technology, Jauch Quartz GmbH & Jauch Battery Solutions GmbH, In der Lache 24, 78056 Villingen-Schwenningen, Germany. www.jauch.com, https://www.jauch.com/downloadfi le/5c5050fa5b6510e9a8ad76299baae4e53/white_paper_introduction_to_lipo_ battery_technology_11-2018_en.pdf
65. Zainuddin NK, Samsudin AS (2018) Investigation on the effect of NH4Br at transport properties in k–carrageenan based biopolymer electrolytes via structural and electrical analysis. *Materials Today Communications* 14:199–209. https://doi.org/10.1016/j.mtco mm.2018.01.004
66. Aziz SB, Mohd MHH, Kadir FZ, Karim WO, Abdullah RM (2019) Development of polymer blend electrolyte membranes based on Chitosan: Dextran with high ion transport properties for EDLC application. *International Journal of Molecular Sciences* 20 (13):3369. doi:10.3390/ijms20133369
67. Aziz SBH, Karim WO, Kadir MFZ, Brza MA, Abdullah OG (2019) High proton conducting polymer blend electrolytes based on Chitosan: Dextran with constant specific capacitance and energy density. *Biomolecules* 9. doi:10.3390/biom9070267
68. Aziz SB, Hamsan MH, Brza MA, Kadir MFZ, Abdulwahid RT, Ghareeb HO, Woo HJ (2019) Fabrication of energy storage EDLC device based on CS:PEO polymer blend electrolytes with high Li+ ion transference number. *Results in Physics* 15:102584. https:// doi.org/10.1016/j.rinp.2019.102584
69. Shukur MF, Hamsan MH, Kadir MFZ (2019) Investigation of plasticized ionic conductor based on chitosan and ammonium bromide for EDLC application. *Materials Today: Proceedings* 17:490–498. https://doi.org/10.1016/j.matpr.2019.06.490
70. Willfahrt A (2019) Printable acid-modified corn starch as non-toxic, disposable hydrogel-polymer electrolyte in supercapacitors. *Applied Physics A* 125:474. doi:10.1007/s00339-019-2767-6
71. Liew C-W, Ramesh S (2015) Electrical, structural, thermal and electrochemical properties of corn starch-based biopolymer electrolytes. *Carbohydrate Polymers* 124:222–228. doi:10.1016/j.carbpol.2015.02.024
72. Teoh KH, Lim C-S, Liew C-W, Ramesh S, Ramesh S (2015) Electric double-layer capacitors with corn starch-based biopolymer electrolytes incorporating silica as filler. *Ionics* 21 (7):2061–2068. doi:10.1007/s11581-014-1359-x
73. Shukur MF, Ithnin R, Kadir MFZ (2014) Electrical characterization of corn starch-LiOAc electrolytes and application in electrochemical double layer capacitor. *Electrochimica Acta* 136:204–216. https://doi.org/10.1016/j.electacta.2014.05.075
74. Majid SR (2007) High Molecular Weight Chitosan as Polymer Electrolyte for Electrochemical Devices, (Publisher) Jabatan Fizik, Fakulti Sains, Universiti Malaya.
75. Ma J, Choudhury NA, Sahai Y, Buchheit RG (2011) A high performance direct borohydride fuel cell employing cross-linked chitosan membrane. *Journal of Power Sources* 196 (20):8257–8264. https://doi.org/10.1016/j.jpowsour.2011.06.009

76. Majedi FS, Hasani-Sadrabadi MM, Emami SH, Taghipoor M, Dashtimoghadam E, Bertsch A, Moaddel H, Renaud P (2012) Microfluidic synthesis of chitosan-based nanoparticles for fuel cell applications. *Chemical Communications* 48 (62):7744–7746. doi:10.1039/C2CC33253A
77. Mohanapriya S, Bhat SD, Sahu AK, Pitchumani S, Sridhar P, Shukla AK (2009) A new mixed-matrix membrane for DMFCs. *Energy & Environmental Science* 2 (11):1210–1216. doi:10.1039/B909451B
78. Meenakshi S, Bhat SD, Sahu AK, Alwin S, Sridhar P, Pitchumani S (2012) Natural and synthetic solid polymer hybrid dual network membranes as electrolytes for direct methanol fuel cells. *Journal of Solid State Electrochemistry* 16 (4):1709–1721. doi:10.1007/s10008-011-1587-1
79. Oliveira PNd, Mendes AMM (2016) Preparation and characterization of an eco-friendly Polymer Electrolyte Membrane (PEM) based in a blend of sulphonated poly(vinyl alcohol)/ chitosan mechanically stabilised by nylon 6,6. *Materials Research* 19:954–962
80. Mohanapriya S, Sahu AK, Bhat SD, Pitchumani S, Sridhar P, George C, Chandrakumar N, Shukla AK (2011) Bio-Composite membrane electrolytes for direct methanol fuel cells. *Journal of the Electrochemical Society* 158 (11):B1319–B1328. doi:10.1149/2.030111jes
81. Tiwari T, Kumar M, Yadav M, Srivastava N (2019) Study of arrowroot starch-based polymer electrolytes and its application in MFC. *Starch - Stärke* 71 (7–8):1800313. doi:10.1002/star.201800313
82. Moniha V, Alagar M, Selvasekarapandian S, Sundaresan B, Hemalatha R, Boopathi G (2018) Synthesis and characterization of bio-polymer electrolyte based on iota-carrageenan with ammonium thiocyanate and its applications. *Journal of Solid State Electrochemistry* 22 (10):3209–3223. doi:10.1007/s10008-018-4028-6

Index

A

Al, 7, 87, 94, 96
  Al-based MOF, 97
  Al_2O_3, 59, 84, 95, 96, 125, 126, 138, 141, 142, 153, 154, 160, 188, 199, 204, 222, 257, 259
  $Al(OH)_n$, 139
Al-1,3,5-benzenetricarboxylate (Al-BTC), 97
  ion battery, 39
  Li-Al-germanium phosphate (LAGP), 88
Al-1,4-benzenedicarboxylate (MIL-53), 97, 98
Al-doped NASICON ceramic (LATP), 89
alkaline battery, 6, 8, 9
Alkaline manganese batteries, *see* alkaline battery
Alkane, 72
Alkyl
  n-alkyl chains, 70
  alumoxanes, 60
  carbonate, 13
  co-crystallization, 72
  poly(2,5,8,11,14-pentaoxapentadecam ethylene-(5-alkyloxy-1,3-phenylene), 70, 72
Allyl
  1-allyl-3-methylimidazolium chloride, [Amim] Cl, 263
  poly[diallyldimethylammonium]bis-trifluoromethane sulfonimide] (PDAD-MATFSI), 239
  poly(diallyldimethylammoniu) (PDADMA), 247
Amino Acid, 101
  substituted, 63

B

Benzene
  1,4-benzenedicarboxylate, (BDC), 98
  4-dodecylbenzenesulfonic acid, 249
  Al-1,3,5-benzenetricarboxylate, (Al-BTC), 97
  Al-1,4-benzenedicarboxylate (MIL-53), 97, 98
  divinylbenzene, 119
Boron
  difluoride diethyl etherate, 139
  hexagonal boron nitrile (h-BN), 192
  nitride, 85

C

Cadmium, 8, 9
  hydroxide, 11
  metallic, 10
Ceramic electrolyte, 6, 58, 84
  perovskite type, 92
  polymer, 94
  quasi, 89

D

Dimethylacetamide (DMAc), 59, 115, 118–121, 124, 136, 143, 191, 192, 263

E

Electrical energy storage (EES), 14
Electric vehicles, 4, 13, 40, 45, 46, 48, 52, 57, 81, 82, 149, 150, 221, 256
Electrochemical, 2–4, 10, 20, 23, 36, 52, 134, 187, 255, 262, 267, 268
  cell, 4, 10, 36, 52
  double layer capacitor (EDLC), 2, 4, 19, 20, 262, 267, 268
  energy storage devices, 2, 3, 23, 36, 134, 187, 255
  redox reaction, 4
Electromotive force (EMF), 13, 42, 238

F

Fluoride
  ion batteries, 12
  ions, 12
  metal fluorides, 13
Fluorine, 13
Fuel cell, 2, 4, 20–29, 36, 52, 256, 268, 269
  alkaline fuel cell (AFC), 26
  bio-polymer fuel cell, 268, 269
  direct methanol fuel cell (DMFC), 25
  microbial fuel cell (MFC), 28
  molten carbonate fuel cell (MCFC), 27
  phosphoric acid fuel cell (PAFC), 27
  polymeric electrolyte membrane fuel cell (PEMFC), 23–25, 29

G

Graphene, 46, 85

oxide (GO), 17, 188, 222, 264
Graphite, 7, 13, 41–44, 46, 47, 51, 57, 196, 220, 223, 229, 264

H

Hexagonal
　alkane layer, 72
　boron nitrile (h-BN), 192
　crystalline layer, 72
　layer, 71
Hollow, 118
　Electrospun fibers, 120
　Fiber membrane, 119

I

Imidazolium, 202, 240, 247
Immiscible blend, 117
Impedance, 70, 89, 223, 267
Inorganic ceramic fillers, 85
Inorganic filler, 84, 159, 188, 222, 257
in-situ preparation, 84
in-situ-generated silica, 222, 226
Intercalated nanocomposites, 189
Intercalation, 41, 47, 49, 51, 188–191, 194, 206, 208, 219, 238
Interfacial contacts, 84, 89
Interfacial resistance, 59, 92, 93, 95, 97, 160, 168, 249
Interfacial stability, 58, 74, 139, 168, 170, 180, 207, 208, 223
Internal resistance, 8, 9, 11, 246
Internal surface, 8
Intrinsic principles, 19
Iodine, 16
Iodine-poly-2-vinyl pyridine, 15
Ion
　ion batteries, 13
　permeable barrier, 8
　permeable membrane, 7
Ion diffusion, 82
Ion-exchange membrane, 3, 22, 25
Ionic
　interaction, 112, 114, 152
　liquid-based gel polymer electrolyte, 225
　liquid electrolyte, 227, 228
　mobility, 84, 93, 150, 151, 168, 179
　point defects, 58
Iron, 7, 12, 37, 42
　disulfide, 8

L

Lanthanum titanate oxide, 159
Lead, 8–11, 37

acid battery, 10, 11, 37, 38, 39
Leak resistance, 8
Leclanché cell, 6, 37
Lewis acid, 97, 98
Lewis acid-base interaction, 95, 96, 117, 139, 199
Li-Al-germanium phosphate (LiAGP), 88, 89, 161, 174, 178, 179, 188
Ligand, 63
Li$^+$-ion
　mobility, 93–95, 247
　superionic conductors (LISICON), 93, 94
Li lanthanum titanate, 90
Linear sweep voltammetry, 261, 265
Li-rich anti-perovskites, 92
Lithiated phosphorus sulfide, 94
Lithium
　air battery, 16, 17
　alloy, 16
　aluminate (LiAlO$_2$), 95, 257
　aluminum titanium anode, 9, 88
　bis(fluorosulfonyl)imide (LiFSI), 242, 247
　bisoxalato borate (LiBOB), 65, 66, 74, 99
　bis(triflourosulfonyl imide) (LiTFSI), 14, 61, 66, 68, 72, 74, 88–90, 92, 94, 97–99, 125, 137, 175, 204, 225, 227, 228, 239–242, 244, 247–249, 259, 263
　chloride, 263, 264, 266
　cobalt oxide (LiCoO$_2$/LCO), 13, 14, 41, 43, 44, 58, 65, 89, 202, 219, 220, 228
　dendrite, 82, 84–86, 92, 170, 236, 240, 244, 248
　difluoride(oxalate)borate, 139
　dioxide, 9
　hexafluorophosphate (LiPF$_6$), 13, 14, 102, 119, 124–127, 144, 154, 155, 159, 171, 199–201, 204, 223, 229, 263, 264, 268
　iodide, 15, 263
　iodide battery, 15, 16
　ion polymer batteries, 15
　ion technology, 9, 15, 44, 264
　iron disulfide battery, 8
　iron oxide, 41
　iron phosphate (LiFePO$_4$/LFP), 43, 89, 91–93, 119, 121, 124, 158–161, 172, 180, 196, 204, 220, 225–227, 240, 241, 243, 245–250
　lanthanum titanate, 86, 90, 92, 160
　manganese oxide, 9, 14, 15, 41, 43, 44, 159, 196, 223, 227
　metal batteries, 8, 14, 264
　methacrylic acid, 65, 74
　nickel oxide, 14
　nitrate, 124
　oxalate, 139
　oxide, 16, 44, 46, 223
　oxide battery, 43

Index

perchlorate, 14, 60, 62, 65, 69, 71, 72, 87, 92, 95, 96, 101, 102, 115–120, 125, 151–153, 155, 156, 170, 171, 174, 179, 195, 205–208, 223–225, 263, 266, 268
phosphate (LATP), 89, 90, 159, 161, 188
polysulfide, 19
polyvinyl alcohol oxalate borate, 125
redox flow battery, 17, 18, 22
salts, 58–61, 63, 68, 72, 74, 82, 98, 100–102, 112, 114, 134, 135, 168, 169, 172, 177, 192, 199, 220, 236
sulfur battery, 18, 19, 45
sulfur dioxide, 9
tetrafluoroborate, 71, 125, 137, 155, 177, 259
thionyl chloride, 9, 14
thionyl chloride battery (LTC), 9
transition metal-oxides, 41
triflate, 60, 71, 125, 137, 205, 206, 225, 259, 263, 264
trifluoromethane sulfonate, 201
Lyophilic, 190

M

Macro-size, 46, 85
Macrovoids, 195
Magnesium (Mg), 7, 12
Magnesium oxide (MgO_2), 257
Magnesium ion batteries (MIBs), 12
Manganese
   lithium manganese dioxide ($LiMnO_2$), 9
   dioxide, 8, 38
   spinel, 41
Mechanical stability, 95, 96, 99, 102, 112, 121, 134, 137, 138, 141, 152, 158, 159, 170, 174, 189, 199, 206–208, 225, 248, 258
Mechano-chemical mixing, 94
Memory effect, 9, 11, 12, 38, 39, 144
Memory protection, 4
Mercury, 7–9
   batteries, 8
Mesocarbon microbead, 202
Mesoporous silica, 229
Metallic cadmium, 10
Metallic Li electrodes, 41
Metal organic frame work, 59, 83, 96–98, 102
Metal oxides, 19, 20, 43
Methyl methacrylate, 62, 65, 73, 159, 170, 207, 223, 244
1-methyl-1-propyl piperidinium chloride, 245
Micelle, 190
Microbial fuel cell, 23, 28, 29
Microcomposite, 190
Microporous polymer electrolyte, 119, 120
Molten carbonate fuel cell, 23–25, 27

Monovalent, 7
Montmorillonite, 125, 188, 189, 191, 193–196, 201–203, 205, 207–209

N

Nafion, 25, 249, 257, 269
Nanocatalyst, 27
Nanocomposite based solid polymer electrolytes, 82–85, 87, 90, 92, 93, 95–97, 102
Nanocomposite nanofiber (NCNFs), 127
Nanoparticles, 59, 84–87, 89, 95, 97, 101, 108, 125, 139, 224, 228–230, 245
Nano silica, *see* Nano silica-based gel polymer electrolyte
Nano silica-based gel polymer electrolyte, 224, 225
Nanowires (NWs), 86, 87, 92, 232
Ni-Cd batteries, *see* Nickel cadmium batteries
Nickel, 26, 27
   electrode, 10
   hydroxide ($Ni(OH)_2$), 10, 11
   metal hydride batteries (NI-MH) batteries, 11, 38, 39, 42, 235
   oxide (NiO_2), 188, 222
   oxide hydroxide (NiOOH), 11, 12
Nickel cadmium batteries, 11, 12, 38, 39
N-methyl pyrrolidone (NMP), 115, 118–120, 141, 169
N-methyl-N-propyl pyrrolidinium bromide (Pyl3Br), 244
Noble-metal catalyst, 23
Nominal voltage, 9, 41
Non rechargeable batteries, 4, 5, 8, 38, *see also* primary, batteries

O

Oligomers, 62, 67, 68, 94
One dimensional (1D), 84
   nano fillers, 85
Ordinary alkaline batteries, 8
Organic clay (OC), 198, 199
Organic ligands, 96
Organo-modified clay (OMC), 197, 198
Organophilic, 189, 190, 201, 206
Overcharge capability, 11
Oxidation, 14, 15, 17, 18, 20, 23, 25, 44, 268
Oxygen, 11, 16, 17, 21–23, 25–27, 32, 40, 46, 47, 60, 96, 98, 100, 102, 127, 137, 188, 222, 239, 261, 268, 269
   electrode, 17

P

Paintable batteries, 4, 36
Passivation layer, 9, 249

PEG-diacrylate-*co*-vinylene carbonate (PEGDA-*co*-VC), 250
Perfluoropolyether (PFPE), 94
Phase inversion, 112, 114, 118, 119, 121, 127, 135, 136, 139–142, 144, 150, 151, 154–157, 161, 162, 167, 169, 171, 194, 195, 199
Phosphazene, 63, 66
Phosphoric acid fuel cell (PAFC), 22, 23, 25, 27
Physicochemical working principles, 23, 24
Plasticizers, 59, 102, 111, 116, 117, 124, 135, 136, 139, 144, 151, 152, 170, 177, 201, 206, 256
Polyacetylene, 41
Polyacrylonitrile (PAN), 14, 31, 59, 72, 79, 82, 86, 87, 112, 121, 124, 134, 137, 140, 141, 143, 150–162, 168, 180, 187, 204–207, 209, 221, 229, 264
Poly(bis-benzoxazole), 25
Polycaprolactone (PCL), 258
Poly(diallyl dimethylammonium) (PDADMA), 247, 248
Poly[diallyldimethylammonium] bis-trifluoromethane sulfonimide], 239
Polydimethylsiloxane (PDMS), 62, 119, 121, 122, 239
Polyether, 61, 70–72
Polyethylene glycol (PEG), 60, 62, 68, 89, 172, 222, 242, 258, 260
 methyl ethyl methacrylate (PPME), 64
 triboron based (BPEG), 89
Polyethylene oxide (PEO), 14, 57–62, 64–74, 82, 86–90, 92, 94–99, 101, 112, 115, 119–121, 125, 134, 137, 140, 150, 153, 158, 159, 161, 168, 170, 172–179, 187, 205, 207, 221, 226, 259, 260, 264, 268
Poly(ethylene oxide-*co*-propylene oxide), 61, 89
Polyethylene terephthalate (PET), 221
Polyethyl methacrylate (PEMA), 137
Polylactic acid (PLA), 258
Polymer electrolyte membrane fuel cell, (PEMFCs), 23, 25, 29
Polymeric electrolyte membrane fuel cell, (PEMFC), 23–25
Polymethyl methacrylate (PMMA), 14, 59, 34, 35, 66, 112, 115, 116, 125, 137, 141, 143, 150, 152, 153, 158, 159, 161, 167–172, 181, 187, 207–209, 221, 223, 225, 226, 264
Poly(methyl methacrylate-*co*-butyl acrylate), (PMMA-*co*-BA), 225, 226
Poly(*N*-vinylimidazole)-*co*-PEG-methyl ether methacrylate, (PVIM-*co*-PPEGMA), 247, 248
Poly[*N*-(1-vinylimidazolium-3-butyl)-ammonium TFSI]-*co*-PEG methyl methacrylate, 247, 248

Poly(oligo-oxyethylene) methacrylate, (POEM), 68, 69
Polystyrene, 22, 64, 65, 112, 150, 159, 168, 257
Polyurethane, 62, 125, 159, 244
Polyvinyl acetate (PVAc), 134, 158, 159, 168
Polyvinyl alcohol (PVA), 115, 116, 153, 161, 196, 225, 259, 264
Polyvinyl chloride (PVC), 112, 115–117, 124–125, 134, 152, 153, 159, 161, 167, 168, 176, 177, 179–181, 223
Poly(vinylidene fluoride) (PVdF), 4, 77, 89, 91, 99, 111–128, 134, 135, 140, 143, 144, 150, 156, 158, 167–171, 173–181, 187, 192, 193, 194–199, 207, 209, 221, 225, 228, 264
Poly(vinylidene fluoride-*co*-hexafluoropropylene) (PVdF-*co*-HFP), 14, 89, 112, 133–144, 150, 156, 158, 167–172, 179, 181, 187, 192, 199, 200–204, 207, 209, 221, 222, 227, 229, 240, 241, 244–247, 249, 250
Potassium carbonate (K_2CO_3), 27
 hydroxide (KOH), 7, 8, 11, 26
 iodide (KI), 152
Power density, 5, 8, 12, 24, 25, 29, 39, 45, 52, 111, 187, 223, 235, 238, 268, 269
Primary
 batteries, 5–10, 14, 15, 36–39, 44
 fuel cell, 24
Propylene carbonate (PC), 96, 100, 115–117, 124, 135–137, 152, 170, 171, 174, 177, 179, 190, 195, 201, 205–208, 223, 224, 242, 243, 250, 256

Q

Quasi
 ceramic electrolyte, 89
 solid-state electrolyte, 192
Quaternization, 247, 261
Quaternary alkyl ammonium salt, 205

R

Radio frequency identification (RFID), 9
Roll to roll printing, 4, 36
Random
 access memory (RAM), 49
 copolymers, 60, 61, 63, 65, 73
 nanowires, 86
Ruthenium, 64
 oxide, 17

S

Silver oxide batteries, 7
Sodium, 13, 235

Index

carbonate, 27
chloride, 238
hydroxide (NaOH), 7, 8, 100, 258
ion batteries, 13, 39
ions (Na$^+$), 13, 238
sulfur battery, 39
Solid polymer electrolyte (SPEs), 6, 47, 57, 58, 60–66, 69, 74, 83, 85, 94, 96, 98, 100–102, 134, 150, 168, 201, 205, 206, 236, 255, 256, 264
   hybrid, 88
   ionic liquid based, 239
   MOF-based, 97
   novel, 82
   PEO based, 59, 73, 99
Solid state electrolyte, 6, 47, 65, 66, 82, 83, 85, 89, 94, 101, 236, 251
   anti-perovskite based, 93
   ceramic composite, 92
   flexible, 87, 95
   hybrid, 86, 96
   nanocomposite (PNSEs), 83
   polymer ceramic composite, 90
   quasi, 192
Sulfur, 46, 199
   anions, 93
   based cathodes, 47
   dioxide, 61
Sulfuric acid, 10, 269
Supercapacitors, 2, 4, 19–21, 29, 36, 134, 187, 235, 256, 260, 264, 267
   flexible, 82
Super ionic conductor (NASICON), 83–85, 88–90, 102

T

Thionyl chloride (SOCl$_2$), 8
Titanium (Ti)
   dioxide (TiO$_2$), titania, 14, 59, 84, 95, 96, 125, 126, 138, 141, 153, 154, 188, 199, 222, 257, 263
   disulfide, 40, 49

Z

Zero dimensional (0D), 84, 85
Zeolite
   fillers, 97
   powders, 257